GATEWAY
Additional Science

FOR OCR

David Acaster | Mary Jones | David Sang

CAMBRIDGE
UNIVERSITY PRESS

CAMBRIDGE UNIVERSITY PRESS
Cambridge, New York, Melbourne, Madrid, Cape Town, Singapore,
São Paulo, Delhi, Dubai, Tokyo

Cambridge University Press
The Edinburgh Building, Cambridge CB2 8RU, UK

www.cambridge.org
Information on this title: www.cambridge.org/9780521685412

First published 2006
Reprinted 2010

Printed in India by Replika Press Pvt. Ltd

Cover and text design by Blue Pig Design Ltd
Page layout and artwork by HL Studios, Long Hanborough

A catalogue record for this publication is available from the British Library

ISBN 978-0-521-68541-2 Paperback

Contents

Introduction   iv

Biology

B3a Molecules of life   1
B3b Diffusion   9
B3c Keeping it moving   17
B3d Divide and rule   26
B3e Growing up   32
B3f Controlling plant growth   39
B3g New genes for old   44
B3h More of the same   51

B4a Who planted that there?   59
B4b Water, water everywhere   63
B4c Transport in plants   68
B4d Plants need minerals too   75
B4e Energy flow   79
B4f Farming   86
B4g Decay   96
B4h Recycling   103

Chemistry

C3a What are atoms like?   108
C3b How atoms combine – ionic bonding   116
C3c Covalent bonding and the structure
      of the Periodic Table   124
C3d The Group 1 elements   132
C3e The Group 7 elements   138
C3f Electrolysis   144
C3g Transition elements   150
C3h Metal structure and properties   156

C4a Acids and bases   161
C4b Reacting masses   170
C4c Fertilisers and crop yield   178
C4d Making ammonia – Haber process
      and costs   185
C4e Detergents   191
C4f Batch or continuous?   198
C4g Nanochemistry   203
C4h How pure is our water?   208

Physics

P3a Speed   215
P3b Changing speed   225
P3c Forces and motion   233
P3d Work and power   242
P3e Energy on the move   248
P3f Crumple zones   253
P3g Falling safely   257
P3h The energy of games and theme
      rides   262

P4a Electrostatics – sparks   269
P4b Uses of electrostatics   275
P4c Safe electricals   279
P4d Ultrasound   288
P4e Treatment   294
P4f What is radioactivity?   298
P4g Uses of radioisotopes   305
P4h Fission   310

Answers to SAQs

Biology   316
Chemistry   323
Physics   329

Glossaries

Biology   333
Chemistry   338
Physics   342

Periodic Table   345

Physics formulae   346

Index   347

Acknowledgements   352

Introduction

To the pupil

This book is divided into three sections – Biology, Chemistry and Physics. Each of these sections is then arranged by Item, as in the exam specification. Each Item has the following features.

- **Self-Assessment Questions (SAQs)** are placed within the text and refer to material that has gone before. Answers to these are provided at the back of the book.
- **End-of-chapter questions**, including exam-style questions. Answers to these are not provided in this book.
- **Summaries** at the end of the chapter that show the information you need to know in order to do well in the exam.
- **Higher-level** text, questions and summaries are shown by a side bar marked with the letter 'H'.
- **Context boxes** that give you the opportunity to read about the history behind the discoveries and to learn about real-world applications.
- **Worked example boxes** (where relevant).

The book also contains glossaries, a Periodic Table, a list of physics formulae and an index at the back of the book.

To the teacher

The *Cambridge Gateway Sciences* series has been written to cover the new Gateway Specification (B) developed by OCR.

This text contains materials for the Additional Science specification. It is accompanied by an *Additional Science Teacher File* CD containing adaptable planning and activity-sheet resources, as well as answers to the end-of-chapter questions, and by *Additional Science* CDs of interactive e-learning resources, including animations and activities for whole-class teaching or independent learning, depending on your needs.

The Science specification is supported by a book and CDs in the same way.

For further information on all accompanying materials, visit the dedicated website at www.cambridge.org/newsciences

Cells and DNA

How is it that you are alive and the book you are reading is not? You and the book are made of very much the same collection of elements – carbon, hydrogen, oxygen, nitrogen and a few others in much smaller quantities, such as iron, sodium and chlorine.

Living things, like you, are alive because of an amazingly complex set of chemical reactions that are constantly happening inside them. Most of the reactions happen inside cells, although there are a few that take place in the space in between them or in blood plasma. We still don't know about all of these reactions. Some of them have been well studied, for example respiration. Others, such as some of the ones that happen in our brains, are still mysteries. It sometimes seems that, every time scientists eventually come to understand a particular set of reactions that has been puzzling them, they uncover a whole new set that they did not even know existed.

These chemical reactions are called **metabolic reactions**. They don't happen randomly. The places where they happen and the times when they happen, how fast they take place and how much they take place, are controlled by **enzymes**. Enzymes are biological catalysts – protein molecules that are needed before metabolic reactions can happen.

Mitochondria

Figure 3a.1 shows a diagram of an animal cell. Not all the cells in your body look like this, because different kinds of cells are adapted for different functions. The cell shown in Figure 3a.1 is similar to the cells in the smooth, slippery layer that lines the inside of your cheeks (Figures 3a.2 and 3a.3).

Metabolic reactions take place in every part of a cell. Many different ones happen in the **cytoplasm**.

Aerobic respiration happens inside the **mitochondria**. These are little, fat sausage-like organelles, usually much too small to be seen clearly with a school microscope. They often just look like small dots. With an electron microscope, more detail of their structure

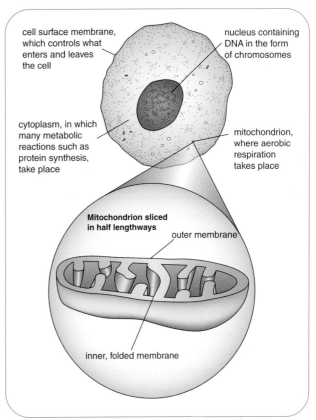

Figure 3a.1 An animal cell.

becomes visible. We can see that they are made of two membranes – an outer smooth one and an inner one that is folded.

Mitochondria are one of the most important parts of the cell, because respiration provides the cell with the energy that it needs to fuel metabolic reactions and so allow living processes to take place. Cells that need a lot of energy, such as muscle cells, have very large numbers of mitochondria. Mitochondria are sometimes called the 'power-houses' of a cell.

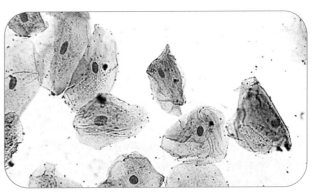

Figure 3a.2 Cheek cells.

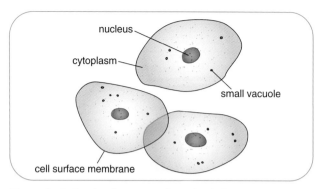

Figure 3a.3 Cheek cells as seen through a light microscope.

SAQ

1   Would you expect plant cells to contain mitochondria? Explain your answer.

The nucleus

Most cells have a **nucleus**, often visible as a dark, roughly circular structure near the centre of the cell. The nucleus contains the **chromosomes**. In a human cell, there are 46 chromosomes. (You cannot usually see them – they only become visible when the cell is about to divide.)

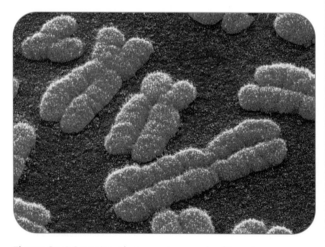

Figure 3a.4 Scanning electron micrograph of human chromosomes.

Each chromosome is an extremely long molecule of **DNA**. DNA controls the production of proteins in the cell. A section of DNA that provides instructions for making one protein is called a **gene**. Each chromosome contains many genes.

We've seen that metabolic reactions are controlled by enzymes, and that enzymes are proteins. Each step of a metabolic reaction needs a different enzyme to make it happen. So, if the production of the enzymes is controlled then the metabolic reactions are controlled, too. DNA controls the manufacture of proteins in the cell, and so DNA controls the metabolic reactions that happen inside it.

The structure of DNA

A DNA molecule is enormous as molecules go. The molecules have a regular, coiled shape, like two helter-skelters wound around each other. The shape is called a **double helix** (Figure 3a.5).

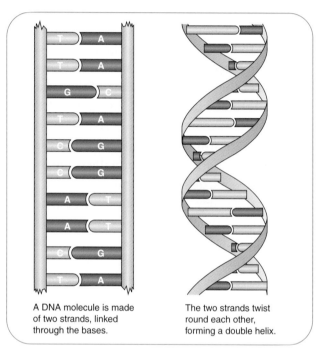

A DNA molecule is made of two strands, linked through the bases.

The two strands twist round each other, forming a double helix.

Figure 3a.5 Part of a DNA molecule.

The two strands of the double helix are held together by pairs of **bases**, which are arranged like rungs on a ladder. The bases on one strand fit perfectly against the bases of the other strand.

There are four different bases in a DNA molecule – adenine, cytosine, thymine and guanine. They are usually known by their abbreviations, **A**, **C**, **T** and **G**. You can see from the diagram that these are always arranged in the same pairs. A is always with T, while C is always with G. This is called **complementary base pairing**.

SAQ

2   A 100 base length of DNA contains 40 T bases.

   a   How many A bases does it have?

   b   How many C bases does it have?

DNA replication

The sequence in which the base pairs are arranged along a DNA molecule is a kind of code. The code carries information telling the cell what proteins to make. It is essential that, when a cell divides to make new cells, all of this coded information is copied perfectly and passed on to the daughter cells.

This means that, before a cell divides, the DNA must divide. This is called **DNA replication** (Figure 3a.6).

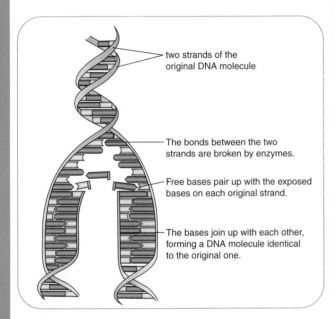

two strands of the original DNA molecule

The bonds between the two strands are broken by enzymes.

Free bases pair up with the exposed bases on each original strand.

The bases join up with each other, forming a DNA molecule identical to the original one.

Figure 3a.6 DNA replication takes place in the nucleus.

First, an enzyme breaks the bonds holding the base pairs together – it 'unzips' the two strands. Next, bases that are already present in the nucleus pair up with the ones that have become exposed on the two DNA strands. The end result is two double-stranded DNA molecules exactly like the original one.

Making proteins

Making proteins is called **protein synthesis**. It happens in the cytoplasm of your cells.

Proteins have many different uses in your body. They are needed for growth and repair of cells. Some of the different types of proteins include:

- enzymes, the biological catalysts that control every metabolic reaction;
- haemoglobin, the red pigment inside red blood cells that transports oxygen;

- antibodies, the molecules that attach to pathogens and help to destroy them;
- keratin, the protein from which hair and nails are made;
- collagen, a slightly stretchy protein that helps to give skin and bones their strength.

A protein molecule is made of a long chain of **amino acids** (Figure 3a.7). There are 20 different amino acids in living things. They can be linked together in any order. The sequence in which they are linked differs in different proteins.

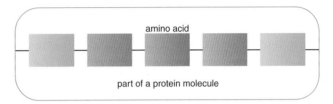

amino acid

part of a protein molecule

Figure 3a.7 A protein molecule.

When you eat proteins, the long chains of amino acids are broken up inside the digestive system. The individual amino acids are absorbed into the blood and carried all over the body so that every cell gets a good supply. A good diet will contain all the different amino acids. If you are short of one particular kind, the liver can usually change one kind of amino acid into another, in a process called **transamination**, but there are a few that can't be made in this way.

When a cell needs to make a protein molecule, it can use the amino acids brought to it in the blood and string them together in the right order to make exactly the kind of protein it needs. So, for example, you might eat an egg containing a protein called albumen. The albumen molecules get split up into individual amino acids and taken to your skin. Here, a cell at the base of your fingernail links the amino acids together in a different order and makes some keratin molecules, which become part of the fingernail.

The DNA in the nucleus contains the information about which order to string the amino acids together. It is in the form of a code, the **genetic code**. The sequence of bases in a DNA molecule is the code. The code tells the cell what sequence of amino acids to join together to make a protein. As we have seen, a length of DNA that carries the code for one type of protein molecule is a gene.

The genetic code

The code in the DNA molecules is a three-letter code. It is read as if each group of three bases is a 'word' standing for one amino acid.

For example, a very short length of a DNA molecule might have this sequence of bases on one of its strands:

A-T-G-C-C-C-G-A-A-G-G-G

If we break this up into three-base groups or **triplets**, it reads:

ATG CCC GAA GGG

Each of these triplets of bases stands for one of the 20 different amino acids that a cell can use to make proteins. The base triplet ATG stands for the amino acid tyrosine; CCC stands for glycine; GAA stands for leucine; GGG stands for proline. So this little stretch of DNA is effectively telling the cell, 'link together tyrosine, glycine, leucine and proline, in that order.'

SAQ

3   Sometimes, a mistake is made when a DNA molecule replicates. This is called **mutation**. For example, in the example on the left, an extra base might get slipped in, so the sequence reads:

A-T-G-G-C-C-C-G-A-A-G-G-G

Explain why this could have a really big effect on the sequence of amino acids in the protein that is made using these instructions.

DNA fingerprinting

Some parts of the DNA molecules in human chromosomes are very variable. They are not the same in any two people – unless they are identical twins. Each person's DNA is unique. This fact is used in a technique called **DNA fingerprinting** or **DNA profiling**.

The DNA is put through a process that leaves a pattern on a gel (jelly). By comparing the patterns

DNA fingerprints and ordinary fingerprints

Fingerprints have been used for many years in criminal cases. A fingerprint is a pattern made by the ridges and grooves in the skin of the fingertips. Figure 3a.8 shows some of the components of these patterns. Police forces in Britain have built up a computer database of fingerprints of people who have been convicted of crimes, and fingerprints left at the scene of a crime can be matched against this database.

It has always been assumed that everyone's fingerprints are unique. But now this has been called into question. And, even if they are unique, the fingerprints found at crime scenes are often only partial, or they may be messy and unclear. No-one has ever done any scientific studies to find out just how reliable fingerprint evidence is. Rather than a zero error rate, which is what people have assumed there was, it seems that error rates could be as high as 4.4%. This could mean that hundreds of people each year are wrongly identified as being present at a crime scene when they were not there at all.

Figure 3a.8 A fingerprint.

Is DNA 'fingerprinting' any better? Yes, definitely. The chance of a random match in the DNA profiles of two unrelated people is less than one in a billion.

made by different samples of DNA, you can tell if two of them came from the same person.

For example, imagine someone has been murdered by stabbing. A suspect has a small bloodstain on his shirt. DNA is extracted from the victim and also from the bloodstain. For comparison, DNA is also taken from the blood of other people, who are known to have had nothing to do with the crime. The patterns made by the different samples of DNA are shown in Figure 3a.9.

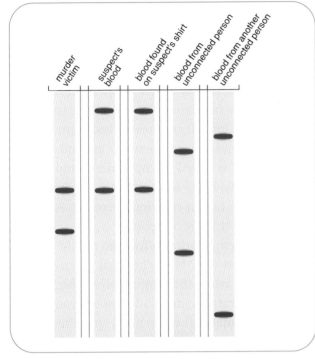

Figure 3a.9 DNA patterns obtained in a crime investigation. Each column is a sample from a person. If the patterns of bands in two columns match, then that could mean those two samples are from the same person. If the bands of two columns don't match, then those are *definitely* not from the same person.

SAQ

4 Does the evidence in Figure 3a.9 support or disprove the hypothesis that the suspect committed the murder, or does it do neither? Explain your answer.

Producing a DNA fingerprint

Figure 3a.10 shows how DNA fingerprinting is carried out.

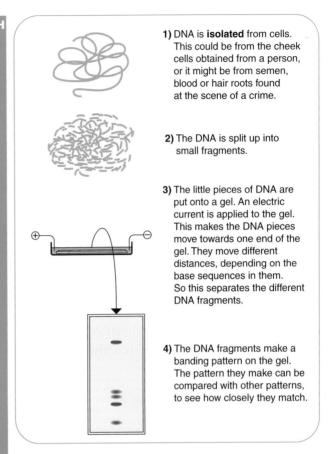

1) DNA is **isolated** from cells. This could be from the cheek cells obtained from a person, or it might be from semen, blood or hair roots found at the scene of a crime.

2) The DNA is split up into small fragments.

3) The little pieces of DNA are put onto a gel. An electric current is applied to the gel. This makes the DNA pieces move towards one end of the gel. They move different distances, depending on the base sequences in them. So this separates the different DNA fragments.

4) The DNA fragments make a banding pattern on the gel. The pattern they make can be compared with other patterns, to see how closely they match.

Figure 3a.10 DNA fingerprinting.

Enzymes

Enzymes are proteins that act as biological catalysts. They speed up metabolic reactions without being changed. Every metabolic reaction – including respiration, photosynthesis and protein synthesis – is catalysed by enzymes.

For example, the enzyme **amylase** catalyses the breakdown of starch to maltose:

$$starch \xrightarrow{amylase} maltose$$

Starch is the **substrate** of the enzyme. Maltose is the **product**. Each enzyme has a very high **specificity** for its substrate – that is, it will work only on that substrate and not on anything else.

How enzymes work

Figure 3a.11 shows how an enzyme works. An enzyme has a particular shape, determined by the sequence of amino acids from which it is made. The long chain of amino acids curls up into a ball. You can't see the chain of amino acids in the enzyme in Figure 3a.11; it just shows the shape of the ball they make when they curl up.

The enzyme has a dent in it called an **active site**. The active site of amylase is exactly the right shape for part of a starch molecule to fit inside it. Once in, the starch molecule is split apart, to form maltose. The maltose molecules move away from the active site, leaving the amylase molecule unchanged and ready to do the same to another starch molecule. This can happen at incredible speeds – hundreds of maltose molecules can be made in one second by one enzyme molecule.

The active site of the enzyme is a very precise shape, exactly the right shape for the starch molecule to fit into. Other molecules won't fit in. So amylase cannot digest proteins, because they are a different shape from starch. The enzyme is like a lock that only a particular key can fit into. Each enzyme can only work on one substrate, because only that substrate can fit into its active site.

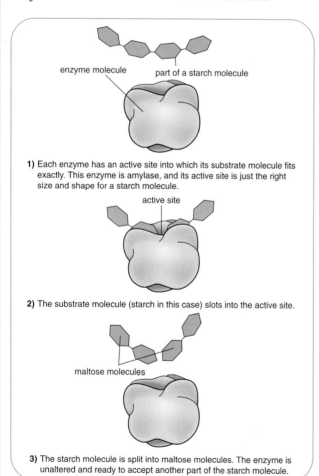

1) Each enzyme has an active site into which its substrate molecule fits exactly. This enzyme is amylase, and its active site is just the right size and shape for a starch molecule.

2) The substrate molecule (starch in this case) slots into the active site.

3) The starch molecule is split into maltose molecules. The enzyme is unaltered and ready to accept another part of the starch molecule.

Figure 3a.11 How an enzyme works.

Enzymes and temperature

Figure 3a.12 shows how the rate of a reaction catalysed by an enzyme varies with temperature.

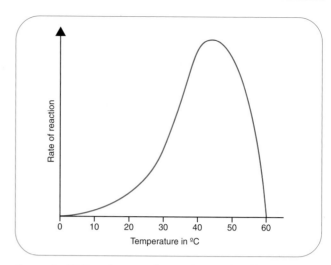

Figure 3a.12 How the rate of a reaction catalysed by an enzyme varies with temperature.

The reaction is slow at low temperatures. If the temperature is increased, the rate of the reaction also increases until it reaches a maximum. For human enzymes, this is usually at around 38–40 °C. This is called the **optimum temperature** for the enzyme. If the temperature is higher than this value, the rate of reaction decreases. At around 60 °C; it stops completely.

Why does temperature affect the rate of reaction in this way? Most chemical reactions happen faster when the temperature is higher. At higher temperatures, molecules move around faster, which makes it more likely that they will hit each other and react together. There is more chance of successful collisions. Usually, a rise of 10 °C doubles the rate of reaction. This happens with an enzyme-controlled reaction, too. As temperature increases, there is more likelihood that a substrate molecule will bump into an enzyme, so the rate of reaction is faster.

However, enzymes are protein molecules. Protein molecules are damaged by high temperatures. At high temperatures, the atoms within the molecule vibrate so energetically that some of the weaker bonds in the molecule break and the shape of the molecule is changed. It is said to be **denatured**. Once the shape of the active site is altered, the substrate can no longer fit into it and the reaction cannot take place. This is why the reaction rate is zero at 60 °C – the enzymes are all denatured and can't work any more.

Once an enzyme is denatured, it can't go back to its original shape. The damage is permanent.

Enzymes and pH

Figure 3a.13 shows how pH affects the rate of an enzyme-catalysed reaction.

SAQ

5 Look at Figure 3a.13.

    a What is the optimum pH for most enzymes?

    b Pepsin is a digestive enzyme which works in the stomach. The stomach makes hydrochloric acid, which kills pathogens that might be present in food. How can this explain the curve for pepsin in Figure 3a.13?

Extremes of pH, whether very high (alkaline) or very low (acidic), can denature enzymes in the same way as high temperatures. The active site of the enzyme molecule loses its shape and can no longer bind with its substrate. Most enzymes have a fairly narrow range of pH in which they can work.

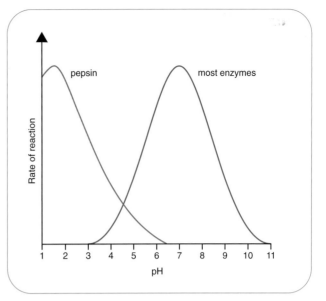

Figure 3a.13 How pH affects the rate of an enzyme-catalysed reaction.

Summary

You should be able to:

- describe the function of a cell membrane, nucleus, cytoplasm and mitochondria

- state the functions of chromosomes and genes

- describe the structure of DNA

- describe complementary base pairing in DNA

- state that DNA replicates before cell division

- describe how DNA replicates

- state that a gene codes for making a particular protein, and know some of the functions of proteins

- describe the three-base code in DNA

- describe the structure of a protein molecule, made from amino acids obtained in the diet

- state that a person's DNA is unique and interpret DNA fingerprinting results

- outline the way in which DNA fingerprinting is done

- describe the structure and function of enzymes, including their specificity

- describe how temperature and pH affect the rate of an enzyme-catalysed reaction

- explain why temperature and pH affect the rate of an enzyme-catalysed reaction

Questions

1 Copy the diagram in Figure 3a.14, and complete it by linking each structure to its function.

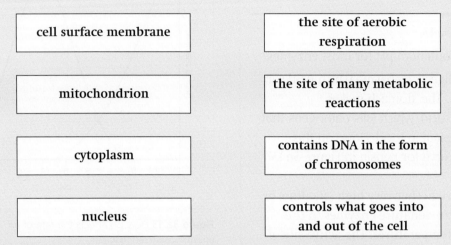

cell surface membrane	the site of aerobic respiration
mitochondrion	the site of many metabolic reactions
cytoplasm	contains DNA in the form of chromosomes
nucleus	controls what goes into and out of the cell

Figure 3a.14

2 If you cut an apple in half and leave it for a while, it goes brown. This happens because an enzyme catalyses a reaction between a colourless substance and oxygen, producing a brown substance.

colourless substance + oxygen ⟶ brown substance

Explain how each of the following can slow down the rate at which the apple goes brown.

a Putting it in the fridge.

b Putting lemon juice onto it.

3 DNA profiles can be used when there is uncertainty about who is the father of a child. A child inherits its genes from its mother and father – it can't get them from anywhere else.

Figure 3a.15 shows a DNA profile obtained from a mother, her child and two possible fathers. Which man could be the father of child A? Explain your answer.

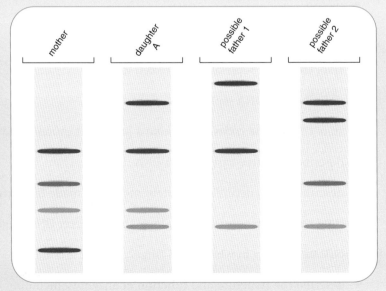

Figure 3a.15 Results of a DNA profiling paternity test.

Figure 3b.1

Diffusion

If someone opened a bottle of perfume in the corner of the room, would everyone be able to smell it straight away? The answer is 'No' – it would take time for the smell to spread across the room. The people close to the perfume bottle would smell it before the people furthest away.

This can happen even if the air in the room is absolutely still. The way in which the perfume spreads through the still air is called **diffusion**. Diffusion is the movement of particles of a substance from a region of high concentration to a region of low concentration. We say that the substance diffuses down a **concentration gradient**.

Diffusion can be explained using the particle theory. When the bottle of perfume is opened, some of the perfume molecules leave the container and go into the air as a gas. These molecules are in constant motion. The directions in which they move are random. Some will be going one way and some another. If they hit another molecule – any of the molecules in the air, such as nitrogen – they bounce off and move in a different direction. Sooner or later, just by chance, the random movement of the molecules will result in them being spread evenly through the air in the room.

Diffusion can happen in gases and in liquids because, in both of these states, the particles can

move freely amongst each other. Diffusion can only happen very slowly, if at all, in solids because the particles in a solid are fixed in place, just vibrating on the spot. However, they can diffuse quite fast in jellies, which aren't really solids. So, for example, substances *can* diffuse through agar jelly in a Petri dish, albeit much more slowly than in a gas or a liquid.

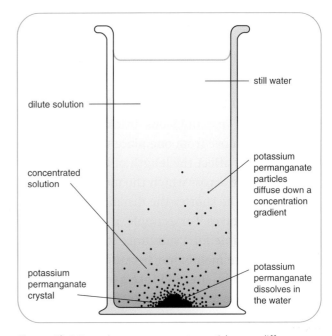

Figure 3b.2 Potassium permanganate particles can diffuse through water, because the particles of solutes move freely and randomly.

Diffusion in and out of cells

Living cells obtain many of their needs by diffusion through their **cell surface membranes**.

- Cells respire, using up oxygen. This makes the concentration of oxygen inside the cell quite low. The concentration of oxygen outside the cell is usually higher. There is a concentration gradient for oxygen. So oxygen diffuses into the cell through the cell surface membrane.

- In respiration, cells produce carbon dioxide. The concentration of carbon dioxide inside the cell is therefore higher than the concentration outside the cell. There is a concentration gradient for carbon dioxide. So carbon dioxide diffuses out of the cell through the cell surface membrane.

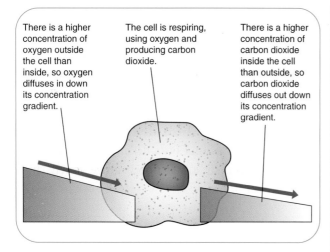

There is a higher concentration of oxygen outside the cell than inside, so oxygen diffuses in down its concentration gradient.

The cell is respiring, using oxygen and producing carbon dioxide.

There is a higher concentration of carbon dioxide inside the cell than outside, so carbon dioxide diffuses out down its concentration gradient.

Figure 3b.3 Diffusion into and out of cells.

Factors affecting the rate of diffusion

Diffusion isn't instantaneous. It takes time for molecules to move from one place to another. Several factors affect the length of this time.

- The distance across which the molecules are travelling. The shorter the distance, the shorter the time it takes. For example, it takes only a very short time for a substance to diffuse across a cell surface membrane, because this is extremely thin. It would take a lot longer for the substance to diffuse right through the cell and out the other side.
- The steepness of the concentration gradient. The greater the difference in concentration, the more quickly the molecules spread out. For example, if a cell is respiring very quickly and making the concentration of oxygen very low inside it, then oxygen will diffuse in more quickly than if the cell was respiring slowly.
- The surface area of the barrier. The more surface there is to diffuse across, the more molecules can be moving across the barrier at the same time. So a larger surface area increases the rate of diffusion. For example, if the cell surface membrane of a cell is folded, rather than just smooth, substances can diffuse across it more quickly.

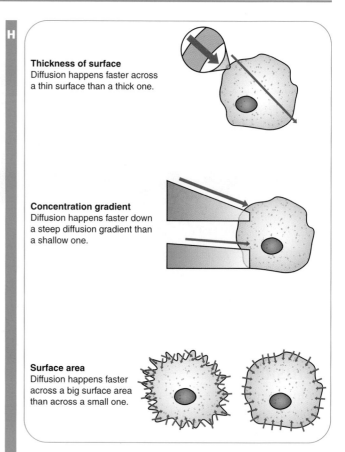

Thickness of surface
Diffusion happens faster across a thin surface than a thick one.

Concentration gradient
Diffusion happens faster down a steep diffusion gradient than a shallow one.

Surface area
Diffusion happens faster across a big surface area than across a small one.

Figure 3b.4 Factors affecting the rate of diffusion.

Gaseous exchange in the lungs

Your body cells are in constant need of oxygen for respiration. They are provided with oxygen by the blood. Oxygen enters the blood in the lungs and leaves it in the body tissues. Carbon dioxide enters the blood in the body tissues and leaves it in the lungs.

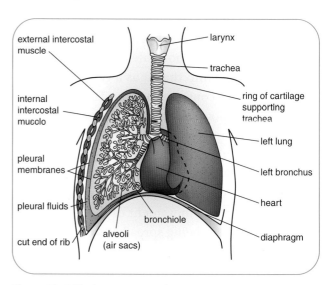

external intercostal muscle

larynx

trachea

ring of cartilage supporting trachea

internal intercostal muscle

left lung

pleural membranes

left bronchus

heart

pleural fluids

bronchiole

cut end of rib

alveoli (air sacs)

diaphragm

Figure 3b.5 The human gas exchange system.

The removal of carbon dioxide and the take-up of oxygen in the lungs are called **gaseous exchange**. The lungs contain many small tubes, called **bronchioles**, leading into blind-ending sacs called **alveoli** (Figure 3b.6). Oxygen and carbon dioxide diffuse between the air inside the alveoli and the blood inside the blood capillaries (Figures 3b.6 and 3b.7).

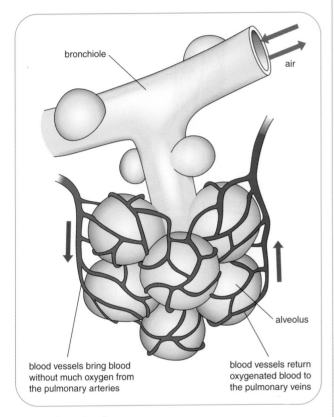

Figure 3b.6 Alveoli.

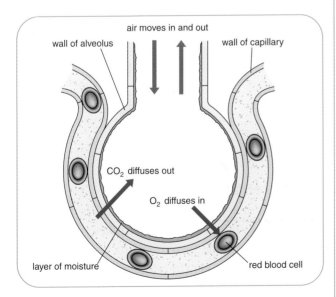

Figure 3b.7 Gaseous exchange in an alveolus.

Diffusion of oxygen and carbon dioxide between the air and the blood needs to happen quickly. The lungs and alveoli have features to help this.

- There is always a steep concentration gradient. Breathing movements bring air into the lungs, ensuring that the concentration of oxygen inside the alveoli is relatively high. Deoxygenated blood constantly flows to the lungs, ensuring that the concentration of oxygen in the blood surrounding the alveoli is always lower.
- The distance across which the oxygen and carbon dioxide diffuse is tiny. The wall of the alveolus and the wall of the capillary are each only one cell thick. What's more, these cells are exceptionally thin.
- The total surface across which the gases diffuse is huge. The total surface area of all the alveoli in your lungs is about $100 \, m^2$.
- The cells are permeable to gases. This means that they let gases pass easily through them.
- The cells are kept moist. This doesn't actually speed up diffusion (gases diffuse faster through air than through liquid) but it does stop the cells in the wall of the alveolus from drying out. There is a danger of this happening because the cells are in direct contact with the air you've breathed in, and if this is dry then water will evaporate from the cells into the air. If the cells did dry out, they would shrivel up and become impermeable to oxygen and carbon dioxide.

SAQ

1 Can you explain why gases diffuse faster through air than through liquid?

Absorption of digested food

In your digestive system, large molecules in the food that you eat are broken down to small ones. These small molecules are able to get across the wall of the intestines and into your blood. Some of them move by diffusion.

Most absorption happens in the **small intestine** (Figure 3b.8). Here, small molecules such as glucose or amino acids diffuse from the intestine and into the blood capillaries. They dissolve in the blood plasma and are carried to all the body tissues in the bloodstream. They diffuse out of the blood and into cells in the body tissues.

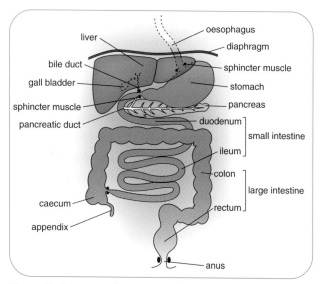

Figure 3b.8 Human alimentary canal.

SAQ

2 Suggest what a body cell might use these substances for:

    a glucose

    b amino acids.

The small intestine has several features that help to speed up diffusion across it.

● It is very long, so food spends a long time travelling through it and has plenty of chance to diffuse across its wall. In an adult, it can be up to 5 m long.

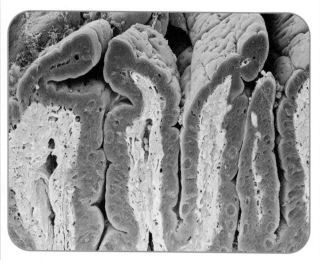

Figure 3b.9 These villi, seen using an electron microscope, have been cut through so that you can see inside them.

● Its lining has a very large surface area. The inner surface of the small intestine is folded and has tiny finger-like projections called **villi** (Figures 3b.9 and 3b.10). What's more, these tiny villi have even tinier folds in their surfaces, called **microvilli** (Figure 3b.11).

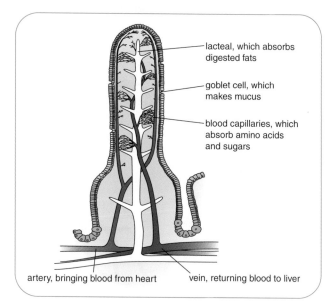

Figure 3b.10 Longitudinal section through a villus.

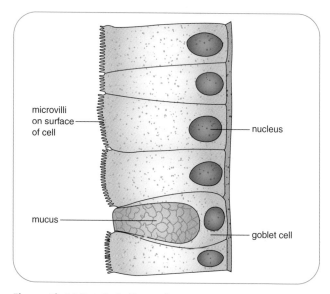

Figure 3b.11 Detail of villus surface.

The placenta

The **placenta** is the organ that connects a foetus to its mother (Figure 3b.12). The placenta is the foetus's life support system. All its requirements are delivered to it via the placenta, and this is also the way its waste products are removed.

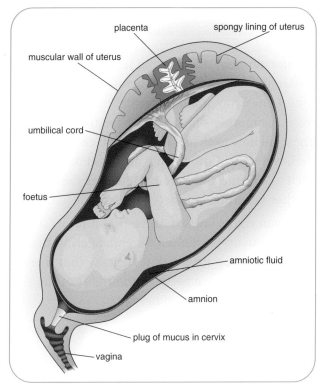

Figure 3b.12 Side view of a developing foetus inside the uterus.

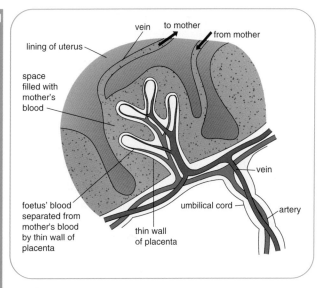

Figure 3b.13 Part of the placenta.

Inside the placenta, the mother's blood and the foetus's blood are brought close together. They do not mix.

Oxygen diffuses from the mother's blood to the foetus's blood, while carbon dioxide diffuses in the other direction. Nutrients, such as glucose and amino acids, also diffuse across from the mother's blood. Urea, a waste product of the foetus, diffuses into its mother's blood.

Like the other structures we have looked at in this Item, the placenta is adapted in ways that allow diffusion to happen quickly across it.

SAQ

3 Look at Figure 3b.13.

   a How does the structure of the placenta provide a large surface area for diffusion?

   b In which direction will concentration gradients exist for:
   i glucose?
   ii carbon dioxide?
   Explain how these concentration gradients are produced.

   c How is the distance across which diffusion has to take place kept small?

Synapses

You may remember that neurones (nerve cells) don't make direct connections with one another. There is always a tiny gap between them, called a **synapse**. When the nerve impulse arrives at the synapse, it causes a **transmitter substance** to be released. This chemical carries the signal across this gap. Transmitter substances are sometimes called **neurotransmitters**.

The transmitter substance crosses the gap by diffusion. The gap is so small – only about 20 nm wide – that it takes no more than a millisecond for this to happen.

SAQ

4 **a** A nanometre (nm) is 1×10^{-9} metres. A millisecond is 1×10^{-3} seconds. Calculate the speed at which the transmitter substance crosses the synapse.

   You will need to think back (or look back) at your earlier work on nerves to answer the next question.

   b Describe how the transmitter substance carries signals from one neurone to the next.

Maternal and foetal rights

In May 1999, in South Carolina in the USA, Regina McKnight gave birth to her third child. The child was stillborn.

Already deeply upset, Regina was plunged into even more despair when she was arrested and given a 12 year sentence for homicide. She was judged to have killed her child by her own actions. Regina was a crack cocaine addict, and the cocaine in her blood had crossed the placenta and affected her growing foetus. Even if the child had survived, it would probably have been badly affected by the drug and could have been born an addict itself.

Unfortunately, the placenta lets many different substances diffuse across it, not only useful ones. Any small molecules can diffuse from the mother's blood to her foetus's blood. These include aspirin, paracetamol, carbon monoxide, nicotine, alcohol and hard drugs such as cocaine and heroin. Whatever the mother is taking, her foetus is taking it too.

Human rights lawyers argue that it is pointless to arrest women for endangering their foetuses by their drug habit. They say that there should be a balance between the rights of the mother and the rights of her baby.

Gas exchange in leaves

The cells inside a plant, just like those inside you, are constantly respiring. They use oxygen and produce carbon dioxide.

glucose + oxygen ⟶ carbon dioxide + water

During the daytime, when light is shining on them, they also photosynthesise. Photosynthesis uses carbon dioxide and produces oxygen.

carbon dioxide + water ⟶ glucose + oxygen

So, at night, when a plant is only respiring, it takes in oxygen and gives out carbon dioxide. In daylight, because there is more photosynthesis than respiration, it takes in carbon dioxide and gives out oxygen.

This exchange of gases happens in the plant's leaves (Figure 3b.14).

Figure 3b.15 shows the cells in a leaf. You can see that there are several layers of them. The gases move into and out of the leaf through tiny holes in the bottom of the leaf, called **stomata**. The gases diffuse between the air outside the leaf and the air spaces inside it.

The shape and construction of a plant's leaves are adapted for the rapid diffusion of carbon dioxide and oxygen.

- The leaves have a very large surface area, because they are wide. The surface area for diffusion is also increased by the presence of air spaces inside the spongy mesophyll layer. This means that many cell surfaces are in direct contact with the air.
- The leaves are thin. This decreases the distance that gases have to diffuse to move between the leaf cells and the air.
- The stomata in the lower epidermis of the leaf allow gases to diffuse freely between the air spaces inside the leaf and the air outside.

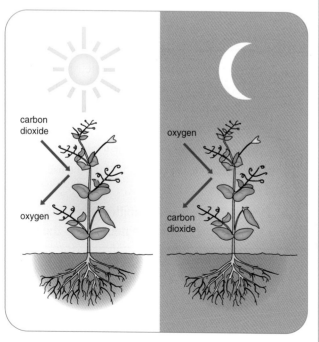

Figure 3b.14 In sunlight, leaves take in carbon dioxide and give out oxygen. The opposite happens at night.

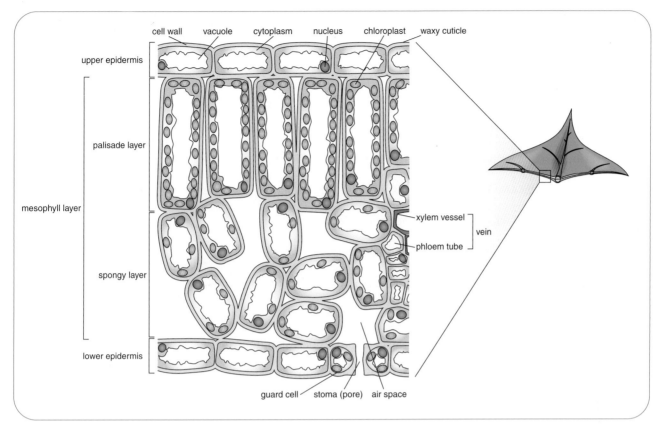

Figure 3b.15 Transverse section through part of a leaf.

Transpiration

Carbon dioxide and oxygen are not the only substances that diffuse between a plant's leaves and the air. Water vapour does this, too. The loss of water vapour from a plant's leaves is called **transpiration**.

Figure 3b.16 shows why this happens. Water seeps out from the cytoplasm of the leaf cells and into their cell walls. You can imagine the cell walls being a bit like wet filter paper, full of water molecules.

Some of this water evaporates into the air spaces inside the leaf. The water molecules are then free to diffuse out of the leaf through the stomata. This happens because there is a concentration gradient for the water. The air spaces inside the leaf are saturated with water vapour, whereas the air outside the leaf is probably drier.

You can read more about transpiration on pages 70–73 in item 4c.

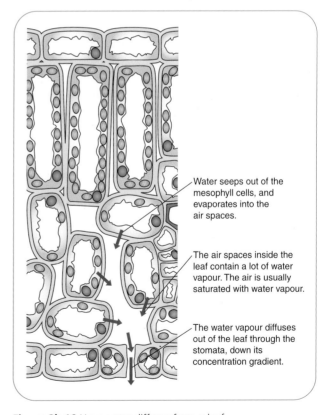

Water seeps out of the mesophyll cells, and evaporates into the air spaces.

The air spaces inside the leaf contain a lot of water vapour. The air is usually saturated with water vapour.

The water vapour diffuses out of the leaf through the stomata, down its concentration gradient.

Figure 3b.16 How water diffuses from a leaf.

SAQ

5   On a really humid day, water vapour does not diffuse out of a plant's leaves. Using what you know about diffusion, explain this.

Summary

You should be able to:

◆ describe what diffusion is, and how molecules enter and leave cells by diffusion

H ◆ explain diffusion in terms of the particle theory and outline factors that affect the rate of diffusion

◆ describe gaseous exchange in the lungs in terms of diffusion

H ◆ explain how the alveoli are adapted for efficient gas exchange

◆ describe how and where digested food molecules are absorbed

H ◆ explain how the small intestine is adapted for absorption

◆ describe how substances are exchanged between mother and foetus in the placenta

H ◆ explain how the placenta is adapted to increase the rate of diffusion

◆ explain how transmitter substances carry impulses between neurones

◆ describe how carbon dioxide, oxygen and water vapour diffuse between leaves and the air

H ◆ explain how leaves are adapted to increase the rate of diffusion

Questions

1   Copy and complete these sentences about diffusion.

Living cells need oxygen for .... The oxygen enters the cell by .... through the cell surface membrane. This happens because the concentration of oxygen is .... inside the cell than outside. The oxygen diffuses down its .... gradient.

2   A pregnant woman eats some food containing glucose.

a   Where and how does the glucose get into her blood?

b   How does the glucose get into her foetus's blood?

3   What are your views on the Regina McKnight conviction, described on page 14? Do you think it is right to convict women who take harmful drugs during their pregnancy? Justify your opinions.

H 4   Copy and complete Table 3b.1 to summarise the ways in which various organs and tissues are adapted to increase the rate of exchange of substances by diffusion.

	Lungs	Small intestine	Placenta	Leaf
What is exchanged				
How surface area is increased				
How diffusion distance is reduced				
How concentration gradient is maintained				

Table 3b.1

Keep it moving

Blood

We are complex organisms, with many different body organs made up of millions of cells. The things that these cells need are brought to them in the blood, and the blood takes away their waste products.

The components of blood

Figure 3c.1 shows what blood looks like under the microscope. The little faintly coloured approximately circular structures are **red blood cells**. These are what make blood look red. They don't appear very red in the photograph, but when you get thousands of them together they look much darker.

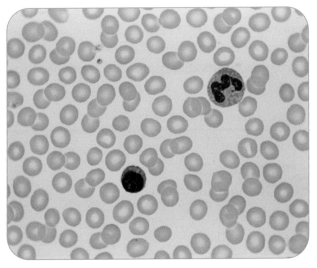

Figure 3c.1 A light micrograph of a stained blood film.

You can also see two cells that look different from the red blood cells. The blood has been stained with a purple dye, and these cells have absorbed the dye so their nuclei look purple. They are **white blood cells**. There are far fewer white blood cells than red blood cells.

If you look very carefully at the photograph, you can see a few pale purple spots. These are tiny cell fragments called **platelets**. Platelets help the blood to clot.

Surrounding the cells is an almost colourless liquid called **blood plasma**. The cells float in the blood plasma and so are carried around as the blood flows.

The functions of red blood cells

Red blood cells transport oxygen from the lungs to the tissues. They contain a red pigment called **haemoglobin**. Haemoglobin easily combines with oxygen when the oxygen is in a high concentration, as in your lungs. It easily releases its oxygen when oxygen is in a low concentration, as in your body tissues.

SAQ

1 Explain why oxygen concentration is usually low in body tissues.

Red blood cells are some of the smallest cells in your body. This allows them to get through tiny blood capillaries, taking oxygen close to every cell in the body. Another unusual thing about them is that they have no nucleus. It is thought that this allows more space inside the cell for haemoglobin, so the cell can carry more oxygen.

A red blood cell is shaped like a disc that has been pressed inwards on both sides (Figure 3c.2). This shape is called a **biconcave disc**. It gives the red blood cell a relatively large surface area, which means that oxygen can diffuse in and out of it quickly. The cell has a large surface area to volume ratio (see page 26).

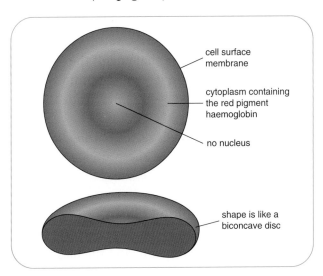

cell surface membrane

cytoplasm containing the red pigment haemoglobin

no nucleus

shape is like a biconcave disc

Figure 3c.2 A red blood cell.

H When a red blood cell is moving through a capillary in the lungs, it is very close to the air inside the alveoli. Oxygen diffuses from the alveoli into the blood, and into the red blood cell. Here it combines with haemoglobin to form a compound called **oxyhaemoglobin**. The blood is said to be oxygenated.

Later, when the red blood cell has travelled to another part of the body, it will pass through a capillary in a tissue that is respiring. Here the oxygen leaves the haemoglobin and diffuses out into the tissue. The blood becomes deoxygenated.

● In the lungs:

haemoglobin + oxygen → oxyhaemoglobin

● In the tissues:

oxyhaemoglobin → haemoglobin + oxygen

SAQ

2 Haemoglobin is a bluish-purple colour, whereas oxyhaemoglobin is bright red. The blood in your veins is deoxygenated but, if you bleed from a vein, the blood may become a bright red soon after leaving your body. Can you suggest why?

The functions of white blood cells

White blood cells help us to fight against pathogens inside the body. They defend us against infectious diseases.

Figure 3c.3 shows a kind of white blood cell that kills pathogens by **phagocytosis**. The cell is able to change its shape easily, putting out projections of cytoplasm that engulf the pathogen and surround it. The bacterium is then digested and destroyed by enzymes inside the white blood cell.

Other white blood cells secrete antibodies that help to destroy pathogens. These are described in Item B1c of Book 1 (*Keeping healthy*).

The functions of blood plasma

Blood plasma is a watery liquid. Many different substances are dissolved in it, and the plasma transports these around the body.

For example, **nutrients** such as glucose, amino acids and vitamins dissolve in the blood plasma after they have been absorbed from the small intestine. The blood carries them from the small

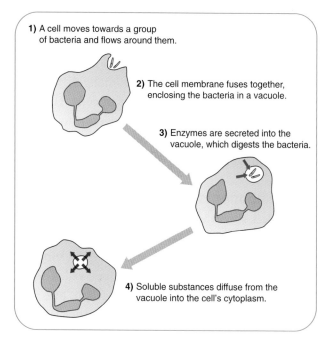

1) A cell moves towards a group of bacteria and flows around them.

2) The cell membrane fuses together, enclosing the bacteria in a vacuole.

3) Enzymes are secreted into the vacuole, which digests the bacteria.

4) Soluble substances diffuse from the vacuole into the cell's cytoplasm.

Figure 3c.3 Phagocytosis.

intestine to the tissues, where they diffuse into cells that need them. The plasma also transports **hormones**, **antibodies**, **water** and **waste products** (especially carbon dioxide and urea).

SAQ

3 Where do each of these substances enter the blood?

a The hormone insulin.

b Water.

c Carbon dioxide.

The circulatory system

The circulatory system is made up of the **blood vessels** and the **heart** (Figures 3c.4 and 3c.5).

SAQ

4 Using Figure 3c.4, describe the pathway that each of these substances would take as they are moved through the circulatory system:

a a glucose molecule that has been absorbed in the small intestine and is taken to the brain;

b a carbon dioxide molecule that has been produced in a leg muscle and is taken to the lungs.

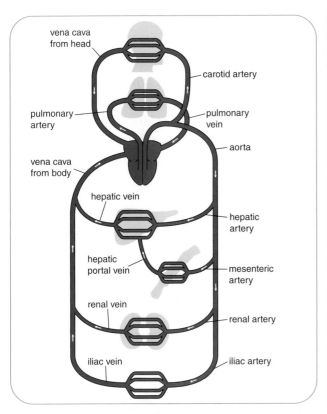

Figure 3c.4 Plan of the main blood vessels in the human body.

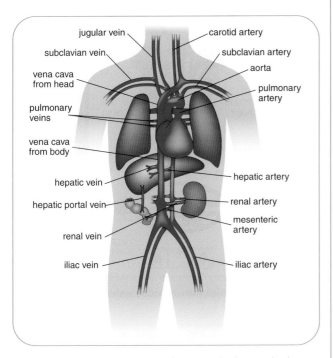

Figure 3c.5 The main arteries and veins in the human body.

Blood vessels

There are three types of blood vessels. **Arteries** carry high-pressure blood away from the heart. They divide again and again, eventually forming tiny vessels called **capillaries**. These join up to form **veins**, which carry blood back to the heart

(Figure 3c.6). The blood in the veins is at a much lower pressure than the blood in the arteries.

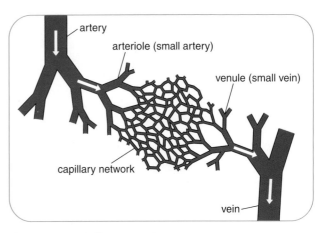

Figure 3c.6 A capillary network

Arteries have very thick, elastic walls (Figure 3c.7). They need these to withstand the high pressure of the blood as it leaves the heart. The blood pulses through the arteries – you can feel it doing this if you place the tips of your fingers against an artery in your neck. Each time the heart beats, the blood surges through the arteries and then slows down as the heart relaxes between beats. The artery walls are stretched outwards with each surge of blood and then recoil back to their normal size as the heart relaxes.

Capillaries have extremely thin and very permeable walls – they are only one cell thick. This allows rapid exchange of substances between the capillaries and the tissues. Oxygen and nutrients can quickly diffuse out of the blood inside the capillaries, while carbon dioxide and other waste material can diffuse in from the body cells. Most capillaries actually have tiny gaps in their walls, in between the cells, and this allows some of the blood plasma to leak out into the tissues, speeding up exchange even more.

By the time blood reaches the veins, it has lost most of its pressure and is flowing gently and smoothly. It doesn't pulse, so there is no need for the walls of the veins to be thick or elastic. Veins have thin walls. They have a larger 'hole' or **lumen** inside them than arteries, so that there is less resistance to the flow of blood.

Because there is so little pressure in the blood in the veins, it is easy for blood just to sit in them and not move back to the heart. This is especially

H

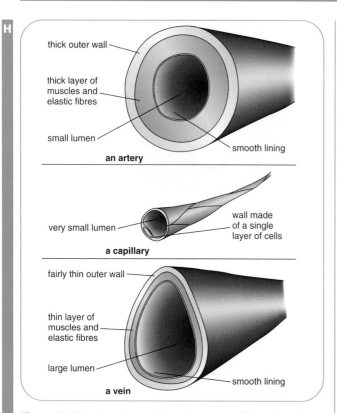

Figure 3c.7 Sections through the three types of blood vessels.

true of the leg veins, where blood has to move upwards against gravity. These veins lie next to some of the muscles that we use when we walk. When we contract the muscles, they squeeze inwards on the veins, pressing the blood along inside them. The blood can only go one way because the veins contain **valves** (Figure 3c.8). If you stay motionless for a long period of time – for example sitting in an aircraft seat on a long-haul flight – these muscles are less active and so the blood can just collect in the leg veins. When this happens, there is a risk that a clot may form, which is very dangerous if the clot reaches the lungs and blocks a blood vessel there.

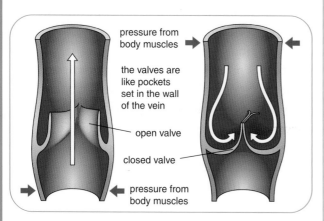

Figure 3c.8 Valves in a vein.

The heart

Your heart is about the same size as your fist. It is made almost entirely of muscle – a special kind of muscle that contracts and relaxes rhythmically throughout your entire life.

Figure 3c.9 shows the relationship between the heart and the rest of the circulatory system. Oxygenated blood is brought to the left side of the heart from the lungs, and is then pumped out of it into the **aorta**. This big artery takes blood towards the body tissues. When the deoxygenated blood leaves the tissues, it travels in the **venae cavae** (*singular*: vena cava) towards the right side of the heart. The heart pumps this blood to the lungs, where it becomes oxygenated once more.

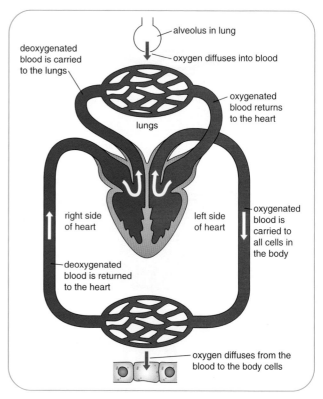

Figure 3c.9 Diagram of the circulatory system of a human.

Figures 3c.10 and 3c.11 show the structure of the heart. The left and right sides (and therefore the oxygenated and deoxygenated blood) are separated by a thick **septum**. On each side, there is an **atrium** at the top and a **ventricle** at the bottom.

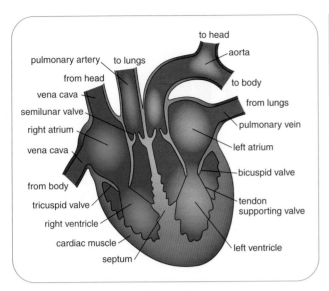

Figure 3c.10 Vertical section through a human heart.

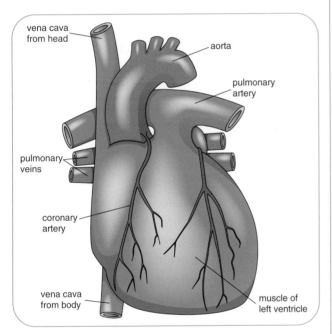

Figure 3c.11 External appearance of a human heart.

The atria receive blood from the veins – the **pulmonary vein** and the venae cavae – that have brought blood back to the heart. The blood flows down through the tricuspid and bicuspid valves into the ventricles. The ventricles then contract, squeezing inwards on the blood and pushing it up and out. The blood cannot go back into the atria, because the blood pushes upwards on the tricuspid and bicuspid valves, and makes them close. It therefore goes upwards into the **pulmonary artery** or the **aorta**.

If you look back at Figure 3c.4, you can see that the left ventricle pumps blood all around the body. The right ventricle only pumps blood to the lungs. The left ventricle therefore needs to have stronger muscle than the right ventricle. You can see in Figure 3c.10 that the wall of the left ventricle is quite a bit thicker than the wall of the right ventricle.

SAQ

5 Explain why the walls of the atria are much thinner than the walls of the ventricles.

The double circulatory system

Look back at Figure 3c.9 and put your finger inside the left ventricle of the heart. Follow the pathway of the blood as it completes one circuit and gets back to where you began. You will find that you have to go through the heart twice on this circuit – once through the left side and then once through the right side.

This arrangement is called a **double circulatory system**. It is found in mammals, but not in fish or amphibians. Why do we have such a system?

When the blood is pumped out of the right side of the heart, it goes to the lungs, where it becomes oxygenated. If we did not have a double circulatory system, the blood would then just carry on to the body tissues. This is what happens in fish. This works well in fish, but in mammals we have an improved version. After picking up oxygen at the lungs, the blood goes *back* to the heart so that it can be pumped once more before it travels on to the body tissues. This means that it travels faster and at higher pressure. It makes a more effective delivery system to the body cells.

Heart disease and treatment

Heart disease is one of the commonest causes of death in the UK. It causes 30% of deaths in men and 22% in women. Although we can never completely eliminate the risk of heart disease, there are things that we can do to reduce that risk. And if a heart does become so badly damaged that it may fail, it is sometimes possible to mend it by surgery, or even to replace it with a transplanted heart.

Cholesterol and heart disease

Cholesterol is a fat-like substance. It is needed in the body to make cell membranes. Nerve cells need especially large amounts of cholesterol.

We get some cholesterol in our diet, especially in foods that have come from animals. The liver can also make cholesterol from other kinds of fats that we eat, especially saturated fats (which again tend to be found in food from animals). So eating large amounts of saturated fats can lead to an increased quantity of cholesterol in the blood. However, the liver usually keeps the amount of cholesterol fairly constant, by making more of it if we don't eat as much as we need and making less if we already have plenty.

Figure 3c.12 These foods contain a lot of saturated fat.

Cholesterol can harm the circulatory system. If there is too much of it in the blood, it can build up on the walls of the arteries and block them. This can happen in the blood vessels that supply the heart muscle with the oxygen and nutrients that it needs. If this happens, the muscle may stop working properly and the heart stops beating. This is a heart attack. Heart attacks can be fatal.

Figure 3c.13 shows how cholesterol can block arteries. The cholesterol collects up in the artery wall, forming a stiff bulge called a **plaque**. The plaque restricts the space through which the blood can flow. This could slow the blood so much that it forms a clot. The plaque also makes the artery wall much stiffer than it should be, so it is less able to expand with each surge of blood and might even burst.

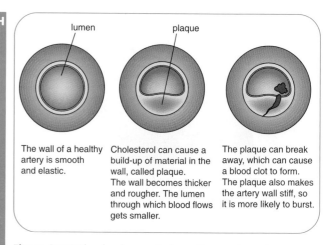

The wall of a healthy artery is smooth and elastic.	Cholesterol can cause a build-up of material in the wall, called plaque. The wall becomes thicker and rougher. The lumen through which blood flows gets smaller.	The plaque can break away, which can cause a blood clot to form. The plaque also makes the artery wall stiff, so it is more likely to burst.

Figure 3c.13 The development of an atheromatous plaque.

Treating a damaged heart

In many cases, surgery can mend damage that has been done to the heart.

Damaged valves Heart valves may become damaged during a person's lifetime. Some children are born with valves that don't work properly right from the start.

A damaged heart valve can be removed and then replaced by a replacement valve. The new valve could have been taken from a human donor or it might be made of metal.

Using a valve from a donor is normally the option that a surgeon prefers, but there are never enough donor hearts for all the people who need them. So, a metal valve could be put into the heart. These work well but a person with one will probably have to take anti-clotting drugs all their life because the blood tends to clot when it comes into contact with metal. And, if the valve has been placed in a child's heart, it may need to be replaced with a larger one as the child grows.

Faulty pacemaker The **pacemaker** is a little patch of muscle in the right atrium of the heart that sets the pace at which the rest of the heart beats. Pacemakers sometimes become faulty. They can be replaced using artificial pacemakers. The new pacemaker doesn't need to be actually in the heart, but can simply be placed under the skin.

One drawback with an artificial pacemaker is that it needs a power supply, and this will need replacing from time to time. Most pacemakers are now powered by lithium batteries, which can last for years before they run out.

Pacemakers used to be quite bulky but now they are only about the size of a small matchbox. All the same, this would be quite large if it needed to be used for a small child.

Badly damaged heart   Sometimes a heart is in such poor condition that the only way a person can be kept alive is to give them a **heart transplant**. This is only done as a last resort, when all other treatment has failed. It is a very big and dangerous operation. Moreover, there is a shortage of donors, so a patient might have to wait for a long time before a suitable heart is available for them. This shortage arises because:

- the heart must come from someone who has just died;
- the heart must come from someone who had a healthy heart when they died and who was not too old;

Figure 3c.14 A heart transplant operation in a young child.

- the tissues of the heart must be a close match to the tissues of the patient, so that the body is less likely to reject them (see Figure 3c.14);
- the heart must be a suitable size to put into the patient.

Surgeons are now able to carry out the actual operation with a high success rate. But problems may arise later. The biggest difficulty is that the body may reject the heart. This happens because the white blood cells recognise the cells in the heart as 'foreign', as though they were a pathogen that has got into the body. They attack the heart cells and destroy them.

To stop this happening, the patient has to take drugs that stop the white blood cells doing this. They are called **immunosuppressant drugs**. They work well but they leave the patient more vulnerable than usual to infectious disease. The patient will need to take them for the rest of his or her life.

H *SAQ*

6 Use the information in the section **Treating a damaged heart** to make a summary of the advantages and disadvantages of a heart transplant compared with replacing or repairing heart valves, or providing an artificial pacemaker.

An impossible choice

Mary and Jodie were twins. They were born in August 2000 to a Maltese couple. Mary and Jodie were conjoined twins – their bodies were joined together and they shared some of their organs.

The hospital in Malta that was looking after the twins' mother recognised early in her pregnancy that her twins were conjoined. They arranged that she would give birth in Manchester, UK, because there was a team of surgeons there who were very experienced at separating conjoined twins.

Mary's and Jodie's bodies were joined at the bases of their spines. They had only one heart between them. The heart was really Jodie's but it did the work of pumping blood to both the girls' bodies.

The surgeons and parents faced a terrible dilemma. Mary could only continue to live if she remained joined to Jodie. But Jodie could possibly live a good life – if she was separated from Mary.

continued on next page

An impossible choice – *continued*

There was a possibility that Mary could be given a heart transplant, and this was carefully considered. But there was only a tiny chance that a suitable heart would be found. There is always a shortage of hearts for transplantations, and this is especially so when tiny babies are concerned, because obviously they cannot be given an adult heart. It would have been difficult to justify giving a heart to Mary, who probably would not survive anyway, rather than to another baby with a better chance of survival. The twins' parents did not want them to be separated. But the doctors felt that this was the wrong decision. They believed that, if the twins were not separated, neither of them would survive for very long. The extra strain on Jodie's heart of doing the work for two bodies was likely to lead to heart failure.

The problem was taken to the High Court and then to the Court of Appeal. Eventually the decision was made that the twins should be separated. This was done and, as expected, Mary died. Jodie survived and now lives with her parents in Malta.

Summary

You should be able to:

- describe the structure of blood and the functions of red cells, white cells, platelets and blood plasma

- explain how the structures of red blood cells and white blood cells are adapted to their functions

- (H) describe where and how haemoglobin is converted to oxyhaemoglobin, and vice versa

- outline the functions of arteries, veins and capillaries

- (H) explain how arteries, veins and capillaries are adapted for their functions

- describe the structure and function of the heart

- explain why the wall of the left ventricle is thicker than the wall of the right ventricle

- (H) explain why we have a double circulatory system

- explain how diet can affect the amount of cholesterol in the blood

- (H) explain how too much cholesterol in the blood can cause damage to arteries

- describe how damaged parts of the heart can be replaced mechanically (using plastic or metal components) or biologically (using donated organs)

- describe the problems associated with heart transplants

- describe the problems associated with using mechanical parts to repair the heart

- (H) describe the advantages and disadvantages of a heart pacemaker or replacement heart valves, compared with a heart transplant

Questions

1   Name the component of human blood that carries out each of the following functions:

    a   transports hormones in solution;

    b   helps with blood clotting;

    c   destroys pathogens;

    d   transports oxygen.

2   a   Describe *four* ways in which a red blood cell is adapted for its function.

    b   Describe *one* way in which a white blood cells is adapted for its function.

3   Copy Figure 3c.15, which is a diagram of a section through a human heart.

    a   Name parts A to F.

    b   Draw arrows on the diagram to show the directions in which blood flows through it.

    c   Lightly colour the parts of the heart that contain oxygenated blood red, and deoxygenated blood blue.

    d   Explain why the wall of the left ventricle needs to be thicker than the wall of the right ventricle.

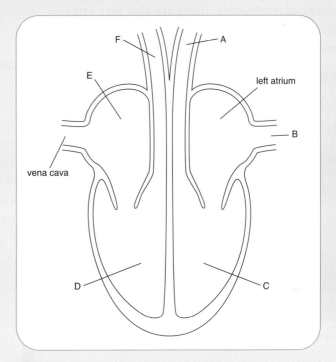

Figure 3c.15

4   Copy and complete Table 3c.1 to compare the structure and function of arteries, veins and capillaries. There will be quite a lot to write in some of the cells in the table. The last row needs a bit more room than the others.

Type of vessel	artery	vein	capillary
Function			
Structure of wall			
Width of lumen			
How it is adapted for its function			

Table 3c.1

Cells and bodies

How many cells are there in a human body? No one has ever counted them all, but we think it is somewhere between 10 000 000 000 000 (10^{13} or 10 trillion) and 50 000 000 000 000 (5×10^{13} or 50 trillion). That's a lot of cells. Why do we have so many and where did they all come from?

Humans are **multi-cellular** animals. 'Multi-cellular' means 'many-celled'. Most of the organisms that we are familiar with are multi-cellular. Plants, insects, worms and birds are all multi-cellular. However, there are also very large numbers of single-celled organisms. Bacteria, yeasts and amoebas are all single-celled.

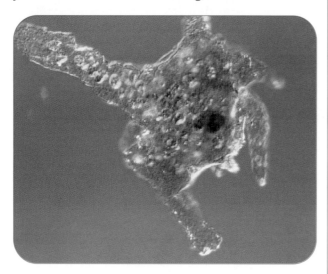

Figure 3d.1 This is a single-celled organism called *Amoeba*. It is about 0.1 mm across. Amoebas live in ponds, where they crawl around and feed on other, even smaller, single-celled organisms.

We are not very aware of all the single-celled organisms around us because they are too small to see with the naked eye. And that is the drawback of being single-celled – they cannot grow very big. Imagine one huge cell the size of your body. How would nutrients get to every part of it without a blood system? How would oxygen get into the middle of it without any lungs? How would it be supported and held in shape without a skeleton?

Being multi-cellular has several advantages.

- The ability to grow large. There is theoretically no limit on how big a multi-cellular organism can get. The more cells there are in the body,

the bigger it can be. (If we have 10 trillion cells in our bodies, imagine how many there must be in a blue whale, which is about 25 m long.)

- Having specialised cells. In a single-celled organism, the one cell has to do everything. In a multi-cellular organism, there can be many different kinds of cells, each specialised to do one thing really well. This is called **cell differentiation**.

- Being more complex. Our bodies are made of many different organs, working together to form many different systems. This couldn't happen if we were made of only one cell.

Perhaps the biggest problem of having just one large cell is related to its surface area to volume ratio.

It's easier to visualise this if we think about a cube-shaped animal (Figure 3d.2). As you increase the size of a cube, both its surface area and its volume increase. But the volume increases much more than the surface area, so the bigger the cube, the smaller the ratio of its surface area to its volume.

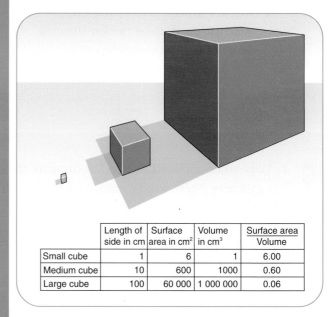

	Length of side in cm	Surface area in cm²	Volume in cm³	Surface area / Volume
Small cube	1	6	1	6.00
Medium cube	10	600	1000	0.60
Large cube	100	60 000	1 000 000	0.06

Figure 3d.2 Surface area to volume ratio.

This is important because living things interact with their environment at their surfaces. In particular, they get oxygen through their surface. If you are very small, this works well – there is plenty of surface available to supply your whole

H volume with oxygen, just by diffusion. But as you get bigger, this doesn't work. The biggest cube in Figure 3d.2 could not supply all of its volume with oxygen through its surface – there's not enough surface and too much volume. However, if this big cube was made up of lots of smaller cubes, each with their own surfaces through which materials such as oxygen could be supplied, all of its body could get what it needs.

That's how our bodies work. We can use some of our trillions of cells to produce extra surfaces in contact with the environment – in particular, the surfaces of the alveoli in the lungs to take up oxygen, and the surfaces of the villi in the small intestine to take up food.

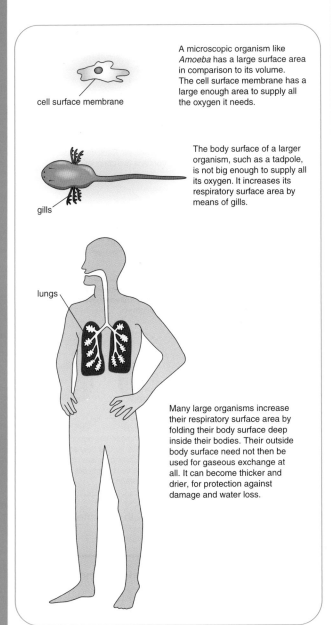

A microscopic organism like *Amoeba* has a large surface area in comparison to its volume. The cell surface membrane has a large enough area to supply all the oxygen it needs.

cell surface membrane

The body surface of a larger organism, such as a tadpole, is not big enough to supply all its oxygen. It increases its respiratory surface area by means of gills.

gills

lungs

Many large organisms increase their respiratory surface area by folding their body surface deep inside their bodies. Their outside body surface need not then be used for gaseous exchange at all. It can become thicker and drier, for protection against damage and water loss.

Figure 3d.3 Gas exchange surfaces.

Mitosis

We each begin life as a single cell but, by the time we are adults, we contain trillions of cells. All of these cells are produced by division of the first cell. The process by which they divide is called **mitosis**.

Human cells contain 46 chromosomes, two sets of 23. A cell with two sets of chromosomes is called a **diploid** cell (Figure 3d.4).

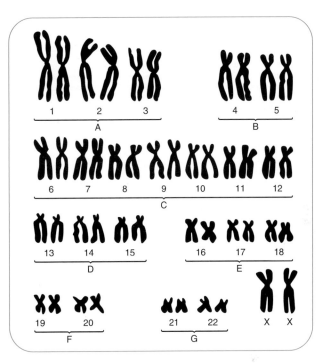

Figure 3d.4 Chromosomes of a normal woman, arranged in order. The two chromosomes in a pair are said to be homologous.

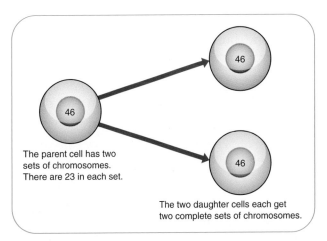

The parent cell has two sets of chromosomes. There are 23 in each set.

The two daughter cells each get two complete sets of chromosomes.

Figure 3d.5 The result of mitosis in a human cell.

When a cell divides by mitosis, each new cell gets the full two sets of chromosomes. The new cells are diploid, just like their parent cell. Mitosis produces new cells that are genetically identical to each other and to their parent cell.

1) This cell has four chromosomes.

2) Before the cell divides an exact copy is made of each chromosome. The two copies called chromatids stay firmly attached to each other at this stage.

3) When the cell divides, the chromatids of each chromosome separate from each other. Each cell gets a complete set of chromosomes.

4) Before the cell can divide again, the chromosomes are copied again.

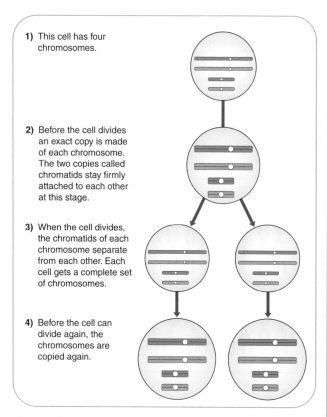

Figure 3d.6 Mitosis.

SAQ

1   Fruitflies have diploid cells, each containing eight chromosomes.

   a   How many sets of chromosomes are there in a fruitfly cell?

   b   How many chromosomes are in each set?

   c   When a fruitfly cell divides by mitosis, how many chromosomes will each new cell get?

Figure 3d.7 *Kalanchoe* can reproduce asexually.

Mitosis is the kind of cell division that our cells undergo whenever we need new cells.

● They are for *growth* of the body.

● They *replace* old cells that have been worn out.

● They *repair* damaged tissues.

Some living organisms also use mitosis to reproduce. For example, the plant *Kalanchoe* produces tiny new plants along the edges of its leaves. The plantlets fall off and grow roots into the soil. These new plants are genetically identical to each other and to the parent plant.

Figure 3d.8 summarises what happens to the chromosomes during mitosis. Only four chromosomes are shown. (If it showed all 46, you would not be able to see what is going on.)

1) When the cell is not dividing, no chromosomes can be seen clearly in the nucleus. They are there, but are so long and thin that they are invisible.

2) The chromosomes get short and fat, so they can now be seen with a light microscope. Each chromosome contains two chromatids.

3) The nuclear membrane vanishes. The chromosomes line up on the equator.

4) The two chromatids separate. The chromatids move away from each other.

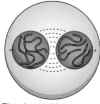
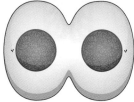

5) The chromatids arrive at opposite ends of the cell, and form into groups. A nuclear membrane appears round each group.

6) The chromosomes become long and thin again, so that they are invisible. The cytoplasm divides, forming two daughter cells. Each cell now goes back to stage 1.

Figure 3d.8 Mitosis in an animal cell with four chromosomes.

SAQ

2   **a**   What substance are chromosomes made of?

   b   What must have happened to this substance before stage 2 in Figure 3d.8?

Sexual reproduction

Figure 3d.9 summarises what happens during sexual reproduction in humans. Two sex cells or **gametes** fuse together to form a **zygote**. The zygote is the first cell of a new human being. It will divide again and again by mitosis to produce all of the cells in the person's body.

The zygote must therefore have the correct number of chromosomes. It must be a diploid cell, with two complete sets of chromosomes.

The two gametes must therefore have only *one* set each. A cell with one set of chromosomes is called a **haploid** cell. Gametes are haploid. When fertilisation happens, the fusion of the two gametes produces a diploid zygote.

SAQ

3 a Fruitflies have eight chromosomes in their body cells. How many chromosomes are there in a fruitfly gamete?

b How many chromosomes are there in a fruitfly zygote?

Meiosis

Sperm are produced when cells in the testis divide. Eggs are produced when cells in the ovary divide. The kind of cell division that is involved is called **meiosis**.

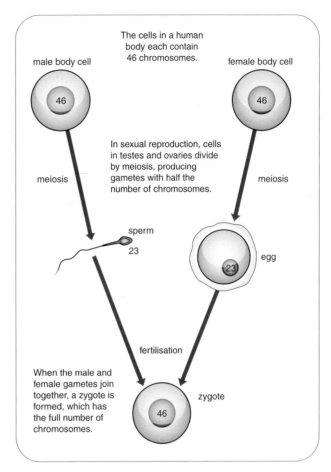

Figure 3d.9 Sexual reproduction.

Triploid zygotes

When fertilisation happens in humans, a sperm containing 23 chromosomes fuses with an egg that also contains 23 chromosomes. The zygote therefore gets 46 chromosomes.

However, sometimes things go wrong. The zygote gets not the two sets of chromosomes it should have but *three* sets. It has 69 chromosomes instead of 46. It is said to be **triploid**.

How can this happen? Sometimes two sperm fertilise an egg at the same time. This isn't meant to happen, but just occasionally it does. So the zygote gets a set of chromosomes from each of the two sperm and another set from the egg – three sets in all.

Another way in which a triploid zygote can be produced is if the sperm or egg is faulty.

Sometimes sperm or eggs are produced that have two sets of chromosomes instead of one. They are diploid when they should be haploid. If a diploid sperm fuses with a haploid egg then the zygote will be triploid. A recent study has found that about 1 in every 250 sperm is diploid.

Scientists think that about 1% of human zygotes are triploid. Most of these never develop, or they are lost as miscarriages very early in the pregnancy. They are thought to account for about one-fifth of all miscarriages. Just occasionally, the foetus does develop right through pregnancy and so a baby with triploid cells is born. These children are severely disabled and die before they are one year old.

Meiosis produces haploid cells from diploid ones. These cells are not the same as each other, or the same as their parent cell. They are genetically different. Not only do they have only half the number of chromosomes as their parent cell but also the genes that are carried on these chromosomes will have got mixed up in different ways. Meiosis produces genetic variation in the new cells that are formed.

Figure 3d.10 shows how a diploid cell divides by meiosis.

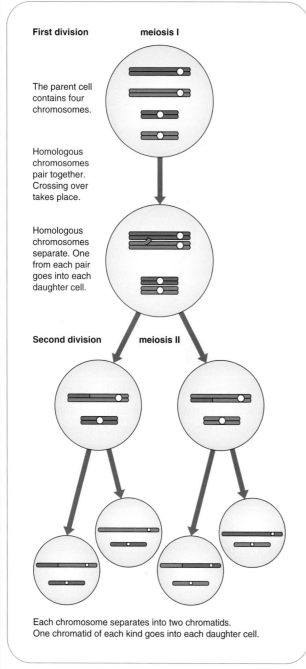

First division   **meiosis I**

The parent cell contains four chromosomes.

Homologous chromosomes pair together. Crossing over takes place.

Homologous chromosomes separate. One from each pair goes into each daughter cell.

Second division   **meiosis II**

Each chromosome separates into two chromatids. One chromatid of each kind goes into each daughter cell.

Figure 3d.10 How a diploid cell divides by meiosis.

H **SAQ** _____

4 In meiosis, the cell divides twice. Which division, the first or the second, is the one that halves the number of chromosomes?

5 Could meiosis take place in a haploid cell? Explain your answer.

Sperm cells

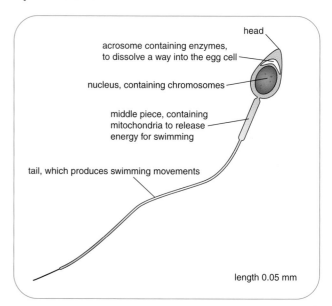

head

acrosome containing enzymes, to dissolve a way into the egg cell

nucleus, containing chromosomes

middle piece, containing mitochondria to release energy for swimming

tail, which produces swimming movements

length 0.05 mm

Figure 3d.11 A sperm cell.

Sperm cells are very tiny, normally only about 0.05 mm long. This makes it easier for them to swim fast and for a long time, because they have less mass to move. It also means that very large numbers of them can be produced, which increases the chance that at least one of them will fertilise an egg.

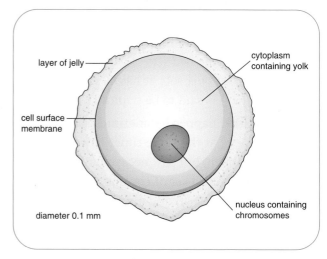

layer of jelly

cytoplasm containing yolk

cell surface membrane

nucleus containing chromosomes

diameter 0.1 mm

Figure 3d.12 An egg cell.

Egg cells

Figure 3d.12 shows an egg cell. Egg cells are much larger than sperm. You can see their relative sizes in Figure 3d.13. The egg cell is large because it contains food stores to provide the zygote with nutrients if the egg cell is fertilised. It takes a few days for the zygote to travel from the oviducts (where fertilisation happens) into the uterus. During that time, it has to depend on its own food stores. Later, when the placenta has formed, the foetus will get its nutrients from its mother's blood.

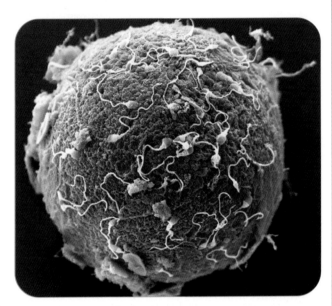

Figure 3d.13 Sperm fertilising an egg.

Summary

You should be able to:

◆ explain the advantages of being multi-cellular

Ⓗ ◆ explain the significance of surface area to volume ratio in single-celled and multi-cellular organisms

◆ state that cells divide by mitosis for growth, replacement and repair

Ⓗ ◆ outline what happens during mitosis, and state that mitosis produces genetically identical cells

◆ state that mammalian body cells are diploid

◆ explain why gametes must be haploid

◆ state that gametes are produced by meiosis, which leads to haploid cells and genetic variation

Ⓗ ◆ outline what happens during meiosis

◆ describe adaptations of sperm and egg for their functions

Questions

1 a State *three* things that a sperm and an egg have in common.

  b State *two* features of a sperm that are not found in an egg. Explain how these features adapt the sperm for its function.

2 a Explain the meaning of the term *multi-cellular*.

  b Give *one* example of a multi-cellular organism.

  c Give *three* advantages of being multi-cellular.

Ⓗ 3 Copy and complete Table 3d.1 to summarise the differences between mitosis and meiosis in humans.

	Mitosis	Meiosis
Where does it takes place?		
What is the effect on chromosome number?		
Do the chromosomes pair up?		
How many times does the cell divide?		
How many cells are produced?		
Are the new cells genetically identical?		

Table 3d.1

Growth

Growth is a permanent increase in size. You have grown since you were born – you have got taller and heavier. You might get a bit fatter for a while and then get slim again. That is not proper growth, because it is not a *permanent* increase in size.

Multi-cellular organisms grow by producing more cells. First a cell gets larger, then it divides to form two new cells. This happens in both plants and animals, although there are some differences in the way that growth takes place in them.

SAQ

1 Name the kind of cell division that takes place in growth.

Animal cells and plant cells

Figures 3a.1 and 3e.1 show the structures of an animal cell and a plant cell.

Both animal and plant cells have:

- a **nucleus**, which contains chromosomes (DNA)
- a **cell surface membrane**, which controls what enters and leaves the cell
- **mitochondria**, where aerobic respiration happens.
- **cytoplasm**, where metabolic reactions take place

Plant cells also have some 'extra' structures.

- **Chloroplasts**, which contain the green pigment **chlorophyll**. Chlorophyll absorbs light energy for photosynthesis. Not all plant cells have chloroplasts.
- A large **vacuole**, which contains a sugary liquid called **cell sap**. When the vacuole is full, it helps to keep the cell firmly in shape and gives it support.
- A **cell wall**, which provides support for the cell.

Looking at onion cells

Cells are very small, so you need a microscope to see them. A good place to find plant cells to look at is inside an onion. You can use the thin tissue inside the onion to make a microscope slide. This is what you do.

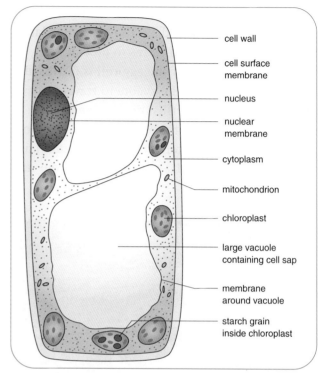

Figure 3e.1 A plant cell.

- Put a little drop of dilute iodine solution into the centre of a clean microscope slide.
- Cut a small piece from an onion bulb. Use forceps (tweezers) to peel away a piece of the very thin skin that you will find on the inside of each layer.
- Carefully place the onion skin in the drop of iodine. Try to keep it flat, and push it down gently so that it goes right into the liquid. Then put a second small drop of iodine solution on top of the onion tissue.
- Gently lower a cover slip (a really thin square piece of glass) onto the piece of onion. Press it down *very* gently until the onion skin is really flat. If iodine solution oozes out around the sides, mop it up with a piece of filter paper.
- Look at your slide under the microscope. You should be able to see cells rather like those in Figure 3e.2. You may be able to see dark spots inside them. These are starch grains that have been stained blue–black by the iodine solution.

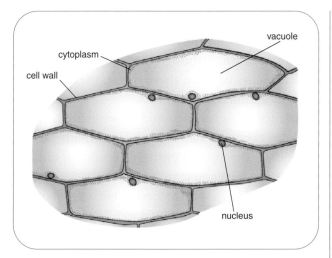

Figure 3e.2 Onion cells as seen through a light microscope.

SAQ

2 Cells in an onion bulb don't have chloroplasts. Why would chloroplasts be of no use to them?

Growth in animals and plants

When a person reaches the age of about 19 or 20, they stop growing in height. Most animals grow to a certain size and then stop growing. Many plants, by contrast, can keep on growing all their lives.

There are some other ways in which growth in plants differs from growth in animals.

We have seen that growth involves cell enlargement followed by cell division. In animals, cell enlargement only happens up to a point. The main way that we get bigger is by making more cells. In plants, cell enlargement is much more significant. A lot of their growth is caused by cell enlargement.

In animals, cells in most parts of the body are able to divide and cause growth. But in plants, only certain parts contain cells that can divide. These places are mostly at the tips of the shoots and the tips of the roots. This means that plant shoots normally grow upwards and in a branching shape. Trees also have a ring of cells that can divide, just beneath the surface of their woody trunks, so that the trunk can grow sideways and get thicker.

Figure 3e.3 shows how a root grows. The place where the cells can divide is just behind the tip. The new cells that have just been made here are small. As they get older, they grow larger. New,

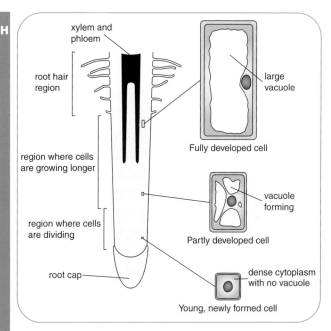

Figure 3e.3 Longitudinal section through a root tip.

small cells are constantly being made near the end of the root, while the ones just above this area grow longer. The overall effect is to make the root get longer.

SAQ

3 **a** Suggest why there is a tough root cap at the tip of the growing root.

   b In which part of the root is cell differentiation taking place?

Differentiation

As you grew from a single-celled zygote, many different kinds of cells developed. The change of a cell from a 'general purpose' one into a cell specialised for a particular function – such as a neurone or a muscle cell – is called **differentiation**.

Most of our cells differentiate early on in our lives. Once they have differentiated, they can't change into any other kind of cell. Once a cell has differentiated into a muscle cell, it will remain a muscle cell to the end of its life. That isn't true of plants. Many plant cells retain the ability to differentiate throughout the plant's life.

A few cells in your body don't differentiate. They keep their options open. They are called **stem cells**. Stem cells are able to divide to produce specialised cells.

H

For example, we have stem cells in our bone marrow. These stem cells divide to form new cells, which then differentiate into different kinds of blood cells.

Stem cells may hold the key to the treatment of a number of diseases. For example, if a person has suffered damage to their spinal cord, the cells in the cord are not able to divide and repair the damage. If some stem cells could be inserted into the place near the damage, maybe they could divide and make new neurones, and mend the cord.

One problem with developing this kind of treatment is getting enough of the right sort of stem cells. The best place to find them is in an early embryo because, at that stage of growth, not many of the cells have differentiated. They are still mostly stem cells. What's more, they are able to produce almost any of the different kinds of specialised cells in the body.

But getting stem cells from early embryos raises a lot of moral and ethical questions. Where do the embryos come from? When a woman is given fertility treatment, especially IVF (*in vitro* fertilisation, which many people know as 'test tube babies'), several eggs are usually fertilised and so several embryos are formed. Only one or two of these will be implanted into her uterus. The rest are all 'spare'. However, many people are uncomfortable with the idea of taking stem cells from these embryos, because they feel that they are already small people and that it is ethically unacceptable to use them in this way.

Another possibility is to find stem cells in the person's own body and use those. This is difficult because (apart from the ones in bone marrow that can make blood cells) we don't have very many stem cells and they are not easy to find and collect.

Despite these problems, stem cell research is moving on fast, and it is probable that stem cells will become an important tool in doctors' and surgeons' armouries, to help to cure currently incurable diseases.

SAQ

4 Explain how a stem cell differs from other body cells.

Growth in humans

We pass through several different stages between birth and death. Although there are no clear cut-off points between these stages, they can be roughly summarised like this.

- **Infancy** This is the period between birth and about two years old. Growth is normally fast in infancy. The child is entirely dependent on adults to feed and care for her.
- **Childhood** This is the period between about 3 years old and 10–11 years old. During this stage, the child's body grows and her brain develops. She learns quickly. She becomes more able to do things for herself.
- **Adolescence** This is the period between about 10–11 years old and about 17. During adolescence, the reproductive system develops. The stage at which the person becomes sexually mature is called **puberty**. Adolescence, like childhood, is a time when the brain is developing fast. Both girls and boys have a growth spurt during adolescence, which tends to happen earlier for girls than for boys.
- **Maturity and old age** Once adolescence is over, growth stops. The person's body is now mature and will not grow in height any more. During the early years of maturity, the body is usually at its peak of health and fitness. As the person ages, body organs gradually become a little less efficient at their tasks. By the time a person has reached 60 or 70, they are usually aware that they cannot do things that they used to be able to do – for example, kicking a football or running to catch the bus. This is old age.

Growth in the uterus

A lot of growth takes place even before a baby is born. The time that it spends in the uterus is called the **gestation period**. In humans this is about 9 months.

Parkinson's disease

It was Russell's friends who noticed it first. His hands always seemed to be trembling slightly and, after a while, he found that he had difficulty in holding a cup still enough to be able to drink from it. Never one to complain, he waited until it was really getting bothersome before he went to see his doctor. By then, his friends had noticed that he didn't seem to be making his normal facial expressions – his smile didn't look right.

Russell was thrown into shock when he was diagnosed with Parkinson's disease. He was only 30. He had always thought of Parkinson's as being an old man's illness. He knew it was a progressive and incurable disease which he would have to battle with for the rest of his life.

Parkinson's disease is caused by the death of cells in the brain. These cells usually produce a transmitter substance called dopamine, which helps brain cells to communicate with one another. Without dopamine, nerve impulses cannot be sent to muscles. The person loses control of their movements.

Parkinson's disease is treated by giving the person a drug called levodopa, which changes into dopamine in the brain. Russell found this helped enormously – he had much better control of his movements. However, it is very difficult to get the dose of dopamine exactly right. If you give the patient too much, their body begins to move with uncontrollable, jerky movements. Moreover, the dosage required gets greater as the person gets older. It may eventually be impossible to find a suitable dose.

A much better way of treating Parkinson's disease would be to replace the lost cells in the brain. Experiments have been tried in mice and monkeys in which embryonic stem cells have been treated so that they differentiate into dopamine-producing cells, which were then successfully transplanted into their brains. Russell is watching the progress of the research on this, and hoping against hope that he might be able to try out the new treatment before his disease has advanced so much that the damage in his brain cannot be repaired.

Figure 3e.4 Mohammed Ali and Michael J. Fox both have Parkinson's disease.

SAQ

5 Table 3e.1 shows the gestation periods of several different animals, and also the average body mass of an adult.

   a Arrange the animals in decreasing order, according to their body mass.

   b Can you see any relationship between the body mass and the gestation period?

   c Suggest reasons for your answer to **b**.

Animal	Gestation period in days	Adult body mass in g
squirrel	30	500
cow	280	800 000
cat	60	3 000
horse	340	700 000
elephant	645	4 000 000
mouse	20	30
lion	110	180 000
rhinoceros	450	2 000 000
goat	150	40 000
chipmunk	30	100

Table 3e.1

Differential growth

As a foetus grows, not all the parts of its body grow at the same rate. This is also true of babies as they grow into children.

Figure 3e.5 shows what happens. Each drawing is drawn to the same size so you can pick out the different proportions of the body parts more easily.

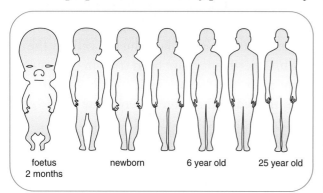

foetus   newborn   6 year old   25 year old
2 months

Figure 3e.5 Differential growth.

SAQ

6   Use Figure 3e.5 to answer these questions.

   a   Which part of the 2-month old foetus has grown most rapidly? Can you suggest why this part of the body develops so early on?

   b   If you could see inside the body, the reproductive system would look really small until the person is an adolescent, when it would begin to look a bit bigger. Can you explain why the reproductive system grows much later than other parts of the body?

7   The graph in Figure 3e.6 shows the body mass of a boy between the ages of 2 years and 20 years.

   a   Give the mass of the boy when he was 10 years old.

   b   By how much did his mass increase between the age of 10 years and 20 years?

   c   During which years was he growing most quickly? Choose from:

      ●   2 to 4 years

      ●   11 to 13 years

      ●   18 to 20 years.

   d   To which life stage does your answer to part **c** correspond?

   e   Does the graph suggest that the boy had finished growing when he was 20 years old? Explain your answer.

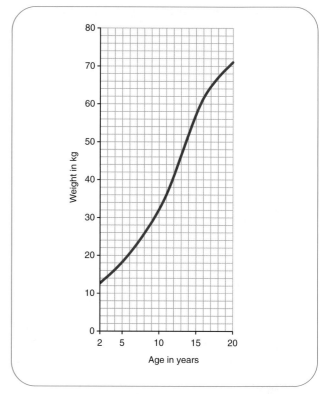

Figure 3e.6 Growth in a boy.

8   Table 3e.2 shows the body mass of a baby girl, and the circumference of her head, between birth and when she was 3 years old.

Age in months	Body mass in kg	Circumference of head in cm
0 (at birth)	3.5	34.5
4	6.2	40.5
8	8.2	43.2
12	9.5	44.6
16	10.6	46.0
20	11.4	46.4
24	12.0	47.1
28	12.7	48.0
32	13.4	48.3
36	13.8	48.4

Table 3e.2

   a   Plot these data as two graphs. (You could draw them on the same graph paper, so that they share the same *x*-axis. Ask for help if you would like to do this and don't know how.) Draw best-fit lines.

b Between which ages did the girl's head grow most rapidly?

c Between which ages did the girl's body mass increase most rapidly?

Making use of growth data

Data about the head size and body mass of a baby can help to check that it is growing normally. All babies tend to follow a similar pattern in their growth even if they were very small or very big at birth.

Babies don't grow as 'smoothly' as the growth curve in Figure 3e.6. The 'ideal' growth curves that you can look up to see how a baby should grow have been drawn using lots of measurements on lots of babies, from which means have been calculated. For any one baby, growth often goes along in fits and starts. They might grow a lot in one month and then not seem to grow at all for the next few. That is normal.

Keeping track of how a baby's body mass changes can give early warning of potential health problems. For example, if a baby is growing unusually quickly, it might simply be that it is eating too much, or it might be that there is a problem that needs to be investigated. A doctor might decide to do a blood test to check that the thyroid gland is working properly, or perhaps a chromosome check to make sure there isn't a genetic disease that requires treatment.

Some babies might have a head size that is increasing unusually quickly or slowly. That could just be something that runs in the family and is not going to cause any problems. But it might be a sign of something going wrong with the development of the brain. Early investigation and diagnosis could make a big difference to the success of treatment.

Summary

You should be able to:

◆ describe the structure of an animal cell and of a plant cell

◆ describe the functions of chloroplasts, vacuole and cell wall in a plant cell

◆ make a stained slide of onion cells

◆ state that animals grow to finite size but plants often grow continuously

◆ describe how plant growth differs from animal growth

◆ state that growth involves cell division and differentiation

◆ explain what stem cells are

◆ discuss issues relating to stem cell research

◆ explain why different animals have different gestation periods

◆ plot data on babies' mass and head size, and interpret these data

◆ explain how these data can be used

◆ state the five main phases of human growth

◆ interpret data on human growth

Questions

1 Give the terms that match these descriptions. Choose from:

adolescence   birth   blood   childhood   chlorophyll   chloroplast
differentiation   expansion   gestation   growth   maturity   stem

a a permanent increase in size

b a cell that is capable of dividing to form specialised cells

c the development of a general-purpose cell into a specialised cell

d a structure in a plant cell where photosynthesis takes place

e the stage of human growth in which the reproductive system matures

f the time that a foetus spends in the uterus.

2 Table 3e.3 shows the body mass of a girl between the ages of 2 years and 20 years.

a Plot these data on a graph, like the one in Figure 3e.6. Draw a best-fit line.

b How are the overall shapes of Figure 3e.6 and your graph similar to one another?

c Describe *two* differences between the graph for the girl and the graph for the boy.

Age in years	Body mass in kg
2	13
4	17
6	22
8	28
10	36
12	45
14	54
16	58
18	62
20	64

Table 3e.3

3 Copy and complete Table 3e.4 to compare plant growth and animal growth.

Feature of growth	Plants	Animals
when do they grow?		
which parts of their bodies grow?		
how important is cell enlargement?		
at which stage of growth can cells differentiate?		

Table 3e.4

4 Use the Internet to research some recent news about the use of stem cells in medicine. Choose *one* issue. Describe the issue and explain the different views that people have about it. What are *your* views? Can you justify them?

Controlling plant growth

Plant hormones

You probably know that hormones are produced in humans. A **hormone** is a chemical that is produced in one part of the body and has an effect on target organs in another part. For example, the hormone insulin is produced in the pancreas and has an effect on the liver, making the liver take up more glucose from the blood.

Plants also have hormones. Like animal hormones, they are usually produced in one part of the plant and have their effects in another part. Like animal hormones, they help to coordinate the actions of different parts of the body. Plant hormones:

● control the growth of shoots and roots;
● make a plant flower or stop it from flowering;
● help fruits to ripen.

Phototropism

If you place a pot plant in a window, it often ends up leaning towards the window, where the light is coming from. This makes sense, because plants need light for photosynthesis. The closer the leaves are to the window, the more light they can get and the more the plant can photosynthesise.

Plants can't get up and move around so they can't actually *move* towards the light. Instead, they *grow* towards it. This response is called a **tropism**. A tropism is a growth response to a stimulus, where the direction of the response is related to the direction of a stimulus.

The growth of a shoot towards the light is called **phototropism**. 'Photo' means 'light'. Because the shoot grows *towards* the light, the response is **positive phototropism**. Some plant roots tend to grow *away* from the direction of any light that is falling on them and this is called **negative phototropism**.

SAQ

1 a i In Figure 3f.1, what variables have been kept the same in this experiment?

   ii Why is it important to keep these variables the same?

b Which variable has been changed in Figure 3f.1? How has this been done?

c Compare the colours of the three sets of seedlings at the end of the experiment. Suggest reasons for the differences.

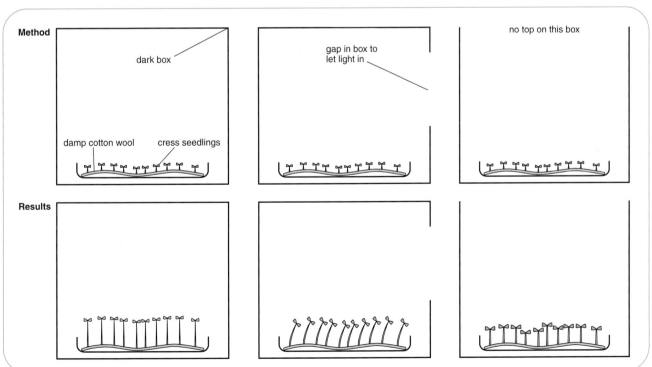

Figure 3f.1 An experiment to show that shoots grow towards the light.

Auxin

Auxin is a plant hormone that is involved in the growth of a shoot towards the light. Figure 3f.2 shows how it works.

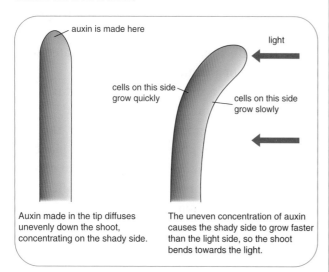

auxin is made here

light

cells on this side grow quickly

cells on this side grow slowly

Auxin made in the tip diffuses unevenly down the shoot, concentrating on the shady side.

The uneven concentration of auxin causes the shady side to grow faster than the light side, so the shoot bends towards the light.

Figure 3f.2 Auxin and phototropism.

Auxin works by making plant cells in a shoot get longer. The auxin loosens the cell walls, so it is easier for a plant cell to stretch lengthways if it takes up extra water, which it does by a process called **osmosis**.

Figure 3f.3 shows an experiment that was done before anyone knew about auxin or how it works. The experiment was carried out using **coleoptiles** – shoots of young wheat plants. Coleoptiles are good for this kind of experiment because they are quite thick and sturdy, and they don't have leaves coming out of them that might confuse the results.

The results for the four different shoots tell us a lot about how phototropism is brought about.

Shoot A did not grow at all. This shows that there is something in the tip of the shoot that is needed to make it grow. We now know that this substance is auxin, which is made in the shoot tip and then diffuses down the shoot. It affects the cells a little way down, making them get longer.

Shoot B grew because it had its tip and so could make auxin. It grew towards the light. The auxin diffused down the shoot, with more of it collecting on the shady side than the light side, making the shoot bend over as it grew.

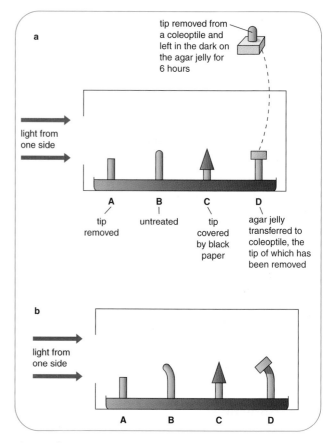

a

tip removed from a coleoptile and left in the dark on the agar jelly for 6 hours

light from one side

A
tip removed

B
untreated

C
tip covered by black paper

D
agar jelly transferred to coleoptile, the tip of which has been removed

b

light from one side

A   B   C   D

Figure 3f.3 An experiment in which coleoptiles have been treated in different ways. **a** The start of the experiment. **b** The results after 24 hours.

Shoot C grew because it had its tip and so could make auxin. But it did not grow towards the light.

Shoot D grew towards the light. Although its tip was removed, the auxin made in the tip was able to diffuse down through the block of jelly and into the shoot.

SAQ

2 What do the results of this experiment suggest about the part of the coleoptile that is able to sense the direction of the light? Explain your answer.

Geotropism

Shoots and roots can also respond to gravity. A growth response to gravity is called **geotropism**. Usually, shoots are negatively geotropic and roots are positively geotropic.

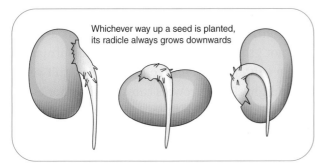

Figure 3f.4 Positive geotropism in roots.

SAQ

3 Why it is useful to the plant if:

   a its shoots are negatively geotropic?

   b its roots are positively geotropic?

Figure 3f.5 shows a method for showing that roots are positively geotropic. It uses a rotating board onto which germinating beans can be pinned. Because the board rotates, the seeds keep experiencing gravity coming from different directions. They grow in no particular direction. If you set up an identical machine, but don't switch on the motor so that it doesn't rotate, the roots all grow downwards, towards the direction from which gravity is pulling on them.

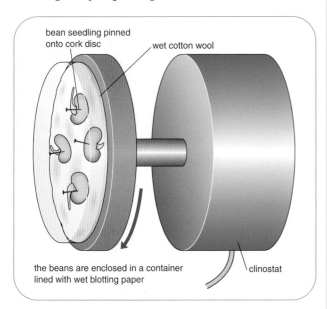

Figure 3f.5 Apparatus to find out how roots respond to gravity.

SAQ

4 How could you use the machine shown in Figure 3f.5 to investigate the response of shoots to light?

Using plant hormones

Plant hormones are used in agriculture and in fruit growing. For example, hormones might be applied to a plant to make it grow faster or more slowly. The poinsettias that you can buy at Christmas have been treated with hormones to keep them short and bushy. If you keep your poinsettia until the next year, it will grow much taller and skinnier, and be less attractive.

Figure 3f.6a Christmas poinsettias. **b** A natural poinsettia.

Plant hormones can be used as **selective weedkillers**. These are weedkillers that kill some plants but not others. Some people spray selective weedkiller onto their lawn, where it kills weeds but not the grass. These weedkillers contain auxin. The auxin makes the weeds grow so much and so quickly that they die.

Rooting powder also contains auxins. Rooting powder is used for taking cuttings (Figure 3f.7). The bottom end of the cutting is dipped in the rooting powder, which makes it grow roots more quickly.

Plant hormones can also be used to make fruit ripen. The hormone that is used is sometimes auxin, but there are other hormones that do this, too. Hormones can be used to make the fruit ripen more quickly, or more slowly.

SAQ

5 Suggest why fruit growers or supermarkets might want to make fruit ripen more slowly.

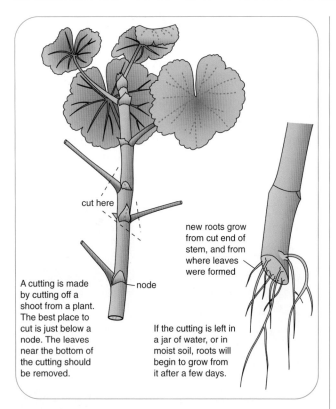

Figure 3f.7 Propagating a geranium plant by cuttings.

cut here

new roots grow from cut end of stem, and from where leaves were formed

A cutting is made by cutting off a shoot from a plant. The best place to cut is just below a node. The leaves near the bottom of the cutting should be removed.

node

If the cutting is left in a jar of water, or in moist soil, roots will begin to grow from it after a few days.

Yet another commercial use of plant hormones is to control **dormancy**. Dormancy means 'resting and not growing'. Many plants become dormant in the winter, when it is cold and there isn't very much light. Plant growers can spray hormones onto them that make them start growing. This is useful if they want to produce plants for people to buy at Christmas, for example.

Seeds can stay dormant for a long time, just resting and not germinating. A hormone called **gibberellin** can be used to make them start into growth. Commercial growers might want to do this so that they can get seeds to germinate at a time of year when they wouldn't normally do it. Some seeds are really difficult to germinate at all, and gibberellin can stimulate them to begin to grow. Brewers sometimes use gibberellin to make barley germinate. Germinated barley contains a sugar called maltose and is an essential ingredient for making beer.

Banana ripening

The bananas we can buy in the UK have all come a long distance to get to us. They have to be picked before they are ripe or they would be well past being edible by the time they got onto the supermarket shelf.

When the bananas arrive here, they are usually put into special 'ripening rooms'. Here, the gas **ethene** is added to the atmosphere. Ethene is a plant hormone – even though it is a gas! Ethene is naturally produced by fruits such as bananas and tomatoes when they are ripening. Other fruits are affected by the ethene and it makes them ripen as well.

Once the bananas are almost ripe, they are taken to shops and put on display. Their shelf life can be extended by using 'modified atmosphere packaging'. They can be wrapped in plastic, with substances called 'ethene scavengers' added to the air inside the plastic. These scavengers soak up all the ethene that the bananas are releasing. This slows down their ripening. Once the bag is opened, the bananas will continue to ripen normally.

Figure 3f.8 Bananas being transported.

Figure 3f.9 Ripe bananas.

Figure 3f.10 Dormant seeds are spread out on a malting floor in a brewery and moistened to encourage them to germinate. Gibberellin may be added to them to speed up their germination.

Summary

You should be able to:

◆ state that plant hormones help to control plant growth, flowering and the ripening of fruits

◆ describe how roots and shoots respond to light (phototropism) and to gravity (geotropism)

◆ explain phototropism in terms of auxin

H ◆ interpret data from phototropism experiments

◆ describe the uses of plant hormones by commercial plant growers, as selective weedkillers, rooting powder, delaying or accelerating fruit ripening and in control of plant or seed dormancy

Questions

1 Copy and complete these sentences.

A tropism is a .... response to a stimulus, in which the .... of the response is related to the direction of the stimulus.

Most shoots are positively .... and negatively .... Most roots are .... and ....

2 Figure 3f.11 shows an experiment investigating tropisms.

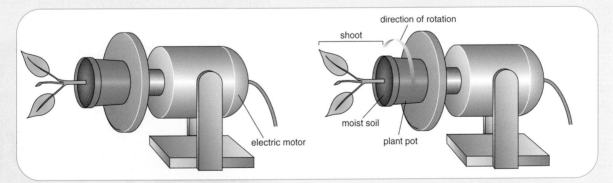

Figure 3f.11 Experiment to investigate tropisms.

a Which kind of tropism is being investigated?

b Predict what the results of the experiment will be. Explain your prediction.

3 Bananas that have been treated with ethene can still be sold as 'organic bananas'. Do you think this is acceptable? Explain the reasons for your opinion.

Mutations

Usually, the characteristics of offspring are very similar to those of their parents, although they may have a mix of features from their father and their mother. But sometimes a new feature suddenly turns up. This is often the result of a **mutation**.

A mutation is a sudden and unpredictable change in an organism's genes. It happens when DNA is being made – the bases get put together into the wrong sequence. So the gene gives

Figure 3g.1 The white coat of this red squirrel is caused by a mutation in its genes.

different instructions and the cells behave in a different way. They make different proteins.

A mutation can happen for no obvious reason – it just happens spontaneously. But the chances of it happening are increased by anything that damages DNA. This includes ionising radiation (such as alpha radiation and X-rays), exposure to ultraviolet light (in sunshine), and various chemicals. Usually, mutations are harmful.

Just occasionally, though, a mutation can be beneficial – the new protein that is made turns out to be even better than the old one.

Selective breeding

For thousands of years, people have been keeping animals for meat, milk and transport and have been growing crops to make food and fibres. Long before we knew anything about genes, farmers knew that the offspring of their animals and plants weren't all the same. They learned to make careful choices about which ones they allowed to breed so that they might get offspring that were even better than their parents.

For example, a farmer might have been keeping cattle to give milk. Some of the cows might give more milk than others. He would choose these higher-milking cows to breed from, hoping that some of their female calves would also give lots of milk when they matured. He would breed from a bull whose mother or sisters gave a lot of milk. If he did this over many generations, he could end up with a herd of cattle which gave especially large amounts of milk.

This process is called **selective breeding**. It is still the main method used today to try to produce animals and plants for farming.

Figure 3g.2 Longhorn cattle are a very old breed. They are thought to be quite similar to the original wild cattle.

Figure 3g.3 Limousin cattle have been bred to have heavy, muscular bodies, to give a high yield of beef. What other differences between the Longhorn and the Limousin do you think might have been produced by natural selection?

SAQ

1 Table 3g.1 shows the average milk yield per cow in the United Kingdom between 1995 and 2003.

Year	Average milk yield per cow in litres
1995	5512
1996	5627
1997	5788
1998	5818
1999	5957
2000	6048
2001	6450
2002	6451
2003	6609

Table 3g.1

a Plot a graph to display these results.

b By how much did milk yield per cow increase between 1995 and 2003?

c Explain how selective breeding could have been used to cause this increase.

d Can you suggest any other reasons for the change in yield over this period?

Selective breeding can also involve crossing two different varieties (Figure 3g.4). For example, there might be a variety of wheat that gave a lot of grain but was very likely to get infected with a disease called rust, where a rusty-coloured fungus grows on the wheat plants. Another variety might be resistant to rust disease but not give so much grain. You could try crossing the two of them together and then choosing the offspring that seemed to have high yields of grain *and* good rust resistance. If you kept doing this for many generations, you could produce a new wheat variety that was both high yielding and very rust resistant.

Reducing the gene pool

Table 3g.1 shows that selective breeding can bring about huge changes in a variety of animal, but this is not always a good thing.

For example, sometimes cattle have been selected to breed just on the basis of their milk yield. Any other features, such as how strong

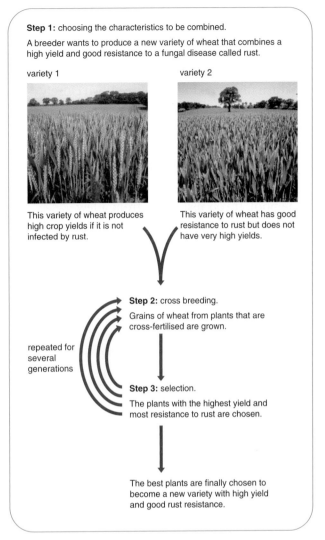

Step 1: choosing the characteristics to be combined.

A breeder wants to produce a new variety of wheat that combines a high yield and good resistance to a fungal disease called rust.

variety 1

variety 2

This variety of wheat produces high crop yields if it is not infected by rust.

This variety of wheat has good resistance to rust but does not have very high yields.

Step 2: cross breeding.

Grains of wheat from plants that are cross-fertilised are grown.

repeated for several generations

Step 3: selection.

The plants with the highest yield and most resistance to rust are chosen.

The best plants are finally chosen to become a new variety with high yield and good rust resistance.

Figure 3g.4 Selective breeding.

their feet were or how good they were at resisting diseases, might have been disregarded. If that goes on for several generations, you run the risk of producing a breed of cattle that is good at producing lots of milk but is always getting foot infections and can only walk with difficulty.

Another problem is that the cattle that are best at producing milk are often all very closely related to each other. That means that they share a lot of genes. Breeding closely related animals or plants with each other is called **inbreeding**. If you pick out closely related animals to breed, generation after generation, you may end up with them all sharing almost identical sets of genes. Some of these genes may be recessive ones for harmful characteristics. Normally, the recessive gene would be rare and usually be present along with a dominant gene in any one animal, so it wouldn't cause any problems. But in these closely related animals, there would be a high chance of *both* parents having the recessive gene, so their offspring could get two of them. This is why it isn't usually a good idea to breed two closely related animals together. It increases the chances of recessive characteristics showing up in the offspring.

Inbreeding also reduces the total number of different alleles in the population – that is, the **gene pool**. If you just breed from a few individuals, you will lose all the different alleles that the others have. You end up with all the animals having practically identical sets of genes. There is less variation.

Figure 3g.5 All the wheat plants are genetically identical and so they are all the same height and all ripen at the same time, which makes harvesting easier.

Why does a reduced gene pool matter? Firstly, farmers might want different characteristics in their animals and plants in the future. For example, they might want to have cattle that can live outside all the time, even in the winter. Perhaps there were cattle with those characteristics years ago, but they were never chosen to breed from, so their particular alleles have now been lost for ever.

Another reason is that all the animals or plants could be wiped out by a disease. If one of them is susceptible to it, then they all will be.

Luckily, many people all over the world are interested in keeping the old breeds of livestock alive (Figure 3g.6). They are often called 'rare breeds'. In future, we may want to use them to begin new selective breeding programmes, trying to produce new breeds with new combinations of characteristics.

Figure 3g.6 Prize-winning rams at a rare-breeds show in Yorkshire.

SAQ

2 In a breed of dog, a recessive gene causes a weakness in the hips.

Using the symbol H for the normal gene and h for the recessive gene, explain why it is not a good idea for two closely related dogs to be bred together. You should include a genetic diagram in your answer.

Gene technology

Selective breeding can eventually produce a population of animals or plants with what we think is the best collection of all the different possible alleles of their genes. But we now have something even more powerful that we can do. We can put *new* genes into an organism. This is called **gene technology**, **genetic engineering** or **genetic modification**. An organism that has been **genetically modified** is often called a GM organism.

For example, the gene for insulin has been taken out of human cells and put into a bacterium. Now, millions of these bacteria are kept in large fermenters – huge metal tanks where they are supplied with the nutrients and oxygen that they need. The bacteria follow the instructions on the gene and make insulin. This is collected and purified and then used to provide people who are diabetic with insulin.

In some countries where rice is the staple food, people eat a diet that is lacking vitamin A. This can cause blindness. Genetic engineering has been used to produce a new variety of rice that is rich in vitamin A. Genes for making vitamin A have been inserted into rice. The new variety of GM rice should help to provide people with plenty of the vitamin. It is called golden rice.

Yet another example involves soya beans. They are a very important source of food in many parts of the world. If weeds are allowed to grow amongst the soya plants, the plants don't produce as many beans. A GM variety of soya has been given a gene that makes it resistant to a particular kind of herbicide. So this herbicide can be sprayed onto the fields of soya, where it will kill the weeds but not the soya bean plants. This means that farmers should be able to produce more soya beans on the same area of land.

Some crop plants – such as potatoes – are killed by frost. If it gets very cold, the water in their cells freezes. The ice crystals break their cell membranes, killing the cells. Other crop plants, such as winter wheat, are able to survive frost. If we can identify and transfer the genes that make them frost resistant into the frost sensitive crops, this would increase the places in the world where that crop could be grown.

Figure 3g.7 Genetically modified golden rice provides extra vitamin A.

The principles of genetic engineering

Genetic engineering isn't quite as easy as it sounds. It involves four main steps.

1 First, you have to choose or **select** the characteristic you want the GM organism to have. In the case of golden rice, the characteristic was a high vitamin A content.

2 Next, you have to find and **isolate** the gene that will produce this characteristic. This is not at all easy. You can't just look down a microscope and

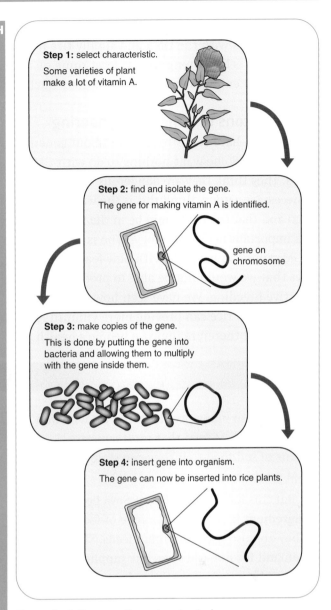

Step 1: select characteristic.
Some varieties of plant make a lot of vitamin A.

Step 2: find and isolate the gene.
The gene for making vitamin A is identified.

gene on chromosome

Step 3: make copies of the gene.
This is done by putting the gene into bacteria and allowing them to multiply with the gene inside them.

Step 4: insert gene into organism.
The gene can now be inserted into rice plants.

Figure 3g.8 How genetic engineering is done.

see the gene. Complex, expensive and sophisticated detection methods have to be used in the laboratory to find the gene amongst all the DNA in the cells, and then to snip it out. In this case, the genes for making vitamin A were taken from daffodils and bacteria.

3 Now you have to make lots of **copies** of the gene. There are several different ways this can be done. One way is to put the gene into a bacterium and let it breed. Every time it divides, it passes on a copy of the gene to each descendant, so after many generations you have millions of copies of the gene.

4 Finally, you have to **insert** the gene into the organism – in this case, the rice. Again, there are several different ways this can be done. One

way is to use a gene gun. You coat tiny silver or gold pellets with the DNA and simply fire it at the plant cells, hoping that at least some of them will end up with the DNA in their nuclei.

Pros and cons of genetic engineering

People often have very strong views about genetic engineering. Some want nothing to do with GM crops. They think they are dangerous and unnecessary. Others think this is the way we ought to go and that the UK should be in the forefront of this important new technology. Who is right?

GM crops can be made with new features – ones that we would not be able to produce just by selective breeding. We have seen how the new variety of GM rice can provide more vitamin A in the diet and therefore improve people's health.

That should be an advantage. But some people would disagree. They say that the real problem is that these people are poor. We ought to be trying to relieve their poverty, not just giving them a different food to eat.

A possible disadvantage of GM organisms is that there could be unexpected harmful effects. For example, how do we know how a gene will behave when it is inserted into a new organism? Perhaps it will have harmful effects that no one could have predicted. And what if the gene for herbicide resistance somehow spread from soya plants into the weeds? Then you might get a population of super weeds that couldn't be killed with herbicides. That hasn't happened yet – a few weeds have been found with the resistance gene in them, but they were all sterile (could not breed).

GM soya

There are so many people in the UK who don't want to eat GM food that most supermarkets want to sell food that they can state contains no GM products.

But this hasn't proved easy. Soya beans are an ingredient in a very wide range of food products. They are also used in animal feeds. There is a big demand for non-GM soya, but supplies of totally non-GM soya are quite limited.

We don't grow soya beans in the UK. Most of what we use comes from the USA or South American countries such as Argentina. Although many farmers there do grow non-GM soya, the factories that process the soya often take in both non-GM and GM soya. Even if they clean out all the equipment between batches, it is still easy for some GM soya to get mixed up with the non-GM soya. On at least one occasion there has been a complete mix-up and soya that was imported to Britain as being non-GM turned out to contain a very high proportion of GM soya.

Argentinian farmers have been growing GM soya since 1997. The soya is resistant to a herbicide, so they can easily get rid of weeds by spraying the crop with it – the weeds die but the soya doesn't. However, a few problems are beginning to emerge. For example, some soya seeds have spread and so soya is growing where

it isn't wanted; because it isn't so easily killed by herbicides, it can be difficult to get rid of it.

GM soya also seems to be less able to deal with hot, dry weather. The gene that confers resistance to the herbicide also appears to make the plants more brittle, so the stems crack open when it is hot. This reduces the yield of the crop.

In Brazil, growing GM soya was illegal for most farmers up until 2005. But, even though the problems were known, Brazilian farmers still wanted to grow GM soya. British retailers tried hard to persuade the Brazilian government not to give in to them. But eventually the government gave in, and now it is legal to grow any GM crop in Brazil. So one of Britain's sources of non-GM soya has dried up.

Figure 3g.9 Campaigners go to great lengths to oppose imports of GM crops.

Moral and ethical issues

The new technology of GM organisms has started up a big debate about how we should use it. Many of the people who pronounce themselves against GM food don't understand what the technology is about, and may have made their decision for unsound reasons. But many others who do understand the science and the technology are still worried about the moral and ethical aspects of it. Here are a few examples of the kinds of questions they are asking. Think about your own views and how you would answer these questions.

- An extra copy of the gene for making growth hormone has been inserted into sheep in Australia. The gene is the sheep's own gene – it hasn't come from a different species. These sheep grow larger and produce more wool and leaner meat. But should we be doing this? How might it affect the general health and well-being of the sheep?

- The gene for a human protein called **antitrypsin** has been inserted into a cow. She produces antitrypsin and secretes it into her milk. Her milk can be used as a source of this protein, which can be used in the treatment of some diseases, including cystic fibrosis. Is this morally and ethically acceptable?

- Faulty genes that cause diseases can be inserted into animals, which can then be used to do research about similar diseases in humans. For example, mice can be given a gene that makes them develop an illness like Parkinson's disease. This kind of research has enabled major breakthroughs in the development of treatments for some devastating human diseases. But is it right to use mice in this way?

- Even if people don't understand about GM food and are completely wrong about what it is, should they still have the right to eat food that is guaranteed GM free, if they want to?

Summary

You should be able to:

- explain what a mutation is and how mutations can happen

- explain how mutations affect the production of a protein

- describe how selective breeding takes place and give an example

- explain how selective breeding can improve agricultural yields

- explain why selective breeding can reduce the gene pool and the problems associated with this

- state what is meant by a *genetically modified (GM) organism* and *genetic engineering*

- describe the principles of genetic engineering

- outline some advantages and disadvantages of genetic engineering

- outline examples of genetic engineering: golden rice, GM bacteria producing human insulin and GM crops that can resist herbicides or are able to survive frost

- discuss some of the moral and ethical issues associated with genetic modification

Questions

1   Copy and complete these sentences. Use some of the words from the list:

genes   protein   useful   harmful   bacteria   radiation   enzymes   chemicals   appearance

A mutation is a change in an organism's .... Most mutations are .... They can be caused by .... or ....

2   Merino sheep are kept for their wool. A selective breeding programme is taking place in Australia to try to increase the quantity of wool that Merino sheep produce.

The mass of the wool obtained from sheep in the flock undergoing the selective breeding programme was measured each year. The same thing was done for a control flock. The results are shown in the graph in Figure 3g.10.

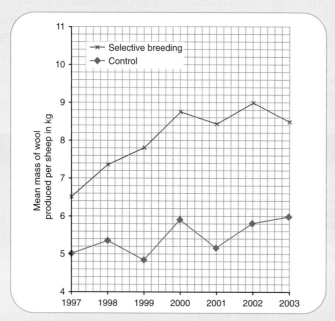

Figure 3g.10 Wool production from selectively bred Merino sheep.   **Figure 3g.11** Merino sheep.

a   Outline how the selective breeding would have been carried out.

b   Suggest what would have been done with the control flock of sheep.

c   Compare the results for the selective breeding flock and the control flock.

d   The breeders are taking care to bring in rams from other flocks to mate with the ewes in the selective breeding flock. Explain why this is a good idea.

H   3   a   Write a paragraph explaining how genetically modified crops could benefit developing countries.

b   Write another paragraph explaining some of the possible dangers associated with growing genetically modified crops.

c   Finally, outline your personal view about genetically modified crops.

Cloning

There are two kinds of reproduction, as we saw in Item B1g of Book 1 (*Who am I*). In asexual reproduction, there is just one parent. The young organisms that are produced are genetically identical with their parent and with each other. They are **clones**.

The process of producing several genetically identical organisms is called **cloning**. Cloning is a kind of asexual reproduction.

Cloning some types of plant is easy, as we shall see later in this Item. Many plants clone themselves, quite naturally – as spider plants do, for example. Cloning animals is more difficult, and it is only recently that ways have been found of cloning mammals, such as sheep, horses or dogs. The very first mammal to be cloned was Dolly the sheep. She was born in 1996 and died at an early age in 2003.

Figure 3h.1 The baby spider plants are being produced by asexual reproduction and are clones of their parent.

Figure 3h.2 Identical twins are naturally occurring clones.

Cloning cows

A top-quality dairy cow produces a lot of milk. She may be quite valuable. If she is used in a selective breeding programme, she could perhaps produce one calf per year.

There is a way in which the number of calves produced from this high-quality cow can be increased. It is called **embryo transplanting**.

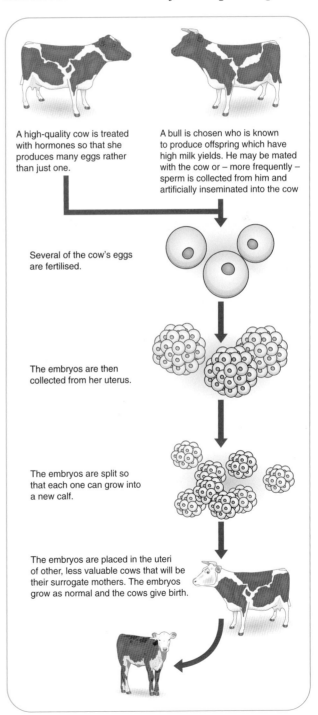

A high-quality cow is treated with hormones so that she produces many eggs rather than just one.

A bull is chosen who is known to produce offspring which have high milk yields. He may be mated with the cow or – more frequently – sperm is collected from him and artificially inseminated into the cow

Several of the cow's eggs are fertilised.

The embryos are then collected from her uterus.

The embryos are split so that each one can grow into a new calf.

The embryos are placed in the uteri of other, less valuable cows that will be their surrogate mothers. The embryos grow as normal and the cows give birth.

Figure 3h.3 Embryo transplants in cows.

SAQ

1 Look at Figure 3h.3.

  a Which is the stage where cloning happens?

  b Are *all* the offspring genetically identical? Explain your answer.

Dolly

Dolly was probably the most famous sheep that ever lived. She was born in Scotland at a research institute where scientists were looking into new ways of producing animals with characteristics that they wanted the animals to have. Dolly was bred using a revolutionary new technique. She was the first cloned mammal.

First, some cells were taken from the udder of a ewe (female sheep) belonging to the Finn Dorset breed. These cells were perfectly normal and their nuclei contained chromosomes that gave the ewe the typical characteristics of a Finn Dorset sheep.

Figure 3h.4 a A Finn Dorset ewe. **b** A Scottish Blackface ewe.

Next, some egg cells were taken from a Scottish Blackface ewe. The nuclei were removed from the eggs – not an easy thing to do with something so tiny. Then one of these egg cells, now minus its nucleus, was made to fuse with a Finn Dorset cell.

The egg cell behaved as though it had been fertilised and was a zygote. It divided repeatedly by mitosis, forming an embryo. The embryo was placed in a Scottish Blackface ewe, where it grew into Dolly.

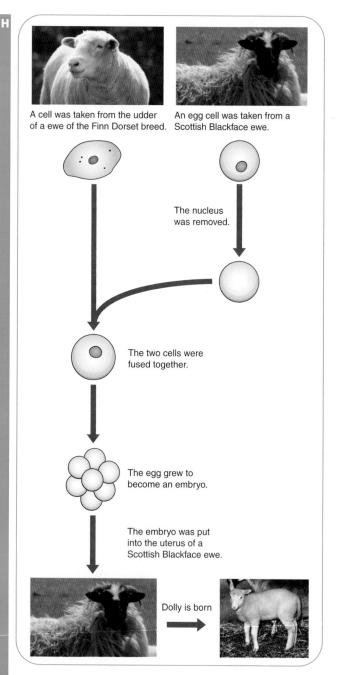

A cell was taken from the udder of a ewe of the Finn Dorset breed.

An egg cell was taken from a Scottish Blackface ewe.

The nucleus was removed.

The two cells were fused together.

The egg grew to become an embryo.

The embryo was put into the uterus of a Scottish Blackface ewe.

Dolly is born

Figure 3h.5 How Dolly was made.

SAQ

2 a Which sheep did all of Dolly's chromosomes come from?

  b Explain why Dolly could be called a *clone*.

Even though Dolly was a success, there were a lot of problems with the procedure and also with her health. The researchers had managed to get 277 pairs of cells to fuse together but only 29 developed into embryos. And only one of these grew into a live and apparently healthy lamb.

Dolly died at a relatively young age for a sheep. She suffered from arthritis. It is thought that cloned mammals are less healthy than ones produced by sexual reproduction, athough we don't yet understand exactly why this is.

Replacement organs

There is always a shortage of organs for human transplants. Each year, many people die because no suitable liver, heart or kidney could be found to give to them. On average, a person needing a kidney transplant has to wait 500 days.

It has been suggested that we could use organs from animals, instead. There were some early experiments involving transplanting baboon hearts into people but none were successful and the trials stopped. More recently, thought has gone into the possibility of using pig organs.

Pigs have organs about the same size as human organs. However, they have 'pig antigens' on their cells and so, if you put them into a human body, the immune system rejects and destroys them. Researchers have been using genetic engineering to put human genes into pig cells or to remove certain pig genes from them, so that the pig cells might be more like human cells and less likely to be rejected. Embryos with altered genes were produced, and then cloned to develop into many more embryos that all contained the altered genes.

Trials carried out in the late 1990s successfully used genetically modified (GM) pig tissue as skin grafts in human patients, and also placed insulin-making pig cells into people with diabetes. Transplanting entire organs, though, is more difficult.

Snuppy

Figure 3h.6 Snuppy, the first cloned dog.

Snuppy stands for 'Seoul National University puppy'. Snuppy is an Afghan hound. He is the first cloned dog.

Snuppy was born in Seoul, South Korea, in 2005. He was produced in a similar way to Dolly. Eggs were taken from an Afghan hound, and their nuclei were removed. Then cells were taken from a male Afghan hound's ear and fused with the eggs. The chromosomes in the 'zygotes' were therefore all from the male hound.

The embryos were transplanted into female dogs, their surrogate mothers.

As with Dolly, Snuppy was a very rare success amongst hundreds of failures. The researchers put 1095 'zygotes' into the surrogate mothers, but only 3 became pregnant. One miscarried, but the other two gave birth to puppies. One of these died at 22 days old. Snuppy was the only survivor.

Should we be doing this? Why would anyone want to clone a dog? Many pet owners say that they would do this if they could; they love their dog so much that they can't bear the thought of it dying before they do. Cloning could keep their pet 'alive' much longer.

But many people think this is not a suitable use of this technology. They cite the huge numbers of eggs and embryos that are lost, and the potential suffering to the dogs involved in the cloning process. Another problem is that, although cloned mammals have identical genes to their parent, they do not actually look identical or behave in an identical way. This is because their environment, including their environment while they are embryos, has an effect on their characteristics. You can be genetically identical, but still not quite the same.

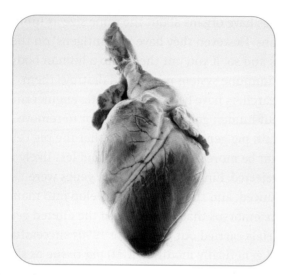

Figure 3h.7 GM pig organs can be used for human treatment.

H SAQ

3   There are a lot of unanswered questions and possible implications of using other animals to produce organs for human transplantation. Write a short paragraph giving your opinions on each of these issues. Try to use your scientific knowledge within your answers, rather than just writing down things that a person who hasn't studied biology might say.

● Is this a morally acceptable way to treat animals such as pigs? Is it right to breed an animal just for the purpose of supplying an organ for a human? Is that any better or worse than breeding it to be eaten?

● What if the pig tissue contains a virus? Might the virus be able to mutate so that it can infect human cells and cause a completely new disease that could spread amongst other people?

● Some cultures and religions see pigs as being unclean animals. How might people belonging to such cultures view the idea of using pig organs for transplants?

Cloning humans

As far as we know, no one has cloned a human yet, though some people claim to have done it. There is a complete ban on this. All the same, once something becomes possible, someone may well do it. Somewhere, some day, a human is likely to be cloned.

For many people, the whole idea of producing genetically identical copies of a human seems bizarre and morally unacceptable. Of course, there are already human clones in existence – identical twins are genetically identical and are naturally occurring human clones. But there is a general feeling that using science and technology to clone a human is just wrong. That isn't a road down which we should go.

To start with, what could possibly be the benefits of doing it? One group of people who might be in favour are infertile couples. If they cannot have a baby from the fusion of their eggs and sperm, perhaps they could have a child that is produced by cloning one of them. But it is possible that such a child might not be healthy, as was seen with Dolly. The child might not like the idea of being a clone of one of its parents, rather than an individual in his or her own right.

At the moment, these arguments against cloning hold sway in most countries. But perhaps, over time, we will come to feel differently about human cloning and it will be made legal.

Cloning plants

Many plants reproduce naturally using asexual reproduction, making clones. You have already seen that spider plants can do this (see Figure 3h.1). Figures 3h.8 and 3h.9 show how two different species of plants produce clones of themselves.

Gardeners often clone plants. If they have a plant that they particularly like, they can take **cuttings** of it. This is a good way to propagate

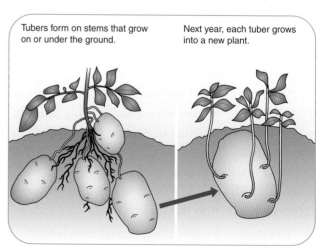

Figure 3h.8 Tuber formation in potatoes.

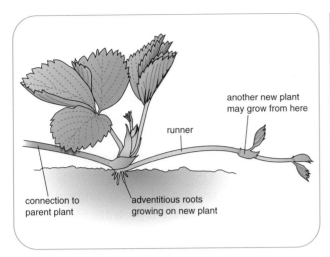

Figure 3h.9 Asexual reproduction of a strawberry plant. In summer, the parent plant grows long stems called runners. New plants grow from the buds along the runner. The new plants get food and water from the parent plant until they have grown their own roots. After several months, the connection to the parent plant dies away. Several new plants may grow from just one runner.

Pelargonium plants. Look back to Figure 3f.7 to see how to do it.

Advantages and disadvantages of cloning plants

Sometimes we don't want new plants that are identical with the old. Sometimes we want to get something new and different. If that is the case then we can breed the plants using sexual reproduction. This involves allowing one plant to produce pollen, which can then pollinate another. Fertilisation takes place and zygotes are produced, all genetically different from their parents. The zygotes grow inside the developing seeds, which may later germinate and grow into new, different plants.

Cloning has some advantages.

- All the new plants are genetically identical. If a commercial plant grower wants to produce lots of rose bushes with exactly the same deep red, scented flowers as one he already has then cloning is the answer.
- Some plants are difficult to produce from seed. For example, bananas have been bred to have no seeds inside their fruits. (Have you ever found a seed in a banana?) So banana trees cannot be propagated using sexual reproduction – they simply cannot produce seeds. They must be propagated by cloning.

Figure 3h.10 A biologist investigates the cloning of plants. Here, he removes a leaf sample for tests.

Cloning also has its disadvantages.

- Imagine a disease gets into a field where a grower has a huge crop of genetically identical potato plants. The plants are all susceptible to the disease, so they all die. If there had been some genetic variation then at least some might have survived. Something like this did happen in Ireland between 1846 and 1850, when a disease called potato blight infected and killed almost all the potato plants, on which people depended for food. If they had been growing many different varieties of potatoes, perhaps this would not have happened and thousands of lives would have been saved.
- If we grow just a few varieties of genetically identical crops then we don't have many possibilities of using selective breeding to produce new varieties. It is important to keep 'banks' of plants and animals that have a wide range of genes, in case we need them in the future.

Tissue culture

Tissue culture is a way of making enormous numbers of genetically identical plants. It involves taking many tiny pieces of the plant you want to replicate and growing them into complete new plants. Because only very small pieces of the plant are needed, it is sometimes called **micropropagation**.

First, you need to take a small piece from the plant you want to replicate. This piece is called an

H **explant**. The explant must contain some cells that have not differentiated and can still divide. As we saw on page 33 (in Item 3e), you can find these at the tips of the roots and shoots. In some plants, they are present in other places, too. The leaves of African violets, for example, have cells that can divide. You can usually get many explants from just one parent plant, because the explants need only contain a few cells.

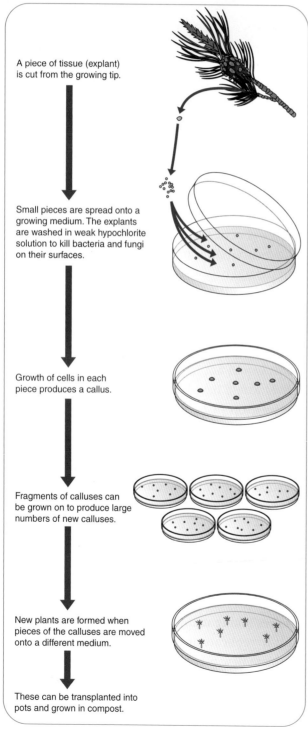

A piece of tissue (explant) is cut from the growing tip.

Small pieces are spread onto a growing medium. The explants are washed in weak hypochlorite solution to kill bacteria and fungi on their surfaces.

Growth of cells in each piece produces a callus.

Fragments of calluses can be grown on to produce large numbers of new calluses.

New plants are formed when pieces of the calluses are moved onto a different medium.

These can be transplanted into pots and grown in compost.

Figure 3h.11 Tissue culture.

H 　Next, the explant is put onto a **medium**. The medium is a nutrient source for the little group of plant cells. It also contains plant hormones. The nutrients are needed because the cells probably cannot photosynthesise at this stage. The hormones are needed to stimulate the cells to grow and divide. They must be kept at a suitable temperature, and provided with plenty of oxygen for respiration.

　Over the next few days or weeks, some of the cells in the explant divide repeatedly, by mitosis. They form a little lump of cells called a **callus**.

　The callus can now be divided into many calluses, and each one placed on a fresh medium with a different mixture of plant hormones. These hormones stimulate the cells in the callus to divide by mitosis and also to differentiate. They become tiny plantlets. When the plantlets have good roots, they can be transplanted into soil and grown in the normal way.

Figure 3h.12 Plantlets propagated by tissue culture.

　One problem that can arise during the tissue culture process is that the tiny plants become infected with fungi, which grow well on the nutrient medium. It is therefore important to keep everything sterile throughout the procedure. Of course, you can't sterilise the explants or you would kill them, but you can at least wash them in sterile water or a dilute solution of a disinfectant like hypochlorite before you put them onto the medium. Everything that you use for handling them (for example, forceps), the medium in which you grow them and the containers in which they are grown must be thoroughly sterilised at the outset, and care must be taken not to let spores of fungi get into the containers at any time. This is called **aseptic technique**.

Plant versus animal cloning

Each day, commercial companies all over the world clone millions of plants using tissue culture and other techniques. Yet scarcely any mammals have been cloned. Why is this?

Many cells in a plant don't differentiate. They remain able to divide and to turn into any kind of specialised cell. If we take just a few of these cells from the plant, they can grow into a complete new plant. The cells divide repeatedly by mitosis and some of them differentiate to form all the different kinds of tissue that are needed to make a new plant body.

In a mammal's body, however, there are far fewer undifferentiated cells. They are called **stem cells**.

And, as far as we know, in an adult mammal there are no stem cells that are capable of producing *all* the different tissues in the body. Stem cells that can do this have so far only been found in embryos. The stem cells that we do have can only differentiate into a few different kinds of cells. For example, the stem cells in the bone marrow can only differentiate into red or white blood cells. So we cannot just take a piece from a mammal's body and grow it into a whole new animal.

There are some animals, though, whose cells do behave rather differently from those of humans. For example, if a salamander (a kind of amphibian) loses a leg, it can grow a complete new leg, with all the different kinds of tissues in all the right places. No matter how many times it loses a leg, it can keep on growing a new one. Researchers are trying to find out how this happens. Perhaps one day it might be possible to use what we learn about regeneration of limbs in salamanders to help parts of human bodies to regrow.

Summary

You should be able to:

◆ explain what is meant by cloning and give some examples

◆ describe how embryo transplants can be used in cattle

H ◆ describe how Dolly the sheep was produced

◆ state that cloning animals could provide organs for human transplants

◆ recall that there are ethical dilemmas concerning human cloning

H ◆ discuss benefits, risks and ethical dilemmas associated with using GM animals to produce organs for humans, and with cloning humans

◆ describe how spider plants, potatoes and strawberries reproduce asexually, and how cuttings can be grown

◆ describe some of the advantages and disadvantages of the commercial use of cloned plants

H ◆ describe how tissue culture is done

◆ explain why cloning plants is easier than cloning animals

Questions

1 Spider plants can reproduce by growing new plants on the end of arching stems.

   a   Is this an example of sexual reproduction or asexual reproduction? Explain your answer.

   b   The new plants are clones. What does this mean?

   c   Give an example of clones that occur naturally in humans.

2 Commercial growers often use cloning to produce plants to sell.

   a   Give *one* example of a method of cloning that the growers might use.

   b   Explain why it is useful to them to be able to produce large numbers of genetically identical plants.

   c   Describe *one* possible disadvantage of propagating plants by cloning.

3 Tissue culture is used commercially to produce very large numbers of genetically identical plants.

   a   Explain the meaning of each of these terms in the context of tissue culture.

      i   Aseptic technique   iii Explant

      ii   Medium   iv   Callus

   b   A grower tested three different growth media to use at different stages of tissue culture of orchid plants. Her results are shown in Table 3h.1.

Growth medium	Time taken for explant to start to grow in days	Percentage of explants that grew	Time taken for callus to form roots in days
A	8.1	71	26
B	10.4	73	30
C	8.6	88	71

Table 3h.1

      i   Suggest how the contents of the three growth media might have differed.

      ii   Which growth medium would you recommend to use for growing the explants? Explain your answer.

Leaves

Leaves are the world's food factories. If it wasn't for plant leaves, we would have no carbohydrates, fats or proteins to eat. All the animals in the world are completely dependent on plant leaves to produce their food.

SAQ

1 Write down the equation for photosynthesis.

Leaf structure

Figure 4a.1 shows the structure of a leaf. The leaf has been cut across so you can see inside it.

The main part of a leaf is usually green. The green colour is caused by the pigment **chlorophyll**.

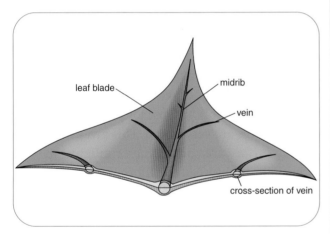

Figure 4a.1 Structure of a leaf.

The leaf has **veins** running through it. The veins bring water and minerals to the leaf, all the way up from the roots. They also take away substances that the leaf cells make, such as sucrose, and deliver them to other parts of the plant.

Figure 4a.2 shows what the cut edge of the leaf would look like if it was magnified. Although the leaf is very thin, you can see that it is actually made up of several layers of cells.

All the cells in the leaf have the same basic structure, like the one shown in Figure 3e.1 (page 32). However, the cells in each layer do each have their own individual characteristics.

Epidermis　The **epidermis** is a thin layer of cells that completely covers the leaf, top and bottom. The cells in the epidermis are different from most of the other leaf cells because they don't have chloroplasts. This means that they cannot photosynthesise. Their function is to protect the cells underneath them. They secrete a layer of wax onto their upper surfaces. This waxy layer is called the **cuticle**, and it forms a waterproofing layer all over the leaf, stopping it from drying out. It also helps to prevent pathogens and parasites from entering the leaf.

Palisade layer　The **palisade layer** is one of two layers of cells that make up the leaf **mesophyll**. 'Meso' means 'middle' and 'phyll' means 'leaf', so the mesophyll is simply the middle of the leaf.

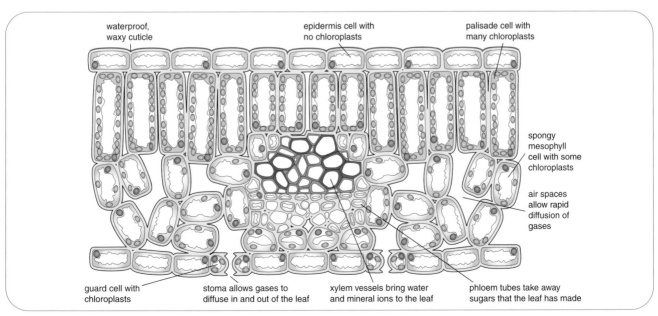

Figure 4a.2 Detailed view of a section through a leaf.

You can see that the cells in the palisade layer are tall and thin. They are stacked close together, with few spaces between. They contain a lot of chloroplasts, which contain chlorophyll. This is where most of the leaf's photosynthesis takes place.

Spongy layer   This is the other layer in the mesophyll. The **spongy layer** also contains cells with lots of chloroplasts, but they don't have quite so many as the cells in the palisade layer. There are big air spaces in between the cells in this layer.

Veins   The **veins** in the leaf lie in between the palisade layer and the spongy layer. They contain **xylem vessels**, which bring water to the leaf, and **phloem tubes**, which carry sucrose (sugar) away

Stomata and guard cells   There are gaps or pores between some of the cells in the lower epidermis. These gaps are called **stomata** (*singular:* **stoma**). They allow air to move into and out of the leaf. Surrounding each stoma is a pair of **guard cells**. The guard cells can open or close the stoma. Guard cells are different from all the other cells in the epidermis because they do have chloroplasts.

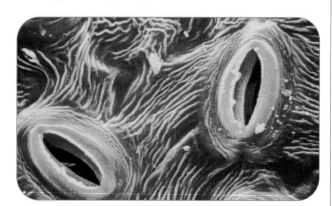

Figure 4a.3 Micrograph of the lower surface of a leaf, showing the closely fitting cells of the epidermis; the oval holes are stomata.

SAQ

2 Which of these structures do *all* the cells in a leaf have?

   chloroplast   cell membrane   cell wall
   nucleus   chromosomes

Supplying the palisade cells

Like any factory, the palisade cells need constant supplies of the raw materials and energy source required for the manufacturing process.

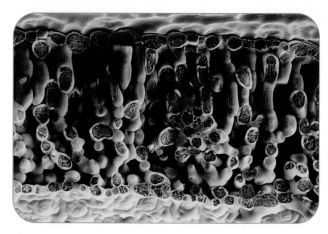

Figure 4a.4 A scanning electron micrograph showing the cells inside a leaf.

These are:
- carbon dioxide
- water
- light.

Figures 4a.5 and 4a.6 show how these supplies get into the leaf and into the palisade cells.

Adaptations for photosynthesis

The structure of plant leaves has evolved over hundreds of millions of years. Natural selection has made them superbly adapted for photosynthesis.
- Most leaves are very **broad**. This gives them a large surface area onto which sunlight can fall. Sunlight is absorbed by chlorophyll, and it provides the energy that drives photosynthesis.

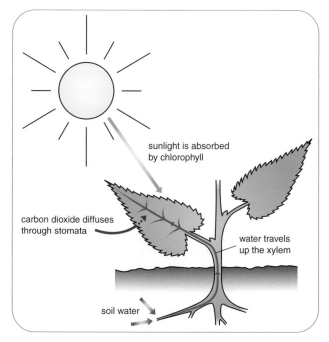

sunlight is absorbed by chlorophyll

carbon dioxide diffuses through stomata

water travels up the xylem

soil water

Figure 4a.5 How the materials for photosynthesis get into a leaf.

- Leaves are usually **thin**. This means that sunlight can reach most of the cells in the mesophyll. It also means that none of the cells are very far away from the air. There is only a short distance for carbon dioxide from the air to diffuse, so the palisade cells are well supplied with it. The oxygen that they produce in photosynthesis has only a short distance to diffuse to pass out of the leaf and into the air.

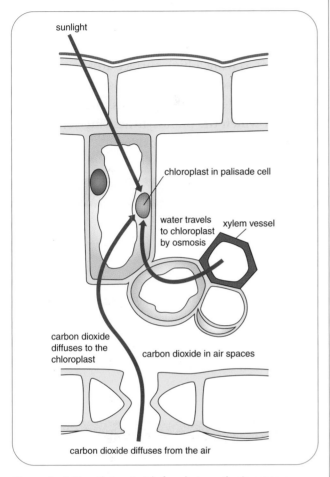

sunlight

chloroplast in palisade cell

water travels to chloroplast by osmosis

xylem vessel

carbon dioxide diffuses to the chloroplast

carbon dioxide in air spaces

carbon dioxide diffuses from the air

Figure 4a.6 How the materials for photosynthesis get to a palisade cell.

- The mesophyll cells contain chloroplasts, which contain the green pigment **chlorophyll**. This absorbs sunlight.
- The leaf has a network of **veins** which transport water to the leaf and take sucrose (made in photosynthesis) away. The veins also help to support the leaf, holding it out flat so that sunlight can fall onto its surface.
- The **stomata** in the underside of the leaf allow gas exchange to take place. When the leaf is photosynthesising, carbon dioxide diffuses into the leaf from the air and oxygen diffuses out.

SAQ

3 Explain why carbon dioxide diffuses into the leaf when it is photosynthesising.

4 a Oxygen diffuses out through the stomata during daylight. Name one other gas that diffuses out of the leaf through the stomata.

  b On very hot, dry days, plants often close the stomata on their leaves. How might this help them to survive?

Leaf pores give clues to the past

Palaeobotanists are people who study fossil plants. They have discovered that we can use plant fossils to tell us about the conditions on Earth millions of years ago.

Plants that grow in an atmosphere where there is a high concentration of carbon dioxide tend to have fewer stomata than if they grow in a low concentration of carbon dioxide. This makes sense – if carbon dioxide is in short supply, it is a good idea to make it easier for it to get into the leaf.

Not many leaf fossils are good enough to be able to see their stomata. However, palaeobotanists have studied the fossils that do allow them to count their stomata, and they have found that plants that grew about 400 million years ago had up to 100 times fewer stomata per cm^2 than modern plants. This matches up with other evidence we have which suggests that the concentration of carbon dioxide in the atmosphere at that time was much higher than it is now.

Now other scientists are using data collected by palaeobotanists to help them to work out how carbon dioxide concentrations have changed over time. This may help us to have a better understanding of global warming.

H
- The **epidermis** is transparent. This allows most of the light that falls onto the leaf to get through the epidermis and into the palisade cells.
- The **palisade layer**, where most of the chloroplasts are, is near to the top of the leaf where it can get most sunlight.
- The **air spaces** in the spongy layer allow diffusion of gases between the stomata and the photosynthesising cells.

H
- The **internal surface** area of the leaf is very large in relation to its volume. The leaf has a high surface area to volume ratio. The large surface area is provided by all the surfaces of the individual cells inside the leaf, many of which are in contact with the air spaces inside the leaf. Having a high surface area to volume ratio increases the rate of diffusion (see pages 26–27 in Item 3d), so this increases the rate at which the photosynthesising cells get supplied with carbon dioxide.

Summary

You should be able to:

◆ describe the structure of a leaf and name the different parts

◆ describe how the materials needed for photosynthesis get into the leaf and how the products of photosynthesis exit the leaf

◆ explain how leaves are adapted for photosynthesis

Questions

1   All of these statements are incorrect. Explain why they are wrong.

   a   Chorophyll inside leaf cells attracts sunlight.

   b   Guard cells are holes in the underside of a leaf.

   c   All the cells in a leaf can photosynthesise.

   d   The only function of veins in a leaf is transport.

2   Think about what an 'ideal' leaf would be like. What features would it have? (Be as inventive as you like, so long as your inventions might work!) Use diagrams, labels, annotations and descriptions to illustrate your design.

H
3   (This is quite a difficult question and there are a lot of possible answers. Concentrate on making sensible suggestions and explaining them, rather than worrying about giving the 'right' answer.)

   In humans, gas exchange takes place through the lungs. In plants, it takes place through the leaves.

   a   How are leaves and lungs similar? Explain why they have these similarities.

   b   Think of at least *three* ways in which the structures of leaves and lungs are different. Why do they have these differences?

Water, water everywhere

Osmosis

All cells need water. The water gets into them through their cell surface membranes. It does this by a special type of diffusion, called **osmosis**.

Partially permeable membranes

You should remember that particles in solids, liquids and gases are in constant motion. Particles in liquids and gases move quite freely. Their movement is random. If there are a lot of particles in one place and fewer particles in another, these random movements cause the particles to end up spread out fairly evenly. This is called **diffusion**.

Figure 4b.1a shows the particles in two solutions, separated by a membrane. On the left, there are a lot of water molecules and just one sugar molecule. This is a **dilute solution**. On the right, there are fewer water molecules and several sugar molecules. This is a **concentrated solution**.

Both the water molecules and the sugar molecules move around freely. They move randomly, bumping into each other and bouncing off in a different direction. If the membrane wasn't there, they would diffuse until there were roughly equal numbers of water and sugar molecules on each side.

However, the membrane stops this happening. It has holes in it that are big enough to let the little water molecules go through but not the much bigger sugar molecules. The membrane is a **partially permeable membrane**. This means that it will let some particles through, but not others.

How osmosis happens

To understand osmosis, you need to think about the different kinds of particles one at a time.

We'll start with the sugar molecules. There is a **concentration gradient** for sugar molecules. There are a lot of them on the right-hand side and fewer on the left. Figure 4b.1b shows their concentration gradient.

However, the sugar molecules cannot diffuse down this gradient. They can't get through the holes. As they move randomly around, they bump into the membrane and just bounce off again.

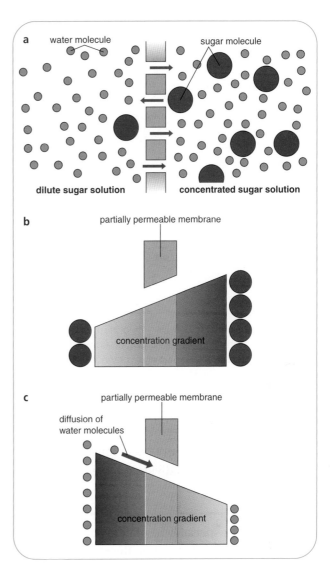

Figure 4b.1 a Principle of osmosis. **b** Concentration gradient for sugar. **c** Concentration gradient for water.

Now think about the water molecules. There is a concentration gradient for them, too. Figure 4b.1c shows their concentration gradient.

Moving randomly, water molecules sometimes hit the membrane and sometimes 'hit' a hole. If they hit a hole, they go through it. More of them go from left to right than from right to left, because there are more of them on the left than on the right. Eventually, so many water molecules will do this that the solution on the left-hand side becomes more concentrated and the solution on the right-hand side more dilute.

This is osmosis. We can define it like this.

Osmosis is the net movement of water molecules from an area of high water concentration to an

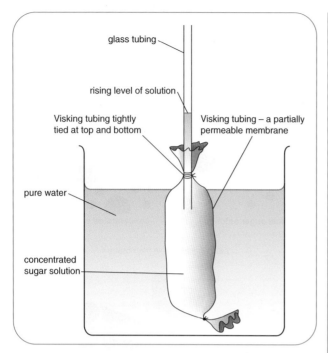

Figure 4b.2 Apparatus to demonstrate osmosis. Water diffuses from the pure water into the sugar solution down its concentration gradient. Sugar molecules cannot diffuse out, because the pores in the Visking tubing are too small for them. Therefore, the level of the sugar solution rises as it is diluted by the water diffusing into it by osmosis.

area of low water concentration across a partially permeable membrane. It is a consequence of the random movement of individual molecules.

SAQ

1 Osmosis is a special kind of diffusion. List three factors that would speed up the rate of osmosis.

Cells and osmosis

The cell surface membrane that surrounds every living cell is a partially permeable membrane. It lets water molecules through, but prevents many others from passing. This means that osmosis happens. Osmosis is very important for living cells.

Osmosis and animal cells

Figure 4b.3 shows what happens if an animal cell is immersed in pure water.

Cytoplasm is a solution of many different substances dissolved in water. Many of these substances cannot get out through the cell surface membrane. Even though there is a concentration gradient for them, higher inside the cell than outside, they cannot move down it.

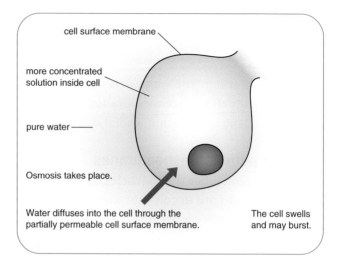

Figure 4b.3 An animal cell in pure water.

There is also a concentration gradient for water. The concentration is higher outside the cell than inside. Water *can* get through the membrane. So water moves, by osmosis, into the cell from the water outside it.

All the extra water going into the cell makes it swell. The cell surface membrane is not very strong, so eventually the cell bursts open. The contents escape and the cell dies. This is called **lysis**.

Figure 4b.4 shows what happens if an animal cell is immersed in a concentrated solution.

Now the concentration gradient for the water is the other way round. There is a greater concentration of water inside the cell than outside. So water moves, by osmosis, out of the cell. The loss of water makes the cell shrink. This is called **crenation**.

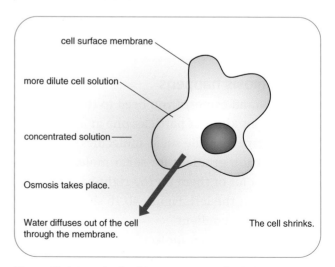

Figure 4b.4 An animal cell in a concentrated solution.

Osmosis and plant cells

Figure 4b.5 shows what happens if a plant cell is immersed in pure water.

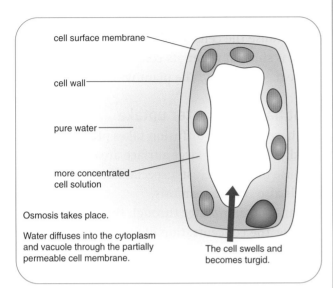

Figure 4b.5 A plant cell in pure water.

Just as with the animal cell, water goes into the plant cell by osmosis and the cell swells. But this time it does not burst. Unlike the animal cell, the plant cell has a strong, inelastic cell wall around it. The cell wall stops the cell from swelling too much.

Figure 4b.6 shows what happens if a plant cell is immersed in a concentrated solution.

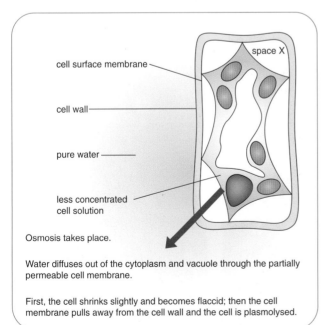

Figure 4b.6 A plant cell in a concentrated solution.

Drowning

If the body of a drowned person is found in suspicious circumstances, forensic scientists may be asked to determine whether the person drowned in salt water or in fresh water.

When a person is drowning, water gets into their lungs. If the water is fresh water, it moves across the walls of the alveoli and into the blood, by osmosis. This happens because there is a higher concentration of water in the lungs than in the blood.

Normally, the concentration of blood plasma and the cytoplasm inside the body cells is about the same. However, in freshwater drowning, the extra water dilutes the blood plasma. Water moves into the cells by osmosis and makes them burst.

The extra water in the blood also increases the total volume of blood in the body. This puts a strain on the heart, which often results in a heart attack.

If a person is drowning in salt water, the salt water in their lungs has a lower water concentration than the blood plasma. Water therefore leaves the blood by osmosis and moves into the alveoli in the lungs. The blood therefore becomes a more concentrated solution than it would normally be. Up to 42% of the water in the blood ends up in the alveoli. The blood cells don't burst and the person usually does not have a heart attack. A person who is submerged in salt water has a much better chance of surviving than someone submerged for the same period of time in fresh water.

The different effects on the body of drowning in fresh and salt water can be detected by a forensic scientist. The scientist will measure the concentration of the blood in the heart and look at blood samples under the microscope to see if the cells have burst.

Just as with the animal cell, water goes out of the plant cell by osmosis and the cell shrinks. The cell wall, however, is quite rigid. As the cell shrinks, the cell surface membrane tears away from the cell wall. A plant cell where this has happened is said to be **plasmolysed**. The cell surface membrane is often so badly damaged when this happens that the cell dies.

SAQ

2 The cell wall of a plant cell is fully permeable – it will let particles of any size go through it. In Figure 4b.6, what do you think fills space X? Explain your answer.

Support in plants

The inelastic cell wall and the water inside the cell play important roles in helping to support a plant.

The cell in Figure 4b.5 contains a lot of water. The contents of the cell are swollen and pushing outwards against the inelastic cell wall. A cell in this condition is said to be **turgid**. It is stiff and held firmly in shape. When there are a lot of cells in this condition close together, they all push on each other, making the tissue firm and strong.

However, if the cells lose water by osmosis, they lose their firmness. As the cytoplasm shrinks, the cell contents stop pressing outwards on the cell walls. The cells become floppy. They are said to be **flaccid**.

Plants rely on **turgor pressure** to support some of their parts. Leaves, for example, are held out flat because the cells in them are turgid. If the plant is short of water, its cells become flaccid and the leaf goes floppy. The plant wilts (Figure 4b.7).

Root hairs and water uptake

You have seen that diffusion takes place more rapidly if there is a large surface area involved. Because osmosis is just a special kind of diffusion, this is also true for osmosis.

Plants take up water through their roots. Usually, the concentration of water in the soil is greater than the concentration of water in the root cells. Water therefore enters the root cells by osmosis, across the partially permeable cell surface membranes.

Some of the cells on the outside of the plant root are specialised to take up water. They are called **root hair cells**. Each root hair cell has a long extension, giving it a large surface area. This speeds up the rate at which water is taken up into the plant's roots.

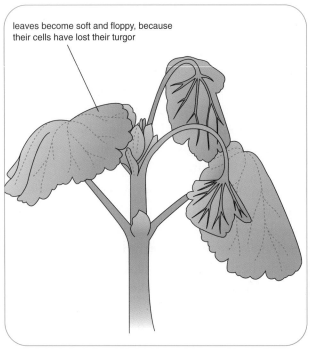

Figure 4b.7 A wilting shoot.

leaves become soft and floppy, because their cells have lost their turgor

Figure 4b.8 You can see the root hairs on the root of this germinating bean.

Summary

You should be able to:

◆ define *osmosis* and explain that it is a special kind of diffusion

◆ explain the term *partially permeable* and state that cell membranes are partially permeable

 ◆ predict the direction of water movement in osmosis

◆ explain what happens when animal and plant cells are immersed in pure water

◆ explain what happens when animal and plant cells are immersed in a concentrated solution

◆ explain how the plant cell wall and water are essential for the support of plants

 ◆ use the terms *flaccid*, *plasmolysed*, *turgid*, *turgor pressure*, *crenation* and *lysis*

◆ explain wilting in terms of a lack of turgor pressure

◆ explain the function of root hairs

Questions

1 Copy and complete these sentences. Use some of these words.

movement   turgor   wall   membrane   sugar   water   diffusion   fully   partially

Osmosis is a special kind of .... In osmosis, ... molecules diffuse through a .... permeable membrane, for example Visking tubing or a cell ....

2 Visking tubing is made of a partially permeable membrane. Starch molecules are much too big to go through the holes in the Visking tubing. Molecules of iodine and water can get through.

a In which direction will the iodine molecules diffuse in Figure 4b.9? Explain your answer.

b In which direction will the water molecules diffuse in Figure 4b.9? Explain your answer.

c Draw and label a diagram to show what the apparatus will look like after an hour. Label the colours of the two solutions. (You will need to think about what colour iodine solution is and what colour it goes when it is mixed with starch.)

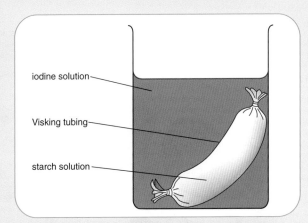

iodine solution

Visking tubing

starch solution

Figure 4b.9

3 Explain why:

a animal cells burst when immersed in pure water but plant cells don't;

b plant leaves wilt when they are short of water.

The plant transport system

Figure 4c.1 shows the main parts of a plant and summarises their functions.

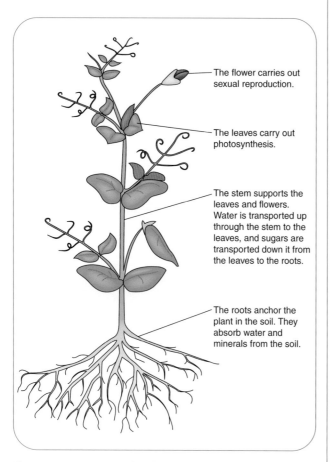

The flower carries out sexual reproduction.

The leaves carry out photosynthesis.

The stem supports the leaves and flowers. Water is transported up through the stem to the leaves, and sugars are transported down it from the leaves to the roots.

The roots anchor the plant in the soil. They absorb water and minerals from the soil.

Figure 4c.1 Main parts of a plant.

Plants have a less complicated transport system than animals. Plants don't have a heart. They rely on other methods to make substances move around inside them. All the parts of a plant need water. The water comes from the soil and is transported to all the cells in the plant.

Across the root

We've already seen how water gets into a plant, through its root hairs by osmosis. What happens next?

Figure 4c.2 shows how water from the soil goes into a root hair by osmosis. Once in the root hair, it travels across into the centre of the root. It may go in and out of the cells as it goes across the root or it might just seep in between them.

Xylem vessels

Once the water has reached the middle of the root, it moves into a dead, empty cell called a **xylem vessel**. These cells are very long and narrow, and they join up with each other end to end. They are completely empty. They are like drainpipes, all linked together to make long, hollow tubes that stretch all the way from the roots, up through the stem and into the leaves and flowers.

Figures 4c.3 and 4c.4 show where the xylem vessels are in a root and in a stem.

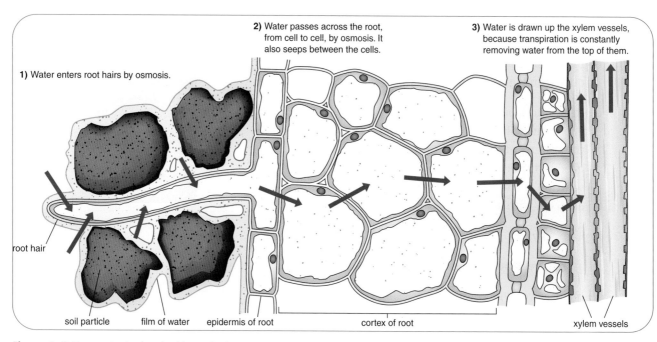

2) Water passes across the root, from cell to cell, by osmosis. It also seeps between the cells.

3) Water is drawn up the xylem vessels, because transpiration is constantly removing water from the top of them.

1) Water enters root hairs by osmosis.

root hair

soil particle   film of water   epidermis of root   cortex of root   xylem vessels

Figure 4c.2 How water is absorbed by a plant.

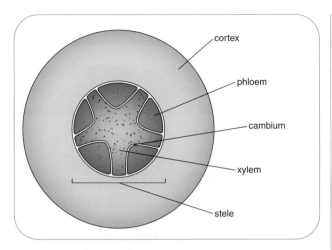

Figure 4c.3 Xylem vessels in a root.

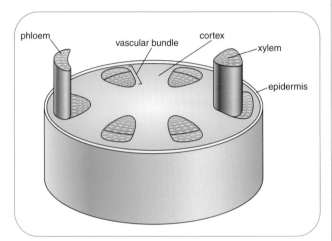

Figure 4c.4 Xylem vessels in a stem.

SAQ

1 Where are the xylem vessels in a leaf?

Figure 4c.5 shows what a xylem vessel would look like if you cut it in half lengthways (a longitudinal section) and crossways (a transverse section). As well as cellulose, the walls contain a very strong, waterproof substance called **lignin**. Lignin stops water leaking out of the xylem vessels. It also helps to support the plant.

If you are sitting at a wooden desk, you can probably see the xylem vessels in it. Wood is made up of xylem vessels.

Another good way of seeing where xylem vessels are is to put the bottom end of a celery stalk into a little bit of ink or blue stain. You can see the blue colour moving up through the stalk, inside little tubes – the xylem vessels. If you cut the celery stalk across, you can see little blue spots where the xylem vessels are.

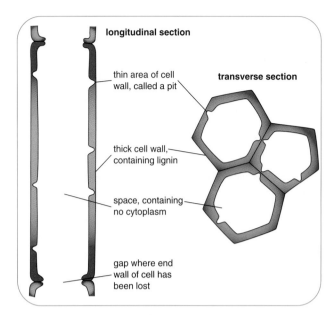

Figure 4c.5 Sections through a xylem vessel.

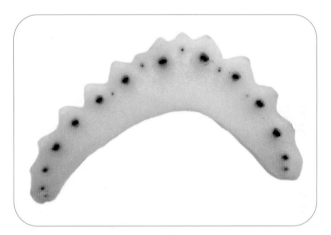

Figure 4c.6 A transverse section of a stained celery stalk.

Phloem tubes

In Figures 4c.3 and 4c.4, you can see that there is a tissue called **phloem** next to the xylem. Phloem, like xylem, is made of columns of long cells stacked end to end to make long tubes. However, in phloem, the cells are alive rather than completely empty, and they don't have lignin in their cell walls.

Phloem tubes transport sugars and other substances that the plant makes by photosynthesis. This is called **translocation**.

Phloem tubes can transport substances in any direction, depending on where they are needed. Sugar can be taken from the leaves down to the roots, where it is stored as starch. Or it could be taken from the stores in the root to other places that need it, such as a growing fruit or a flower.

Using plant fibres

You may be wearing some bits of vascular bundles. Several different kinds of fabrics are made from plant cells that are found in the vascular bundles, alongside the xylem vessels and phloem tubes.

For example, linen is made from long fibres that are part of the vascular bundles in flax plants. The fibres are very strong. To get the fibres out, the stems of the flax plants are allowed to rot away, leaving the fibres behind.

Sisal is made from the vascular bundles of agave plants. Sisal fibres are even stronger than linen, and they are used for making ropes and hard-wearing matting. Figure 4c.7 shows how the fibres are extracted and processed.

Figure 4c.7b Men feed the sisal leaves into a crushing machine. It is dangerous and noisy work – health and safety standards are not high in Madagascar. The machine squeezes all the soft tissues out of the leaves, so only the tough fibres are left. The fibres that come out of the machine are very wet. They are hung up to dry.

Figure 4c.7a Sisal comes from agave plants. These are being farmed in Madagascar. The leaves are very thick and tough. They contain thick, strong fibres in their vascular bundles.

Figure 4c.7c The dried fibres are packed into bales, ready to go off to rope makers or carpet manufacturers.

SAQ

2 Suggest why a growing fruit or a flower might need supplies of sugar.

Phloem and xylem tubes are often found together in the plant. The bundles of xylem and phloem tubes are called **vascular bundles**.

Transpiration

What makes water move up xylem vessels? Plants don't have a heart to pump water around their bodies.

Water moves up through the plant because it is continually evaporating from its leaves, a process called **transpiration**.

The transpiration stream

Figure 4c.8 summarises how water moves through a plant. A large, flat leaf spreading out in the air is a bit like washing hanging on a line. Water evaporates from the wet clothes and then diffuses into the air. As we saw on page 15, the cells in a leaf are wet, so water evaporates from them into the air spaces inside the leaf. The water vapour then diffuses out into the air, through the stomata.

As the water is lost from the leaf cells, they end up with a lower water concentration inside them. This causes water to flow into them from the nearby xylem vessels, by osmosis. It is a bit like sucking on a straw in a glass of water. As you suck, you reduce the pressure at the top of the straw. The pressure at the bottom is now higher than at

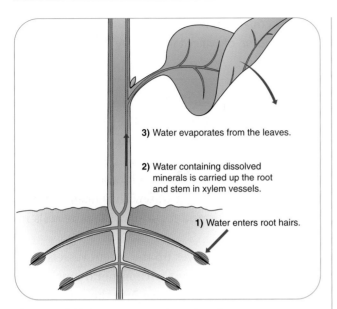

Figure 4c.8 How water moves through a plant.

3) Water evaporates from the leaves.

2) Water containing dissolved minerals is carried up the root and stem in xylem vessels.

1) Water enters root hairs.

the top and this pushes the water up the straw. You can think of xylem vessels as being like a straw and transpiration as being like someone sucking on the top of the xylem vessels in the leaf.

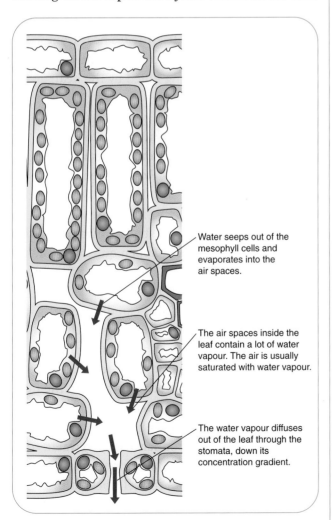

Water seeps out of the mesophyll cells and evaporates into the air spaces.

The air spaces inside the leaf contain a lot of water vapour. The air is usually saturated with water vapour.

The water vapour diffuses out of the leaf through the stomata, down its concentration gradient.

Figure 4c.9 How water diffuses from a leaf.

Factors affecting the rate of transpiration

There are several factors that affect the rate of evaporation of water from the cells in a leaf, or the rate of diffusion of water vapour out of the leaf and into the air.

Temperature On a hot day, water evaporates more quickly from the leaf cells. Diffusion also happens faster. So the transpiration rate is greater on a hot day than on a cold day.

SAQ

3 Why does diffusion happen faster at a higher temperature?

Light intensity When there is plenty of light, a plant can photosynthesise more quickly. To do this, it needs plenty of carbon dioxide, so its stomata open more and allow more carbon dioxide to diffuse in from the air. If the stomata are open then water vapour can diffuse out. So bright light can increase the rate of transpiration.

Air movement On a windy day, water vapour is quickly removed after it has diffused out of the leaf. This means that there is a greater diffusion gradient for the water vapour. Transpiration is greater on windy days than on still days.

Humidity On a humid day, the concentration of water vapour in the air outside the leaf is much higher than on a dry day. So transpiration happens more slowly on a humid day, because the concentration gradient for water vapour is very small.

Investigating transpiration

Figure 4c.10 shows some apparatus you can use to investigate the factors that affect transpiration.

This apparatus is called a **potometer**. 'Poto' means to do with water. As water evaporates from the leaves of the plant, water is drawn into the stem of the plant from the tube. You can watch the **meniscus** moving along inside the capillary tube, towards the plant. The faster the plant is transpiring, the faster the water moves along.

You can take readings of where the meniscus is every minute and then draw a graph like the one in Figure 4c.11.

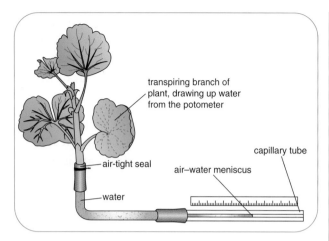

Figure 4c.10 A potometer is used to investigate the factors that affect transpiration.

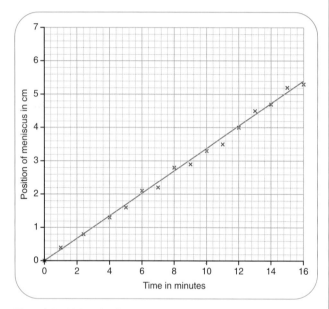

Figure 4c.11 Results from a potometer experiment.

If you refill the tube with water, you can then try the same thing again, but this time changing a variable. For example, you could put the plant in a dark cupboard (you'd have to look in quickly to take readings and shut the door as fast as you could) or use a fan to simulate a windy day.

SAQ

4 How would you expect the graph to differ from the one in Figure 4c.11 if:

   a the plant is kept in a dark cupboard?

   b a fan is blown onto the plant?

5 How could you investigate the effect of these factors on the rate of transpiration?

   a Temperature

   b Humidity

6 How could you use this apparatus to work out the *volume* of water taken up by the plant in one hour?

How transpiration helps the plant

If it wasn't for transpiration, water would not move up xylem vessels. Plants would not be able to grow tall, because they would not be able to supply their leaves with water. They need water for photosynthesis and for support for the cells in the leaves.

The water that moves up the xylem vessels also helps to supply the plant with mineral ions. These come from the soil and move into the root hairs by diffusion or active transport. They dissolve in the water in the xylem vessels and are transported all over the plant. Without transpiration, this would not happen.

Transpiration also helps to cool the plant. It is a bit like sweating in humans. When water evaporates, it absorbs heat energy. When water evaporates from leaf cells, it takes heat from them. On hot days, this can protect the leaves and prevent them from being killed by the heat.

Reducing water loss

A plant must balance its water uptake by the roots with the water loss from the leaves. If it loses more water than it takes in, then the cells become short of water and the plant wilts.

Plant leaves can't help losing some water by transpiration. Their whole structure has evolved to allow gas exchange to happen quickly, because they need to allow carbon dioxide to diffuse into them for photosynthesis. If carbon dioxide can diffuse in then water vapour can diffuse out. The large surface area and presence of stomata, which have evolved as adaptations for photosynthesis, also speed up water loss from the leaf.

Most plants have ways of cutting down the rate of transpiration from their leaves.

- They have a waxy cuticle on the top surface of the leaf. The wax is not permeable to water, so it acts as a waterproofing layer, preventing water from evaporating from the cells beneath it.
- Most of the stomata are on the lower surface of the leaf. This surface doesn't get sunlight

shining directly onto it, so it stays a little cooler. If there were stomata on the upper surface, a lot of water vapour would diffuse out through them. (Look back to Figure 4a.3 on page 60 to see the stomata.)

- The stomata can be closed if the plant is in danger of losing too much water. The guard cells change shape, pushing closely against each other and closing the stoma between them. Figure 4c.12 shows how they do this.

SAQ

7 How would you expect light to affect the guard cells? (Think about why a plant needs to have its stomata open.)

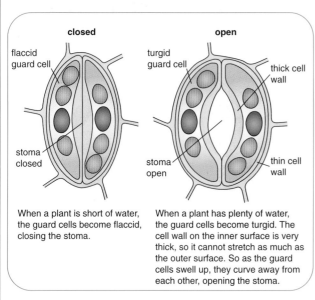

Figure 4c.12 How stomata open and close.

- Plants that live in hot, dry places generally have fewer stomata than plants that live where there is plenty of water. The stomata are often tucked away inside grooves or other places that reduce their contact with the air. This means that the plant can't photosynthesise as quickly, but conserving water takes precedence. This doesn't only apply to desert plants. Figure 4c.13 shows a section cut across a leaf of marram grass, which grows in sand dunes near the sea. The leaf can roll up tightly, with its stomata tucked inside. The stomata are sunk down into little grooves. Water vapour that diffuses out of them collects inside the rolled-up leaf, rather than diffusing out into the air around the leaf. When it rains or when it is cooler, the leaf can quickly unroll.

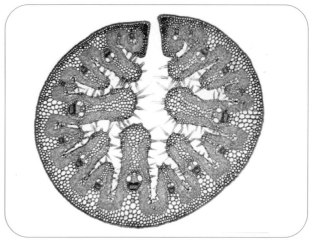

Figure 4c.13 Section through a rolled-up marram grass leaf. The leaf has been stained with a blue dye and photographed using a light microscope. The leaf is actually about 10 mm across.

Summary

You should be able to:

◆ explain what *transpiration* is and why it happens

◆ describe how water moves from the roots to the leaves in xylem vessels

◆ explain how transpiration helps water to move through the plant

◆ describe the function of phloem

◆ describe where xylem and phloem are found in a root, stem and leaf

◆ outline the cellular structure of xylem and phloem

continued on next page

Summary - *continued*

◆ describe experiments to investigate the effect of temperature, humidity, wind and light on transpiration rate

◆ describe how transpiration rate is affected by temperature, humidity, wind and light

Ⓗ ◆ explain why transpiration rate is affected by temperature, humidity, wind and light

◆ state that water uptake should balance water loss

◆ explain how a waxy cuticle and having the stomata on the underside of a leaf reduce transpiration

Ⓗ ◆ explain how guard cells can open and close stomata, and how the number, distribution, position and size of stomata can change the rate of transpiration

Questions

1 Figure 4c.14 shows an experiment to measure the rate of water loss from a plant's leaves.

The plants are weighed at the beginning of the experiment and then again every day for two weeks.

a Which variables should be kept the same in this experiment?

b Which plant would you expect to lose mass? Explain your answer.

2 Copy and complete Table 4c.1.

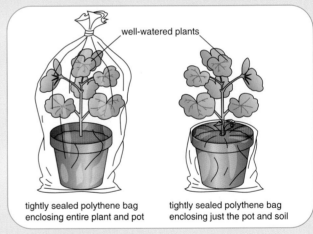

Figure 4c.14 Experiment to measure the rate of transpiration of a potted plant.

	Xylem vessels	Phloem tubes
What do they transport?		
In which direction is the transport?		
What are they made of?		

Table 4c.1

Ⓗ 3 Compare the transport systems of a mammal and a plant. Think about these features and any others you can think of.

● Both systems use tubes. What are the tubes? How are they similar? How are they different?

● Both systems transport substances as liquids. What are the liquids? What do they contain?

● What makes the liquid move around the body in mammal? What makes it move in a plant?

Plants need minerals too

Uses of minerals

You probably know that you need minerals in your diet. You need iron to make the haemoglobin in your blood, and calcium for making bones and teeth. Plants also need minerals. Like us, they need them to make structures and chemicals in their bodies.

Plants absorb minerals from the soil, through their roots. They need many different kinds of minerals, but four are especially important.

- Nitrate ions, NO_3^-, are needed for cell growth.
- Phosphate ions, PO_4^-, are needed for respiration and growth.
- Potassium ions, K^+, are needed for respiration and photosynthesis.
- Magnesium ions, Mg^{2+}, are needed for photosynthesis.

H More about minerals

Nitrate Plants use nitrates for making proteins. Proteins are made of long chains of amino acids, and each amino acid contains at least one nitrogen atom. Plants can make amino acids. They do this by using glucose (made in photosynthesis) and nitrogen (from the nitrates).

Plants can't use nitrogen from the air, because it is much too unreactive. They can only use it when it is already combined with something else. They normally use nitrates, NO_3^-, as a nitrogen source, but they can also use ammonium ions, NH_4^+.

Phosphate Phosphorus (found in phosphate ions) is needed to make DNA. It is an essential part of the 'backbone' of the DNA molecule, which holds the bases in place. Phosphorus is also needed to make cell membranes.

Potassium Many enzymes can only work if they have potassium ions around them. For example, the enzymes needed in photosynthesis and respiration must have potassium ions available.

Magnesium Chlorophyll molecules have magnesium ions in their centres. The magnesium ions are essential for the chlorophyll to absorb energy from light and pass it on so that it can be used in photosynthesis.

Fertilisers

If the soil in which a plant is growing doesn't have enough of a mineral that a plant needs, then the plant will not grow well. Gardeners and farmers can add **fertilisers** to the soil to add minerals and help their plants to grow better.

Many fertilisers contain nitrogen, phosphate and potassium. They are called **NPK fertilisers**. (See also Figure C4c.7 on page 180.)

It's a good idea to test the soil first, to find out what minerals need to be added. It is very wasteful to add a mineral if there is already plenty in the soil. Adding extra minerals to soil can also lead to problems like eutrophication (see page 179 in Item C4c). You can buy fertilisers with different quantities of minerals, so you can get one that fits your plants' needs closely.

Many farmers now use GPS (global positioning system) technology to help them to apply the right balance of fertilisers to the right part of the field. Samples are taken from the soil in different parts of the field and tested. A computer then draws a map of the field, showing what fertilisers are needed in different parts. The farmer feeds this information into a computer in his tractor cab. He loads the fertiliser into a machine that is pulled behind the tractor. As he drives around the field, the computer receives signals from GPS satellites, which tell it exactly where the tractor is. It combines this information with the information about the soil, and allows exactly the right amount of fertiliser to be released onto each part of the field.

Figure 4d.1 Some farmers now use GPS technology.

SAQ

1 How can the use of GPS technology help a farmer to make more profits?

2 How might this technology help the environment?

Mineral deficiencies

What happens to a plant that is short of a particular mineral? Sometimes it will just die, but often it manages to survive, even though it may look very unhealthy.

Figure 4d.2 shows some apparatus you can use to find out what happens to plants that are lacking a particular mineral. The culture solution can be made up to have different amounts of each kind of mineral you are testing. For example, you could grow one plant in a solution with all the minerals and another in identical conditions but in a solution without nitrate.

SAQ

3 **a** Why is the glass vessel in Figure 4d.2 blackened?

   b Why is air bubbled into the culture solution?

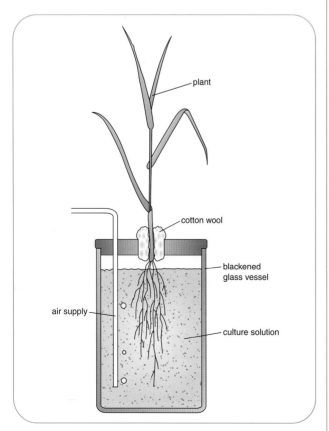

Figure 4d.2 Apparatus for testing mineral needs.

Millionaires and bird droppings

Many people have made their fortunes out of bird droppings. Some of the big houses in Britain today were built from the proceeds of collecting bird droppings in distant parts of the world and transporting them back to England.

The bird dropping boom started in 1835, when a merchant imported some from Peru to Liverpool. The bird droppings are called guano. The waters off the coast of Peru are very rich in food for fish and this supports huge numbers of birds that eat the fish. The birds roost on islands and, over hundreds of years, the guano built up to huge depths. Guano was also collected from islands off the coast of Africa.

By 1847, 220 000 tonnes of guano were being imported each year. The guano was used as fertiliser. It contains a lot of ammonia and a substance called uric acid. Both contain nitrogen in a form that plants can use. Phosphate and potassium ions are also present in guano.

Guano was also used in the chemical industry, to make a dye called murexide. The dye was a rich purple – a colour that had only recently been discovered. Everyone wanted the rich, purple fabrics that could now be made. In 1857, an industrialist in Manchester was spending £6000 a year on guano – an enormous sum of money in those days – but selling the murexide that he made for £100 000.

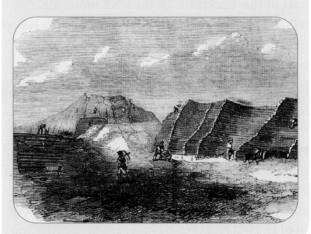

Figure 4d.3 Collecting guano.

Figure 4d.4 Symptoms of different mineral deficiencies. **a** This plant is lacking nitrogen. It is very small and has pale leaves. Its old leaves have gone yellow and died. **b** This plant is lacking phosphorus. Its leaves are dull and discoloured, not the bright green that they should be. **c** This plant has not had enough magnesium. Its leaves are yellow and brown rather than a bright, healthy green. **d** This plant is short of potassium. It has discoloured leaves. Plants that lack potassium usually have poor flower and fruit growth.

How plants absorb minerals

Minerals are present in the soil, dissolved in water. Plants take them up into their root hairs. They are usually present in the soil in very low concentrations, which is why farmers often need to add fertilisers to the soil.

Very often the concentration of a particular mineral, for example nitrate, in the soil is less than its concentration inside the root hair cells. If left alone, the nitrate ions could diffuse *out* of the cell and *into* the water in the soil.

However, plants can make the mineral ions move into the root hair cells, even though their diffusion gradient is going the other way. The process by which they do this is called **active transport**.

In active transport, the plant has to use energy to make the nitrate ions move against their concentration gradient. The energy comes from respiration. Roots need a good supply of oxygen so that they can respire and provide the energy for active transport.

Summary

You should be able to:

- state that plants need nitrate, phosphate, magnesium and potassium for healthy growth, and that these can be supplied as fertilisers

- state why plants need these ions

- explain why plants need these ions

- state that mineral deficiency results in poor plant growth

- describe the symptoms of deficiency of nitrate, phosphate, magnesium and potassium

- state that minerals are usually in relatively low concentrations in the soil and are taken into the plant through its root hairs

- explain how plants take up minerals against their concentration gradient by active transport

Questions

1   Copy and complete Table 4d.1.

Mineral	Why it is needed	Deficiency symptoms
nitrate		
phosphate		
potassium		
magnesium		

Table 4d.1

2   The diagram shows a sack of fertiliser that a farmer is going to use on his fields.

a   What does NPK stand for?

b   Fertilisers are expensive. Explain why it is often worth while for a farmer to add fertiliser to a field.

c   Explain why the farmer might want to use different fertilisers in different fields.

Figure 4d.7

3   a   A gardener grows tomato plants. The fertiliser she uses for her plants is high in potassium. Explain why this is a good fertiliser to use for growing tomatoes.

b   Name the part of the plant:

i   through which the potassium ions get into the roots;

ii   through which the potassium ions are transported up to the leaves and flowers.

c   i   Name *one* other mineral that the growing tomato plants need.

ii   Describe what the plants will look like if they are short of this mineral.

4   Table 4d.2 shows the concentrations of three ions, both inside a plant's root hair cells and in the soil around the roots.

Ion	Concentration inside roots in arbitrary units	Concentration in the soil in arbitrary units
nitrate	10	5
potassium	5	4
chloride	6	6

Table 4d.2

a   Which ion or ions could be moving into the root hair by diffusion? Explain your answer.

b   Which ion or ions must have been taken into the root by active transport? Explain your answer.

c   When it rains very hard, soil can become waterlogged. All the air spaces in the soil get filled up with water. Explain why this can kill plants growing in the soil.

Energy and living organisms

Many of the metabolic reactions in living things need energy. Energy is needed to maintain life. Animals such as ourselves get all of our energy from the food that we eat. Plants get all of their energy from sunlight.

Food chains

You'll remember that food chains show how energy is transferred from one organism to another. For example, in South America, energy could be transferred to anacondas (very large snakes) like this:

grass   →   guinea pig   →   anaconda

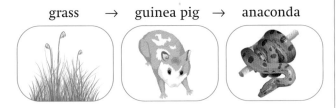

The arrows show the direction of energy flow. Energy flows from the grass to the guinea pig when it eats the grass, and then from the guinea pig to the anaconda when the snake eats the guinea pig.

We can also add the energy in sunlight to this food chain. Now it is an 'energy chain'.

Sun → grass → guinea pig → anaconda

In this energy chain, the grass is a **producer**. It produces food by photosynthesis. Producers are at the beginning of every food chain. Without them, no other living things could exist.

The guinea pig and the anaconda are **consumers**. They consume the food that has been made by producers. All animals are consumers. The guinea pig is a primary consumer and the anaconda is a secondary consumer.

Pyramids of numbers

Within a food chain, we normally find that the number of organisms in the early stages of the chain is much greater than towards the end of it. For example, in the guinea pig → anaconda chain, one guinea pig will need to feed on many grass plants to stay alive, and one anaconda will

need to feed on many guinea pigs. If we counted the numbers of these organisms that lived in a certain area, we would find that the number of grass plants was much greater than the number of guinea pigs, which in turn would be much greater than the number of anacondas.

This pattern can be shown in a kind of diagram called a **pyramid of numbers** (Figure 4e.1)

Each level in the pyramid is a **trophic level**. The producers are on the first trophic level, the primary consumers on the second and the secondary consumers on the third.

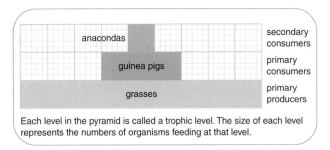

Each level in the pyramid is called a trophic level. The size of each level represents the numbers of organisms feeding at that level.

Figure 4e.1 A pyramid of numbers.

SAQ

1 From the pyramid in Figure 4e.1, work out how many grass plants are needed to support:

  a one guinea pig

  b one anaconda.

2 This is a food chain in a garden pond. The numbers below each organism are the numbers of them in the pond.

  pond weeds → tadpoles → pond skaters

    110   50   10

  Draw a pyramid of numbers to show this information. Use 1 cm^2 to represent 10 organisms.

Biomass

When plants photosynthesise, they take carbon dioxide from the air and water from the soil, and use the energy in sunlight to make them react together to make carbohydrates, such as glucose.

Word equation for photosynthesis:

carbon dioxide + water → glucose + oxygen

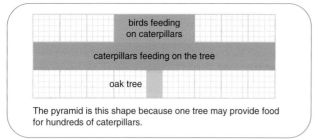

The pyramid is this shape because one tree may provide food for hundreds of caterpillars.

Figure 4e.2 An inverted pyramid of numbers.

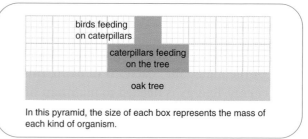

In this pyramid, the size of each box represents the mass of each kind of organism.

Figure 4e.3 A pyramid of biomass.

The glucose can then be used to make substances for building new plant cells. Some of it is converted into cellulose, to make new cell walls. Some is converted to starch and stored inside cells in the leaves and roots of the plant. Some is changed to proteins, which are used to build new cells.

All this new material means that the mass of the plant increases. The mass of living material is called **biomass**.

We can draw pyramids of biomass in the same way as pyramids of numbers, only in this case each bar shows the mass of all the organisms in the food chain, not their numbers.

For example, Figure 4e.2 shows a pyramid of numbers for a food chain in a British oak wood.

The pyramid of numbers is this shape because one oak tree is large and can supply enough food for hundreds (or even thousands) of caterpillars.

If, however, we draw a pyramid of biomass for this food chain, we get a 'normal' shaped pyramid (Figure 4e.3).

Energy losses

Every time energy is transferred from one place to another, or transformed from one kind to another, some of it is wasted. This happens as energy is transferred along food chains.

There are several different ways in which the energy is lost. For example, think about a food chain with grass, rabbits and foxes (Figure 4e.4).

- The rabbits' digestive systems cannot digest all of the grass. Some of it passes through and is lost in their faeces, in the process of **egestion**.
- The rabbits break down some of the glucose from the grass by **respiration** in their cells. Some of the energy in the glucose is lost as heat from their bodies.

So, by the time the energy gets to the foxes, there is only a very small proportion of it left. On average, only about 10% of the energy in the organisms at one level in a food chain gets passed on to the next.

SAQ

3 A farmer grows maize (sweetcorn) and uses it to feed his beef cattle. The food chain in Figure 4e.5 shows how much energy there is in a small patch of maize, and how much of this energy is passed on to the cattle and people.

   a How is energy from the maize transferred into the cattle?

   b What proportion of the energy in the maize is transferred into the cattle's bodies?

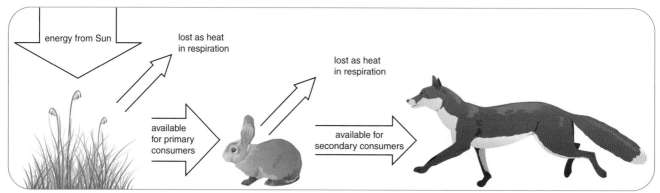

Figure 4e.4 Energy losses in a food chain.

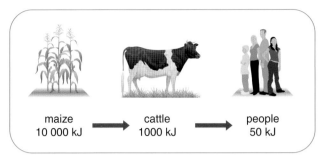

Figure 4e.5

c Explain why only some of the energy in the maize is transferred to the cattle.

d Explain why only some of the energy in the cattle is transferred into the people.

Efficiency of energy transfer

We can calculate the efficiency of energy transfer from one step in the food chain to another like this:

$$\text{efficiency of energy transfer} = \frac{\text{energy transferred to next level}}{\text{energy in the previous level}} \times 100\%$$

Worked example 1

In the food chain in Figure 4e.5, what is the efficiency of energy transfer from the maize to the cattle?

$$\text{Efficiency of energy transfer} = \frac{1000}{10\,000} \times 100\%$$
$$= 10\%$$

SAQ

4 Calculate the efficiency of energy transfer from the cattle to the humans in the food chain shown in Figure 4e.5. Show your working.

Energy losses and pyramids

As energy is lost along a food chain, there is less energy available at each step in the chain. In the 'grass → guinea pig → anaconda' food chain, there is a great deal of sunlight energy available

for the grass but, by the time the energy gets to the guinea pigs, there is less. By the time it gets to the anacondas, there is less still.

The energy is needed to make new biomass and to keep the living cells alive. If there is less energy then there must also be less biomass. This is why pyramids of biomass are the shape that they are. The further you go along the food chain, the smaller the biomass of the animal populations.

SAQ

5 Explain why pyramids of numbers don't always look like a 'pyramid', whereas pyramids of biomass do.

If you go more than about five steps along a food chain, there is so little energy left that there probably isn't enough to support any organisms at all. It is rare to find a food chain that has more than five links in it.

SAQ

6 Write down a real-life food chain that has five links in it. Which animals in the food chain are likely to be the rarest? Why?

Biomass as fuel

The food that we eat is 'biomass'. It has all been made from living organisms. All the energy that we use in our bodies comes from biomass. Biomass can also be used as fuel to generate heat or electricity. Biomass contains energy that originally came from the Sun. It is **renewable** energy – so long as plants keep on photosynthesising, they will keep making more biomass.

Electricity generation from biomass

The UK government has set a target of generating at least 15% of its electricity using renewable sources by 2015. This is a tough target. If we are going to reach it, then biomass could be a major contributor. At the moment, only a small amount of biomass is used in power stations to generate electricity.

One type of biomass that can be used in power stations is wood. A good way of providing the wood is to grow fast-growing trees such as willow or poplar. They can be harvested by coppicing – cutting them down to the ground – and then allowing them to regrow for about three years before harvesting them again.

Figure 4e.6 Fast-growing trees can provide biomass for electricity generation.

One problem with this is that transport costs can get so high that it is hardly worth transporting the biomass to the power station. Not all power stations can use biomass as a fuel – they have to have special burners – so it is important to make sure that the crops are grown close to a power station that can use them, to minimise transport costs.

As well as being renewable, energy from burning biomass has another ecological advantage. It is **carbon neutral**.

There is great concern that burning fossil fuels is increasing the quantity of carbon dioxide in the atmosphere, leading to global warming. The fossil fuels were formed millions of years ago and, when we release the carbon dioxide from them, it just goes into the air. We can't put it back into more fossil fuels.

However, when we burn biomass, the situation is rather different. Carbon dioxide is released when the biomass burns. But that carbon dioxide was only taken from the atmosphere last year or a couple of years ago, when the trees were growing. As the trees regrow, they take more carbon dioxide from the air. So the quantity of carbon dioxide that is released into the air when we burn the biomass is no greater than the quantity that is being taken out of the air by the growing trees.

H

SAQ
7 Explain how the advantage of carbon neutrality when using biomass as fuel could be wiped out if the power station is a long way from the place where the biomass is grown.

Yet another advantage of using biomass as a fuel is that we can keep growing our own. We don't need to import it from another country. This isn't true for fossil fuels. Although we have had natural gas from the North Sea for many years, this is now running out and we are importing oil from other countries in order to generate electricity. With biomass, we could be self-reliant.

Biogas

Biomass can be used for generating a fuel called **biogas**. Biogas contains methane, which provides heat when it is burnt.

Biomass can be changed into biogas by **fermentation**. This is done by bacteria and yeasts. Usually, the biomass is put inside a **digester** (Figure 4e.7), where the bacteria and yeasts break it down.

In some parts of the world, for example rural areas of China, people have small biogas generators close to their houses. All their waste goes into it – their own faeces and urine, the waste from their animals, waste food and so on. They can tap off the biogas and use it as a fuel for heating their houses. The solid waste inside the fermenter breaks down into a harmless sludge that can be used as a fertiliser for their crops.

Alcohol

Alcohol can be used as a fuel. A type of alcohol called **bioethanol** can be made by allowing yeasts to ferment biomass. (Bioethanol is just ethanol that has been made by living organisms.)

For example, Brazil began a programme to make bioethanol from sugar cane as long ago as 1975. The bioethanol was mixed with petrol and used to fuel cars. Car engines can also be made to run on bioethanol alone. By 1988, one third of the cars in Brazil were fuelled by bioethanol. Some bioethanol is now being imported into the UK from Brazil and added to petrol for use in cars.

Train fuelled by cow's intestines

In 2005, the world's first biogas-fuelled train began to take passengers. The train runs in Sweden and the biogas that fuels it comes from the intestines of cows.

The cows are killed for meat in a town called Linkoping. Their intestines cannot be used for food and are not easy to dispose of hygienically. Now they are being used to make biogas. They are taken in a tanker to a nearby biogas factory, where they are put inside a huge tank and left for one month, while bacteria break them down and produce biogas.

The biogas is used instead of diesel in the trains. At the moment, it is more expensive than diesel – but with world fuel prices going up, this probably will not be true for long. It takes the intestines of one cow to move the train about 4 km.

The biogas is also sold at pumps like petrol pumps. The 65 buses in Linkoping are also fuelled by it and some people use it in their cars.

SAQ

8 Unlike coal, biogas and bioethanol do not contain much sulfur. Explain how using biogas and bioethanol could produce less atmospheric pollution than burning coal.

Choices

You have seen that we have a wide range of choices about what to do with biomass that farmers can produce on their land. We can:

- eat it ourselves as food, for example as bread (made from wheat), potatoes, or fruit;
- feed it to animals and then eat those;
- burn it as a fuel.

With cereal crops such as wheat, there is another choice. Instead of eating the wheat grains (seeds) or feeding them to animals, we can save them to plant to produce another crop the next year.

Which is the best thing to do? It depends on the circumstances – all of them are good choices at different times and in different places.

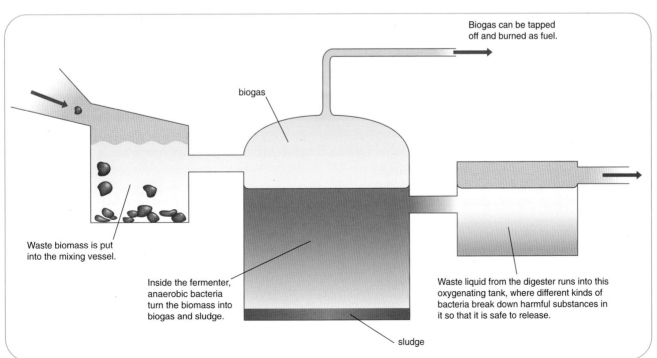

Waste biomass is put into the mixing vessel.

biogas

Biogas can be tapped off and burned as fuel.

Inside the fermenter, anaerobic bacteria turn the biomass into biogas and sludge.

Waste liquid from the digester runs into this oxygenating tank, where different kinds of bacteria break down harmful substances in it so that it is safe to release.

sludge

Figure 4e.7 A biogas digester.

H *SAQ*

9 A farmer has a field of maize. Which of the following would give us the most energy from the maize?

● Eating the maize ourselves.

● Feeding the maize to cows and then drinking their milk.

Explain your answer.

Summary

You should be able to:

◆ explain the terms *producer* and *consumer*

◆ draw pyramids of numbers and pyramids of biomass, and explain the difference between them

◆ explain how energy flows through food chains and interpret data about this

◆ explain how energy is lost from a food chain as heat and through egestion

H ◆ calculate efficiency of energy transfer in a food chain

◆ explain the shape of pyramids of biomass and the limited length of food chains in terms of energy losses

◆ explain what is meant by *biomass* and give examples

◆ explain how energy can be transferred by burning fast-growing trees or by generating biogas

H ◆ explain that biofuels are renewable, how they make us self-reliant and why they are non-polluting

◆ discuss the choices about how to use biomass

Questions

1 Write down what you had, or will have, for lunch today. Are you a primary consumer, a secondary consumer or both? Explain your answer.

2 a Explain what is meant by a renewable fuel.   c Explain why wood is a renewable fuel.

  b What is biomass?   d How can biomass be used to generate electricity?

3 Figure 4e.8 shows a food web in a wood.

  a Name one producer, one primary consumer and one secondary consumer in this food web.

  b What do the arrows in the food web mean?

continued on next page

Questions - *continued*

c   What is the longest food chain in the food web?

d   Sketch a pyramid of numbers for this food chain. (You cannot draw it accurately because you don't have any figures – just show the approximate shape you would expect it to be.)

e   Would a pyramid of biomass be the same shape? Explain your answer.

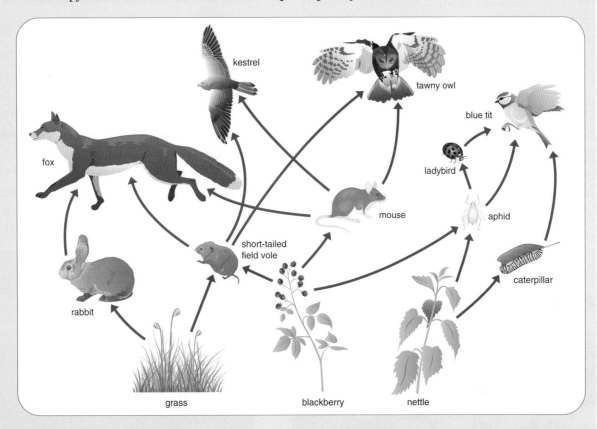

Figure 4e.8 A food web in a wood.

H   **4**   This family is trying to survive through a drought year. They have a difficult decision to make.
Do they:

●   eat their tiny crop of grain;

●   save the seeds so that they will be able to grow a crop next year;

●   feed them to their mule so they have transport to the town where they can get food and healthcare from a charity?

Imagine that three members of the family all have different views about this. Write a short paragraph explaining each person's point of view.

Figure 4e.9

Intensive farming

Farming involves growing crops and keeping animals to produce food. Before the Second World War, most farming in Britain was extensive farming. Crops were grown without the use of very much fertiliser or other chemicals. Animals spent most of their time outside, grazing on grass in open fields.

During World War II, the UK ran very short of food because ships could not get here to bring imports of food from other countries. Farmers were encouraged to get as much food as possible from their land. This continued all through the second half of the 20th century. The government encouraged farmers to get the biggest yields they could.

To get high yields from crops or farm animals, you have to have high inputs. This means that you use chemicals, such as fertiliser, to help crops

Figure 4f.1 Extensive farming.

Figure 4f.2 Intensive farming.

grow well. Herbicides and pesticides are also used. Animals might be kept indoors instead of outside, and fed on high-energy diets. Although this means that the farmer has to spend more, in the end he may make more profit because he gets more produce to sell. This is **intensive farming**.

Pesticides and herbicides

Wheat is a major crop grown by British farmers. Wheat is a cereal – a kind of grass – that produces quite large seeds, full of starch, protein and vitamins. Wheat seeds are sown in the ground in the autumn. They germinate and grow very slowly through the winter, speeding up in spring. In summer, they flower and the flowers develop seeds (grain). The ripe wheat is harvested in August. The grains are sold and are used to make flour for making bread, biscuits or breakfast cereals.

Figure 4f.3 Only wheat grows in this field.

If you look at Figure 4f.3 you will see that the whole field is full of wheat and nothing else. No other plants grow there. If the farmer allowed other plants to grow, they would compete with the wheat for light, water and minerals from the soil. To prevent this competition, the weeds in the field are killed. This is usually done using chemicals called **weedkillers** or **herbicides**. The farmer will normally use a **selective weedkiller**, which kills the weeds but not the wheat. We saw earlier that plant hormones can be used as selective weedkillers (see page 41 in Item 3f). Using herbicides helps the farmer to get a higher yield of wheat from the field.

Other chemicals called **pesticides** may also be sprayed onto the crop. Wheat can get quite a few different diseases. For example, a fungus can cause 'rust' disease. The fungus grows on the wheat leaves, taking nutrients out of them and weakening the plant so that it does not produce as much grain. The farmer can stop this happening by spraying a type of pesticide called a **fungicide** onto the wheat. He may also spray **insecticides**, to kill insects such as aphids, which feed on the wheat plants.

Figure 4f.4 Poppies and other weeds are growing in this wheat field, because the farmer has not used herbicides.

SAQ

1 A farmer grows a crop of wheat on 1 hectare of land. If the crop grows well, he can sell it for £500. If a weed called wild oats grows in the crop, he gets a lower yield of wheat. Table 4f.1 shows how the wild oats affect his profits.

Number of wild oats per m^2	Price he gets for the crop if he does not kill the wild oats
35	£182
80	£84

Table 4f.1

a Explain why the wheat yields are lower if wild oats grow in amongst the wheat.

b Calculate how much money the farmer loses if there are 35 weeds per m^2 in the crop and he does not spray it with a herbicide.

c It costs him £20 to spray the field of wheat with a herbicide that kills all the wild oats. If he does this, what return does he get on each £1 that he spends?

Problems with pesticides and herbicides

Pesticides and herbicides are unpleasant and dangerous chemicals. They have to be – they are used to kill animals, plants or fungi. If not used carefully, they can harm the environment or people's health.

● Pesticides may kill animals other than the ones you want to kill. For example, if you spray a pesticide onto a crop to kill aphids (greenfly) you could also kill ladybirds. Ladybirds are helpful insects – they don't hurt the crop at all and they eat the aphids that are doing harm. Or you might kill honeybees, which help to pollinate the flowers of many crop plants, such as tomatoes and apples. Moreover, killing one organism removes it from a food web, and this can affect other organisms in the web.

Figure 4f.5 Ladybird eating an aphid.

● Pesticides and herbicides can be harmful to humans. Many people think that they have been harmed by breathing in pesticides or herbicides that have been sprayed onto crops near their homes or when they were walking through a field. It has been very difficult to prove that their ill health really has been caused by the pesticides, but there is certainly a possibility that this has happened. People are also worried that some of the food we eat, especially fresh fruit and vegetables, may still have traces of pesticides or herbicides on it. There are strict limits on the quantity of pesticide that can be present in a food before it is sold, but not every individual consignment is tested so it is always possible that some badly contaminated food might go on sale.

Figure 4f.6 The numbers show the amount of DDT in each kind of animal in this food chain. An investigation in the 1970s showed that, although the DDT levels in the water of an estuary were only 0.000 05 parts per million (ppm), the amounts in the animals feeding in the estuary were much greater.

- Some of the pesticides that were used in the past, especially DDT, caused a lot of harm to animals near the end of food chains. DDT is a **persistent** insecticide. This means that it does not break down easily. The DDT stayed in the bodies of insects that were killed by it. Birds or other animals ate the insects and the DDT stayed in their bodies, too. By the end of the food chain, the concentration of DDT had built up so much that it killed birds of prey or weakened their egg shells so much that they could not hatch their young. Once DDT and other persistent pesticides were banned, this problem stopped. We now use **biodegradable** pesticides – ones that can be quickly broken down by micro-organisms in the soil. However, some countries still allow DDT to be used.

DDT can accumulate in food chains because it is not broken down in an animal's body. It dissolves in fat stores in the animal and just stays there. If the animal is not killed by it then every time it eats food that contains DDT, more and more DDT builds up in its fat stores.

If a predator then eats that animal, it is eating all the DDT that its prey ate during its lifetime. If it eats a lot of prey then it eats a lot of DDT – and all of it stays in its body. This carries on all the way up the food chain. The concentrations of DDT that were found in some dead birds of prey were up to 30 parts per million – much greater than the concentration needed to kill an insect or other pest.

H Why intensive farming works

We have seen that the yield of a crop in which no weeds are growing is greater than the yield from a crop that is full of weeds. If weeds are growing, they compete with the crop plants for light, water and minerals. Some of the energy in the sunlight that falls onto the field is trapped and used by the weeds, rather than by the crop plants. If the weeds are killed, then the crop plants have all the light energy to themselves. They can photosynthesise more and therefore grow more.

Pests such as insects feed on the crop plants. They take in energy-containing foods from the plants. This means that some energy from the plants is transferred into the pests, rather than being used for growth of the plants. By spraying the crop with pesticides, the farmer reduces the amount of energy that is being transferred to pests.

Crops in glasshouses

In Britain, some crops are grown under glass. This includes tomatoes and cucumbers. These plants need warm conditions to grow well. They can grow outside in summer, but yields are usually better if they are grown in glasshouses.

In a glasshouse, the grower can control:
- the temperature – even if it gets very cold outside, heaters can be used to keep the temperature just right for the plants to grow well;

Figure 4f.7 Hydroponic cultivation of lettuce plants.

- the lighting – even when days are quite short (in autumn, winter and spring) extra lighting can give the plants long days in which to photosynthesise and grow.

This increases the **productivity** of the plants – how much they grow and how much fruit they produce.

Growing in a glasshouse can also reduce problems with pests and weeds.

SAQ

2   Suggest why it is easier to control pests and weeds in a glasshouse than it is outside in a field.

3   An experiment was done to investigate the best temperature for growing cucumbers in a glasshouse. Table 4f.2 shows some of the results.

Temperature in °C	Mean number of cucumbers per plant	Mean mass of cucumbers per plant in kg
23	9.9	4.48
25	11.4	5.20
27	11.1	5.14

Table 4f.2

  a Calculate the mean mass of each cucumber when the temperature was 23 °C.

  b Which temperature produced the highest yield?

  c Suggest why the other two temperatures did not produce such high yields. (Deal with them one at a time, because the reasons are not the same for both of them.)

Hydroponics

Hydroponics is a way of growing plants without soil. Quite a large proportion of the UK tomato crop is grown like this. Hydroponics is also used in countries like Saudi Arabia, where the soil is much too dry and barren to support crops.

To understand how hydroponics works, you need to think about why plants need soil. Soil provides plants with:

- something for their roots to anchor them, so they don't fall over or blow away;
- water;
- mineral ions;
- oxygen.

Hydroponics provides all these things without using soil. There are many different ways of growing plants hydroponically. Figure 4f.8 shows one way in which it can be done.

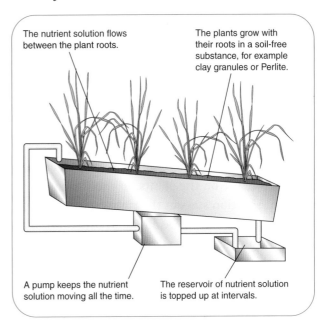

The nutrient solution flows between the plant roots.

The plants grow with their roots in a soil-free substance, for example clay granules or Perlite.

A pump keeps the nutrient solution moving all the time.

The reservoir of nutrient solution is topped up at intervals.

Figure 4f.8 Growing plants hydroponically.

Plants are grown in a substance such as clay pebbles, which helps to support them. They are supplied with water and minerals by providing them with a solution containing all the substances that they need, in just the right concentrations. This is called a **nutrient solution**. In some kinds of hydroponics, the plants just grow in pots and are watered with the solution. In others, such as the one shown in Figure 4f.8, the nutrient solution is pumped over their roots.

It is important to make sure that there is plenty of oxygen in the solution, so that the plant roots can respire. This can be done by bubbling air through the nutrient solution.

There are many different reasons for growing plants hydroponically rather than in soil.

- Being able to grow plants in places where there is not any soil, for example in a desert. In 1997, NASA experimented with growing plants hydroponically in the Space Shuttle Columbia as it orbited the Earth. If long manned space journeys ever happen, growing plants in the spacecraft could help to provide oxygen and possibly even food for the astronauts.
- Having complete control over the nutrients supplied to the plants. You can make up the nutrient solution to exactly match the requirements of the plants. None of the minerals are wasted; if the plants do not take them up, they just stay in the solution and go round the system again.
- Making it easy to control diseases. Soil can harbour pests and diseases, but this is less of a problem with hydroponics. It is relatively easy to control these when plants are grown hydroponically.

On the downside, hydroponics can be an expensive way of growing plants. Although the system can be automated, someone does need to keep a close eye on the plants, to check that the nutrient solution stays at the right concentration. Fresh minerals need to be added to the solution at regular intervals. If you were growing the plants in soil, the soil could hold enough minerals to keep the plants going for a much longer time.

Battery farming

Battery farming is an intensive way of farming animals. It involves keeping them indoors in a fairly small space. In Britain, chickens are kept like this to produce eggs (Figure 4f.9).

Keeping chickens penned like this can increase their productivity. They are not wasting energy by walking around, so more of their energy can go into making eggs. They can be kept warm and free of pests and diseases. They don't peck each other as they sometimes do when they can interact with one another.

The downside is that the chickens have lost their freedom. They cannot walk around and

Figure 4f.9 Battery hens provide cheap eggs.

behave in their normal manner. Many people question whether it is right to keep animals like this. They prefer to buy eggs from chickens that can roam outside in a large pen (free-range eggs) or from chickens that live in a large building (barn eggs). These eggs cost more, because the farmer does not get as many eggs per chicken as he would if he used battery farming.

Cattle can also be kept indoors, rather than outside in fields. In Britain, most farmers bring their cattle inside during the winter. If they are left outside, they need a lot to eat, so they can produce heat by respiration and keep their body temperature high enough. If they are kept inside, they can be kept warm. More of the food that they eat can be used to make new biomass in their bodies. So they grow faster, or make more milk, than they would outside.

SAQ

4   An investigation was done into the use of energy by dairy cows. The amount of energy in their food was measured, and also the amount of energy that their bodies used for various purposes per day. The mean values per cow were then calculated. The results are shown in Table 4f.3.

	Energy in MJ per day
Energy in cows' food	500
Energy in faeces	115
Energy in urine	20
Energy in heat produced	170
Energy in milk	120

Table 4f.3

H

a Which types of nutrients in the food given to the cattle would contain energy?

b Explain how faeces and urine can contain energy.

c Calculate the amount of energy per day that can be used to produce new tissues in the cow's body.

d Use the data to explain how keeping cows indoors in a warm environment could increase their milk yield or their rate of growth.

Fish farming

Many species of fish that have traditionally been caught in the seas around Britain are now endangered. We have caught so many of them that there are not enough left to reproduce and keep the populations at a high level. The European Union imposes limits on how many fish of each species can be caught, as well as other measures to try to conserve the remaining fish stocks. But these measures have not been strict enough and the populations of cod and other fish have fallen to dangerously low levels.

Fish is a good food, so it would be a pity if we could no longer buy it or eat it. One way to try to deal with the problem is to farm fish. This is done on quite a large scale around the coast of Scotland.

Many of the fish that are farmed in Scotland are salmon. The salmon are kept in underwater cages in sheltered sea lochs. They are fed well and grow quickly. They can be treated with pesticides and antibiotics to keep them free of disease.

Figure 4f.10 The cages at a salmon farm in Scotland.

However, there have been environmental problems with fish farming.

● Some of their food, and all of their faeces, drop to the bottom of the water, causing pollution.

● The pesticides used to treat the fish can kill other animals that live in the loch.

Organic farming

Organic farming is a way of growing crops or raising animals without using:

● artificial fertilisers
● herbicides
● most kinds of pesticides.

Organic farming techniques

The aim of organic farming is to farm in a way that works with the natural environment rather than harming it. (The word 'organic' doesn't mean the same as it does in chemistry.)

Organic farmers do not use artificial fertilisers, such as ammonium nitrate, on the land. Instead they use natural fertilisers such as manure from cattle or compost made from plant remains. They might also grow crops such as beans, which have **nitrogen-fixing bacteria** in their roots that convert nitrogen in the air into a form that the plant can use (see pages 105–106 in Item B4h). They are **nitrogen-fixing** crops. Some of this fixed nitrogen stays in the ground and can be used by the crop that grows in that field next year.

Instead of using herbicides, they can just allow weeds to grow in amongst the crop. They may also use machinery to hoe the weeds (cut them off at ground level).

Organic farmers try not to use pesticides, but if they really must then there are a few that are allowable. There are more than 500 synthetic chemicals that 'conventional' farmers are allowed to use, but only seven that are licensed for use in organic farming.

Organic farmers usually use **crop rotation** to help to keep pests down to a manageable level. Crop rotation involves growing different crops in a field each year. Organic farmers may also 'stagger' the sowing of a crop so that the plants are not all in the same stage of growth at the same time. This can mean that some of them may escape the attentions of pests.

Figure 4f.11 An organic farmer stacking a tray of onions.

Organic farming does not produce such high yields as conventional farming. This means that the crops are more expensive to buy. Many people are prepared to pay the higher prices because they like the idea of eating food that has been produced without using pesticides or herbicides.

However, there is no evidence at all that organically produced food is better for you than other food. The nutrients in it are about the same. In controlled tests, people could not tell any difference between the flavours of organic food and non-organic food. There is, however, less of a risk that organic food will have traces of pesticides on it, but organic farmers can still use unpleasant substances such as copper sulfate to kill pests if necessary.

But there *is* evidence that organic farming can be good for the environment. Biodiversity on organic farms is often higher than on conventional farms. In a big study carried out in 2005, researchers found that organic crops contained 85% more plant species than conventionally grown crops. There were 17% more spiders, 5% more birds and 33% more bats living on organic farms.

SAQ

5 Suggest why there are more spiders, birds and bats on organic farms.

6 An investigation was carried out into the yields of wheat, oats and beans that could be grown on similar land, either organically or conventionally. Table 4f.4 shows some of the results.

Farming method	Wheat in tonnes per hectare	Oats in tonnes per hectare	Beans in tonnes per hectare
organic	4.30	4.30	3.60
conventional	7.12	5.75	3.60

Table 4f.4 Yields for three crops.

a Compare the yields of the two cereal crops when grown organically or conventionally.

b Suggest at least *two* reasons for these differences.

c Suggest reasons for the results for beans.

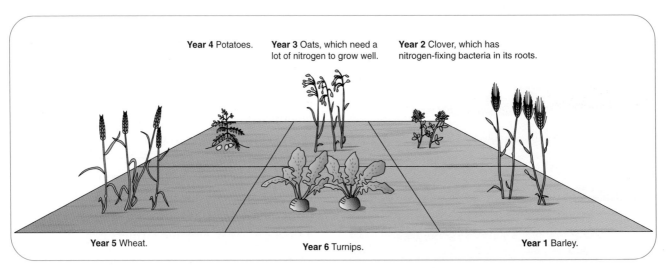

Year 4 Potatoes. **Year 3** Oats, which need a lot of nitrogen to grow well. **Year 2** Clover, which has nitrogen-fixing bacteria in its roots.

Year 5 Wheat. **Year 6** Turnips. **Year 1** Barley.

Figure 4f.12 One type of crop rotation. By growing different crops on the same piece of land in successive years, a farmer can gain several benefits. For example, crops like clover may provide nitrogen for the following crop. Also, a disease of one crop will not get the chance to infect that crop in the following year, and may die out before the same crop comes around again.

Biological control

Biological control is a way of keeping pests under control without using pesticides. It usually involves using a predator of the pest to kill and eat it.

For example, tomato plants can be infested with a tiny pest called whitefly. The whitefly feed on the sap in the veins of the plant's leaves. They breed quickly and can rapidly weaken the tomato plants and reduce their productivity.

If the plants are being grown in a glasshouse then a tiny wasp called *Encarsia* can be introduced into the glasshouse. The wasps lay their eggs in the whitefly larvae. Each egg hatches into a wasp larva, which eats the whitefly larva from inside. The whitefly larva is killed.

Figure 4f.13 *Encarsia* wasps (on the right of this photo) are used for biological control of whitefly.

The *Encarsia* wasps can keep the whitefly numbers down to such a low level that they don't cause a problem. Usually, they don't get rid of the whitefly completely, but that is acceptable so long as the few remaining ones are not enough to reduce the yield of tomatoes.

Advantages of using biological control

- The predator will usually only attack one kind of prey, so there is no risk of killing other beneficial organisms.
- Once introduced, the predator can stay around for a long time. The grower doesn't have to keep on putting new ones in. If she sprayed a pesticide onto the crop, she would probably need to keep on doing it at regular intervals.
- The predator will not harm the health of anyone. Pesticides can be harmful to humans if they breathe them in or swallow them.

Disadvantages of biological control

- The predator does not completely eliminate the pest.
- The predator needs to be introduced before the pest population has got too high or it may not be able to kill enough of the pest to bring the population down to an acceptable level.
- It doesn't work very well on outdoor crops, because the predator is likely to fly away and new pests can fly in at any time.
- The predator can sometimes become a pest itself.

Cane toads

Figure 4f.14 Biological control gone wrong.

In the 1930s, Australian farmers had a huge problem with beetles feeding on their sugar cane crops. Someone decided it would be a good idea to introduce a large toad, called a cane toad, from South America as a biological control agent.

The cane toads did well. They ate the beetles and improved yields from the sugar cane crops. However, the toads bred rapidly. Now there are millions of them in Australia. With not many beetles left to eat, the toads eat almost any small animal that they come across. Several of Australia's rare small mammals have become in danger of extinction because of the cane toad.

continued on next page

Cane toads *continued*

But the problem is even worse than that. Cane toads secrete a heart and nerve poison on their skin. If another animal tries to bite the toad, it can be poisoned. Several dogs in Australia have died because of this.

People are trying to get rid of the toads but it is very difficult. Anyone who comes across a cane toad is asked to pick it up using gloves and then destroy it. The recommended method is to double wrap it in freezer bags and put it in the freezer over night. Unfortunately, many of the 'cane toads' that people destroy turn out to be native frogs, some of which are endangered species.

Summary

You should be able to:

- explain what is meant by *intensive farming* and give some examples of farming crops and animals intensively

- explain why using pesticides and herbicides (including insecticides and fungicides) can increase yields

- explain why pesticides and herbicides can cause harm to the environment and human health

H
- explain how intensive food production improves the efficiency of energy transfer

- explain why some pesticides may accumulate in food chains

- describe how glasshouses, hydroponics, fish farming and battery farming can increase productivity

- describe how plants can be grown hydroponically, and outline some of the uses of this technique

H
- explain the advantages and disadvantages of hydroponics

- describe what is meant by organic farming

- describe techniques used in organic farming

H
- discuss the advantages and disadvantages of organic farming techniques

- describe how pests can be controlled biologically by introducing predators

- explain the advantages and disadvantages of biological control

Questions

1 Copy Table 4f.5 and complete it.

Farming technique	Advantage	Disadvantage
using pesticides		
using biological control		
keeping battery hens		
fish farming		

Table 4f.5

2 Write down the definitions of:

a pesticide   b herbicide   H c fungicide

d insecticide.

3 Barley is a cereal crop. As the plants grow, they put up new shoots called tillers. The graph in Figure 4f.15 shows the effect of weeds on the number of tillers that barley plants grew.

a How many tillers did the barley plants produce when there were no weeds?

b Calculate the difference in number of tillers if there were 800 weeds per square metre.

c Explain the reasons for this difference.

d How could the farmer get rid of the weeds by:

i farming conventionally?

ii farming organically?

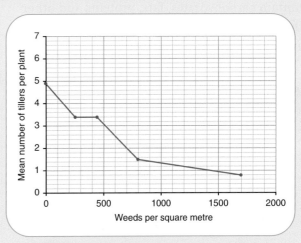

Figure 4f.15 Growth of barley.

4 If everyone in the UK farmed organically, we would need to use more land to get the same amount of food from it.

Put forward arguments for and against growing food organically.

5 DDT is used in developing tropical and subtropical countries to kill the mosquitoes that spread malaria. Malaria kills 1.5 million people, many of them children, each year. There is no vaccine for malaria.

The World Health Organization wants to keep on using DDT, because it is a cheap and very effective insecticide. There are no others that work so well. The World Wide Fund for Nature wants to ban DDT completely.

What are your opinions on this debate? Use scientific facts to support your suggestions.

Micro-organisms and decay

You already know that some micro-organisms – which include bacteria and fungi – can cause disease. But there are millions of others that do not cause harm to humans. Some of these are very useful because they help to decay waste products, making them harmless and in some cases useful. However, decay can sometimes be a nuisance, for example if food decays or wet leather shoes go mouldy.

How decay happens

Food, bodies, faeces and paper are examples of materials that can decay. They are all substances that were once part of a living organism. They contain carbohydrates, fats, proteins and other chemicals that can be broken down by micro-organisms. They are **biodegradable**.

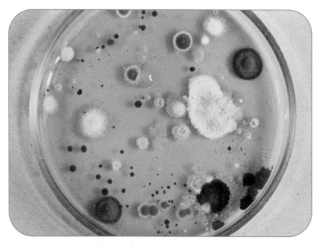

Figure 4g.1 Bacteria and fungi growing on agar jelly in a Petri dish, which had been left open to the air. The 'furry' colonies are fungi and the smoother ones are bacteria. Each colony probably grew from a single organism which landed on the agar.

The micro-organisms in Figure 4g.1 are breaking down the nutrients in the agar jelly in the Petri dish. They secrete enzymes, which digest the substances in the jelly. The jelly gradually decays.

SAQ

1 Which of these things can decay?
   a plastic box; a banana skin; a brown paper bag; a drinks can; toe nail clippings; a glass bottle.

If you leave food around in a warm place, it will probably decay. It will decay most quickly when the conditions are best for the micro-organisms to grow quickly. These include:
- water – if there is only a little water, decay will be very slow;
- oxygen – if there is no oxygen, decay will be slow;
- a suitable temperature – if it is too hot or too cold, decay will be slow.

Figure 4g.2 Bacteria and fungi cause food to decay.

Micro-organisms, like all living things, need water to survive. We make use of this fact by drying some foods so they won't decay. Dried fruit, dried mushrooms and dried soups all last for a long time without decaying.

Oxygen is needed for the micro-organisms to respire. If food is packed in a container without any oxygen, it won't decay, or will only do so slowly. For example, vacuum-packed bacon lasts much longer than 'loose' bacon. However, there are some micro-organisms that can live without oxygen, and they can cause slow decay even when no oxygen is present.

The temperature affects the rate at which the micro-organisms grow. At very low temperatures, metabolic reactions take place only slowly, so bacteria and fungi don't grow fast and they don't break down the food quickly. This is why keeping food in a freezer (at a temperature below −10 °C) or a fridge (at about 4 °C) makes it last longer. The micro-organisms aren't killed and, once the food is warmed up, they will start to grow again.

H Micro-organisms are killed at high temperatures. Boiling will kill most of them, but some have tough spores that need to be heated to over 100 °C to kill them.

SAQ

2 Explain why you should not refreeze food once it has thawed.

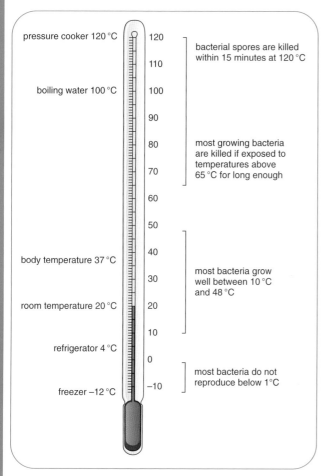

Figure 4g.3 How temperature affects bacteria.

pressure cooker 120 °C

bacterial spores are killed within 15 minutes at 120 °C

boiling water 100 °C

most growing bacteria are killed if exposed to temperatures above 65 °C for long enough

body temperature 37 °C

room temperature 20 °C

most bacteria grow well between 10 °C and 48 °C

refrigerator 4 °C

freezer −12 °C

most bacteria do not reproduce below 1°C

Preserving food

We can make food last longer by taking away one or more of the things that micro-organisms need. This will prevent the food decaying.

Canning involves packing the food into a metal can. The can is heated to a temperature high enough to kill all the micro-organisms in it, and then sealed so no more can get in. Canned food can last for a very long time. Captain Scott and his team left some cans of food at a camp when they tried to reach the South Pole. The cans were found 47 years later and the food was still good to eat.

Cooling food in the fridge, as we have seen, slows down metabolic reactions in micro-organisms. You can keep foods like fresh meat in the fridge for a few days, whereas it would go bad quite quickly if you left it in a warm place.

Freezing slows the micro-organisms down even more. Foods such as frozen vegetables and meat can be kept in a freezer for several months before there is any sign of decay.

Drying deprives micro-organisms of water, so they cannot grow. Dried foods such as raisins can keep for a very long time.

Adding salt or sugar is another way of depriving micro-organisms of water. The concentrated solution of salt or sugar causes water to move out of the micro-organisms by osmosis. We use this when making jam or salted meats.

Adding vinegar makes the environment too acidic for micro-organisms to grow. You will remember that enzymes work best at a particular pH. The low pH of vinegar stops the enzymes in most micro-organisms from working. We use this when making pickles and chutneys.

Figure 4g.4 a Fish kept fresh by cooling them on ice. **b** Ham being salted to preserve it.

Detritivores and decomposers

Organisms that cause decay are called **decomposers**, because they decompose (break down) biomass. Most decomposers are microscopic bacteria and fungi. However, some larger organisms also feed on biomatter and help to break it down. They are called **detritivores**. Earthworms, maggots and woodlice are detritivores.

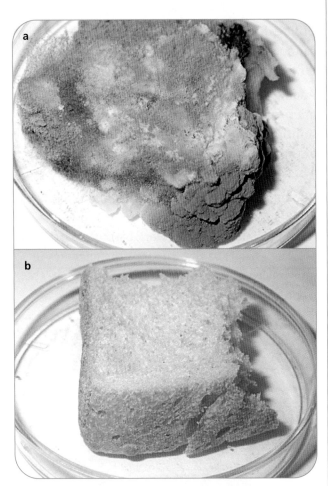

Figure 4g.5 a This mouldy bread was kept moist in a warm place. **b** This bread was dipped in a weak hypochlorite solution to kill bacteria and fungi.

Detritivores help to break down large pieces of biomatter – for example, leaves – into smaller pieces. This increases their surface area, so the decomposers can digest the biomatter more quickly. Detritivores speed up the rate of decay.

Organisms that feed by secreting enzymes onto their food are called **saprophytes**. Fungi feed like this and so do many other micro-organisms, including bacteria.

Figure 4g.7 shows the structure of a fungus called *Mucor* and also how it feeds. *Mucor* feeds on

bread and other kinds of food. It is made up of long, thin threads called **hyphae**. Enzymes are secreted from the tips of the hyphae. The enzymes digest the large molecules in the bread. The enzymes are the same kind as the ones in our digestive system – proteases, lipases and carbohydrases (including amylase).

SAQ

3 Name the substrates and products of these three kinds of enzymes.

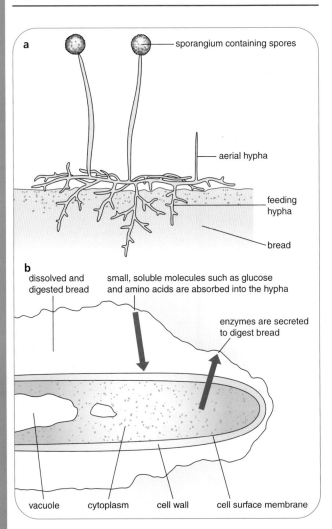

Figure 4g.6 a *Mucor*: the network of hyphae is called a mycelium. **b** The tip of a hypha growing through bread.

Making compost

Many people with a garden have a compost heap. They put garden waste and also food waste – such as leftovers or potato peelings – onto the heap. Micro-organisms in the heap gradually feed on the waste and break it down. Figure 4g.8 shows how to make a good compost heap.

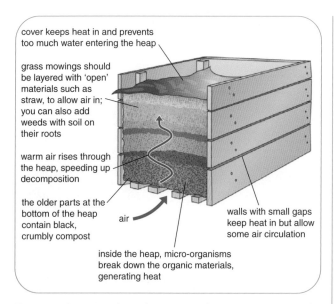

cover keeps heat in and prevents too much water entering the heap

grass mowings should be layered with 'open' materials such as straw, to allow air in; you can also add weeds with soil on their roots

warm air rises through the heap, speeding up decomposition

the older parts at the bottom of the heap contain black, crumbly compost

air

walls with small gaps keep heat in but allow some air circulation

inside the heap, micro-organisms break down the organic materials, generating heat

Figure 4g.7 Section through a compost heap.

The micro-organisms in the heap produce enzymes that digest the proteins, fats and carbohydrates in the waste. They use these nutrients in respiration, to provide energy for themselves. Eventually, after a few months, the waste material is unrecognisable. The micro-organisms turn it into a dark brown, crumbly material, excellent for putting onto the garden to improve the growth of plants.

To do this, the micro-organisms need:

● a suitable temperature – the compost heap stays warm because some of the energy released by the micro-organisms in respiration is lost as heat;

● plenty of oxygen – the compost heap should be constructed so that air can move through it;

● some water, but not too much – the waste material will probably have quite a lot of water in it to start with and this will be enough; a cover on the heap stops too much rain water getting in.

Treating sewage

Sewage is waste liquids from homes and industry. Sewage should not be allowed just to flow into rivers or the sea.

● It may contain pathogenic micro-organisms, such as the bacteria that cause cholera or typhoid.

● It contains material that provides food for bacteria in the river. This can make the

Forensic entomology

Entomology is the study of insects. A forensic entomologist uses knowledge of insects and other invertebrates to help to answer questions about possible crimes.

All dead bodies, whether plant or animal, are broken down by detritovores and then decay organisms. This happens in a predictable sequence.

Firstly, certain species of flies arrive and lay their eggs on the body. The eggs hatch into maggots, which feed on the dead tissue. They moult several times over the next few days. By identifying the stage of growth of the maggots on the body, the entomologist can work out how long ago the eggs were laid and therefore the probable time of the person's death. To do this with any accuracy, he has to take into account the temperature of the body. If the environment is cold, the maggots take longer to develop than if it is warm.

Other organisms appear later on. Beetles that feed on fly maggots have to wait until there are maggots to eat, whereas beetles that eat bone have to wait until some bone is exposed.

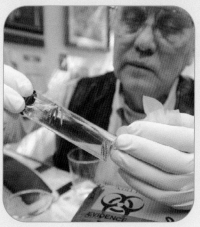

Figure 4g.8 This forensic entomologist is examining insect larvae from a murder victim. This helps to work out when the person died.

population of bacteria in the river increase. As they respire aerobically, they use up oxygen from the water. There is not enough oxygen left for fish, so the fish must move away or die.

● It contains chemicals that can harm living organisms, such as detergents and chemicals that are used in factories.

Most of the sewage produced in the UK is treated to remove all of these things before it is allowed to flow into rivers or the sea. The treatment is done by micro-organisms, including many different kinds of bacteria and microscopic fungi.

Figure 4g.11 shows one way in which sewage is treated. It is quite similar to the way in which biogas is produced (page 82–83). The raw sewage is changed into sludge, methane and much cleaner water called effluent. The sludge can be used as fertiliser on fields. The methane can be used as a fuel. The effluent can safely flow into rivers or the sea.

Figure 4g.9 Nutrients in untreated sewage encourage the growth of bacteria and other organisms.

grit and stones are removed by screens | primary Ⓐ settlement tank | aeration tank Ⓒ | secondary settlement tank Ⓓ | boiler house | effluent Ⓔ

raw sewage

solid organic wastes

activated sludge return

anaerobic digester Ⓑ

methane

sludge

Ⓐ **Primary settlement tank**
Solid wastes sink to the bottom and are sent to the anaerobic digester.

Ⓑ **Anaerobic digester**
There is no air here so all the bacteria which need oxygen are killed. Other bacteria which grow well in anaerobic conditions feed on the sludge, digesting it and producing methane gas. The methane can be used as a fuel. The digested sludge contains no harmful bacteria, and can be used as fertiliser.

Ⓒ **Aeration tank**
The liquid from the top of the primary settlement tank flows into here. Air is bubbled into it, so aerobic microorganisms grow, breaking down any remaining organic matter.

Ⓓ **Secondary settlement tank**
Here, many of the microorganisms from the aeration tank sink to the bottom. The material that sinks is called activated sludge. It is piped back to the aeration tank, where the microorganisms help to digest the organic material.

Ⓔ **Effluent**
The liquid at the top of the secondary settlement tank is quite clear, does not smell, and has no pathogenic organisms in it. It can be released into rivers or the sea.

Figure 4g.10 One way in which sewage is treated.

Summary

You should be able to:

♦ recognise materials that can decay (biodegradable materials)

♦ state that micro-organisms, a suitable temperature, oxygen and moisture (water) are needed for decay to happen

♦ describe how temperature, amount of oxygen and amount of water affect the rate of decay

H ♦ explain why temperature, amount of oxygen and amount of water affect the rate of decay

♦ describe food preservation processes and explain why they reduce the rate of decay

♦ describe an experiment to show that decay is caused by decomposers

♦ describe what detritivores do and give some examples

♦ explain how detritivores increase the rate of decay

H ♦ explain how decay involves saprophytic nutrition by bacteria and fungi

♦ state that micro-organisms are used to break down sewage and produce compost

Questions

1 a List *three* environmental factors that affect the rate of decay.

  b For *each* of the factors you have listed, briefly describe *one* method of food preservation that works by affecting that factor.

2 Peter left some strawberries on a kitchen worktop for two days. They went soft and mouldy. Peter wanted to find out what made that happen.

He bought some more strawberries. He divided them up into three groups.

● A were kept in the fridge

● B were left on the worktop

● C were dipped into hypochlorite (to kill any micro-organisms) and then left on the worktop

Table 4g.1 shows Peter's results.

Group	Time taken to start to decay in days
A	8
B	2
C	6

a Which group of strawberries took longest to start to decay? Explain why this happened.

b Explain the results for group C.

Table 4g.1

continued on next page

Questions - *continued*

Peter also wanted to know what kind of micro-organism was causing the decay. He took a sample from one of the decaying strawberries and spread it onto some agar jelly in a Petri dish. He covered the dish to stop it drying out and left it in a warm place. After two days, the dish looked like this (Figure 4g.12).

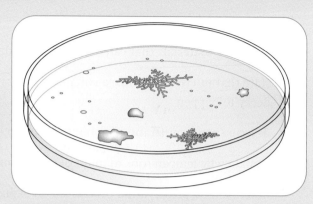

Figure 4g.11

c   Explain what Peter can conclude from these results.

H   3   a   Define the term *saprophyte*.

b   Explain how saprophytes are involved in the process of decay.

Detritivores and decomposers in a compost heap release heat as they respire. Some kinds of bacteria are able to survive in much higher temperatures than others. A bacterium called *Thermus* lives in hot compost heaps.

Figure 4g.13 shows the temperature changes in a newly made compost heap over 40 days. It also shows the population size of *Thermus* bacteria.

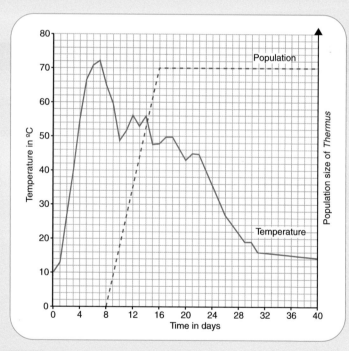

Figure 4g.12 Data for a compost heap.

c   Describe what happens to the temperature of the compost heap over these 40 days.

d   Explain why this happens.

e   Explain the results for the population of *Thermus* between 0 and 16 days.

f   Suggest why the population of *Thermus* does not drop significantly between 16 and 40 days.

Cycles

Not much is wasted in the living world. Plants and animals are made up of chemicals containing many different elements, including carbon and nitrogen. These are passed around between different organisms in an ecosystem. The carbon and nitrogen atoms in your body will have been part of someone else's body in the past and perhaps even part of the body of a dinosaur or ancient bacterium.

As organisms grow, they take more and more of these elements into their bodies. When they die and decay, the elements are recycled.

The carbon cycle

Figure 4h.1 shows the carbon cycle. As you read the next few paragraphs, keep an eye on the diagram.

Carbon dioxide is present in the air, but only in a very low concentration. Only about 0.04% of the air is carbon dioxide. Plants take in carbon dioxide during photosynthesis and use it to make carbohydrates. Some of the carbohydrates are used to make new biomass. The carbon in the carbon dioxide becomes part of the plant's cells.

The plant uses some of the carbohydrates it has made to provide energy for its cells. It does this by **respiration**. In respiration, glucose is broken down and carbon dioxide is released back into the air.

Animals get their carbon by eating plants or by eating other animals. Once again, some of the carbon is used to make new biomass and some of it is released back into the air as carbon dioxide, through respiration.

Waste products from the animals and plants, or their dead bodies, are broken down by **detritivores** and **decomposers**. These respire, releasing more of the carbon back into the air as carbon dioxide.

Millions of years ago, in the Carboniferous period, many places on Earth were so wet that there was not enough oxygen in the soil for the decomposers to be able to break down dead bodies and waste materials fully. Instead, the partly broken down substances got buried in the soil. As more and more material accumulated on

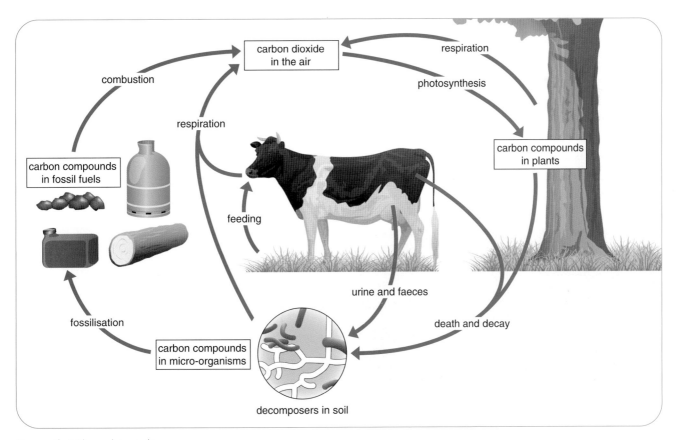

Figure 4h.1 The carbon cycle.

top of them, they became buried in the rocks and changed to form oil, or squashed to form coal. Now we are extracting these fossil fuels from the rocks and burning them. The carbon in them is being returned to the air as carbon dioxide.

SAQ

1 Use Figure 4h.1 to outline how an atom of carbon in a plant that grew in the Carboniferous period could become part of a molecule in your brain.

In the sea, the producers are tiny, floating plants called **phytoplankton**. They photosynthesise, incorporating carbon from carbon dioxide into their bodies. Floating animals called **zooplankton** feed on the phytoplankton.

Many phytoplankton and zooplankton have tiny shells containing calcium carbonate. When they die, these shells fall to the sea floor. Over hundreds of thousands – or millions – of years, they can build up enormously thick layers. As more and more shells pile up, they compress the ones below them. A rock called **limestone** is formed.

As the Earth's tectonic plates move around, the limestone may become buried deep under the surface or lifted high to become mountains.

If it is buried deep, it may be melted and perhaps erupt out onto the surface through a volcano. The calcium carbonate is broken down in this process and released into the air as carbon dioxide.

If the rock is lifted high then it will be exposed to the rain. The limestone will be weathered. Normal rain is slightly acidic and this weak acid reacts with the calcium carbonate, releasing carbon dioxide from it.

Figure 4h.2 Limestone is made from calcium carbonate, from the shells of tiny marine organisms.

Figure 4h.3 Volcanic eruptions release carbon dioxide from the molten rock.

SAQ

2 Use Figure 4h.1 to draw an outline of the carbon cycle (you do not need to illustrate it). Then add the formation and breakdown of limestone to the diagram.

The nitrogen cycle

Nitrogen is needed for making proteins and DNA. Figure 4h.4 shows how nitrogen atoms are cycled around within an ecosystem.

Unlike carbon, there is plenty of nitrogen in the air: 78% of the air is nitrogen. But this nitrogen gas is so unreactive that plants and animals cannot use it.

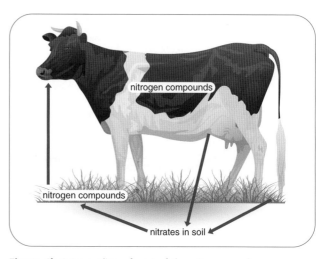

Figure 4h.4 An outline of part of the nitrogen cycle.

Instead, plants get their nitrogen in the form of **nitrate** ions from the soil, through their roots. They use the nitrates to make amino acids and these are used to make **proteins**. The proteins help the plant to make new cells and grow.

When animals eat the plants, they digest the proteins and use them to build new proteins in their own bodies. This is how animals get their nitrogen – in the form of proteins from plants or from other animals.

When plants and animals produce waste materials or die, they are broken down by decomposers. The decomposers use some of the nitrogen to make proteins in their own bodies. They release some of it into the soil as nitrates.

SAQ

3 Explain how an atom of nitrogen in a bacterium in the soil could become part of a cell in your body.

4 Have you eaten anything containing nitrogen today? How did the nitrogen get into the food?

Nitrogen fixation

We have seen that plants and animals cannot use nitrogen gas from the air. But there are some bacteria that can do this. They are called **nitrogen-fixing bacteria**. They change nitrogen gas, N_2, into ammonium ions, NH_4^+. Changing unreactive nitrogen gas into a more reactive compound of nitrogen is called **nitrogen fixation**.

Many nitrogen-fixing bacteria live inside little swellings called **nodules** on the roots of plants belonging to the pea and bean family. The bacteria use nitrogen from the air spaces in the soil to make ammonium ions that the plant can use for making proteins. The plant provides carbohydrates for the bacteria. This is an example of **symbiosis** – two different species of organisms living in close association, each one benefiting from the relationship.

Lightning can also fix nitrogen. It causes nitrogen gas to combine with oxygen in the air, forming **nitrate ions**, NO_3^-. As it rains, the nitrates dissolve in the raindrops and are carried to the ground.

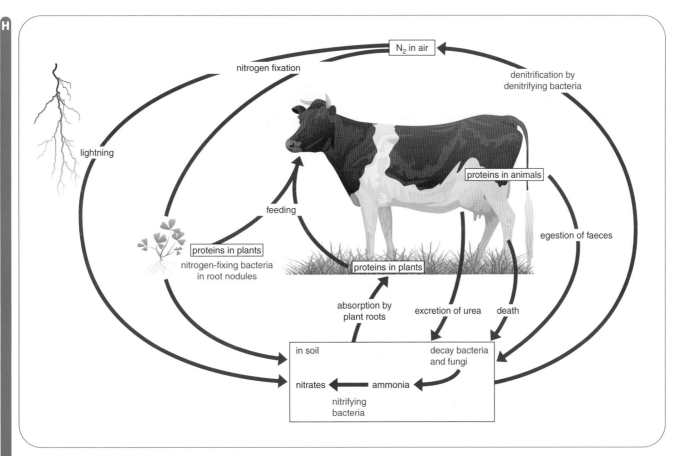

Figure 4h.5 The nitrogen cycle.

Figure 4h.6 Root nodules on a pea plant contain nitrogen-fixing bacteria.

Nitrification

Decomposers have a very important role to play in the nitrogen cycle. They convert proteins and urea into nitrates. Urea is a nitrogen-containing compound that is excreted in the urine of animals.

First, the proteins and urea are broken down by bacteria and fungi and converted into ammonia, NH_3. If you ever visit a farm where animals are kept indoors, you may smell the ammonia that has been produced from their urine.

Next, **nitrifying bacteria** in the soil convert ammonia into nitrates. Now the nitrogen is in a form that plants can take in and use.

Some bacteria take the process a step further. They convert the nitrates into nitrogen, which goes back into the air. These are called **denitrifying bacteria**.

Carnivorous plants

Most plants get their nitrogen from the soil, in the form of nitrates or ammonium ions. But in some soils, these ions are in very short supply. This happens in soils that are almost always waterlogged, such as on wet moorlands. Here, some of the plants have evolved a different way of getting nitrogen.

Carnivorous plants such as sundews, pitcher plants and Venus flytraps eat insects. Sundews trap them on their sticky leaves, while pitcher plants drown them in liquid at the bottom of slippery tubes made from their leaves. Venus flytraps tempt the insects into their leaves with sweet nectar, then slam the leaf shut like a prison door, trapping the insect inside.

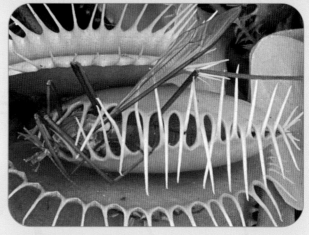

Figure 4h.8 Venus flytrap.

The plants secrete enzymes onto the trapped insects, which digest them in a very similar way to digestion inside your alimentary canal. Proteins are broken down into amino acids and absorbed into the plant's cells. Sundews can get as much as half of their nitrogen in this way, while Venus flytraps get up to 80% of theirs.

Some huge pitcher plants that live in Borneo are said to be able to trap and digest rats, but probably any rats that weren't able to scramble out of the pitcher would have been fairly ill anyway, even before they fell in.

Figure 4h.7 Sundew.

Summary

You should be able to:

◆ state that animals and plants incorporate carbon and nitrogen atoms into their bodies as they grow and that these elements are recycled

◆ explain how carbon is recycled, including the roles of plants, animals and decomposers

H ◆ explain how carbon is recycled in the sea, and by volcanic eruption and weathering

◆ explain how nitrogen is recycled in nature, including the roles of plants, animals and decomposers

H ◆ explain how decomposers, nitrifying bacteria and denitrifying bacteria are involved in the nitrogen cycle

Questions

1 a Name the only process by which carbon atoms are removed from the air and incorporated into living organisms.

b Describe how these organisms return some of the carbon back to the air.

c Explain why fossil fuels contain carbon and how this carbon can be returned to the air.

2 a Explain why nitrogen is important to living organisms.

b State the form in which each of these organisms obtains its nitrogen.

  i A green plant. ii Nitrogen-fixing bacteria. iii A person.

H 3 A fish tank was filled with water and some bacteria were added. Some phytoplankton (microscopic floating plants) were then introduced. The tank was put into a dark place and left for eight months.

At intervals, the water was tested to find out what it contained. The results are shown in Figure 4h.9.

a Explain why the phytoplankton died so quickly.

b The phytoplankton contain nitrogen in their cells. In what form is most of this nitrogen?

c Explain why the quantity of dead phytoplankton decreased during the first two months of the experiment.

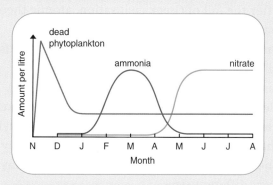

Figure 4h.9 Experimental results for phytoplankton.

d Explain where the ammonia came from and name the type of bacteria that produced it.

e State the time at which nitrate begins to appear in the water.

f What kind of bacteria are responsible for its production?

Figure 3a.1 a This coin is made of silver atoms.

Figure 3a.1 b We cannot see the atoms using a microscope.

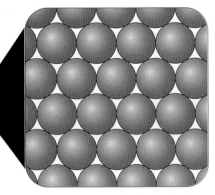

Figure 3a.1 c If we could see the silver atoms, they would look like this.

Atomic theory

Everything is made from tiny particles called **atoms**. Here 'tiny' means *really* tiny – the full stop at the end of this sentence is about one million atoms wide. The existence of atoms was first suggested 2400 years ago by Greek philosophers.

If you look at a piece of copper, or a stone, or your fingernail, or anything else, you can't see the atoms. They are too small. Even if you get close up, the atoms are too small to see. We cannot see atoms even with the best microscopes. We believe that atoms exist because they are an excellent explanation of all the things we can see and measure.

John Dalton

After the fall of Greek civilisation, atomic theory was more or less forgotten for centuries. The theory was revived in 1803 by a British scientist, John Dalton. By this time, scientists had discovered over 30 **elements**. They defined an element as a substance which cannot be broken down chemically. Dalton explained this by saying that an element was a substance made of only one type of atom. He said that different atoms were different sizes.

We have now discovered over 100 different elements. The original definition of an element, and John Dalton's explanation, are still very useful.

● An element is a substance that cannot be broken down chemically.

● An element is a substance made of only one type of atom.

Figure 3a.2 a The outside of this 2p coin is made of copper. Copper is an element.

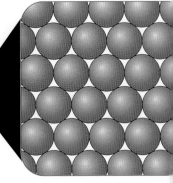

Figure 3a.2 b If we could see th copper atoms, they would look like this.

Figure 3a.2 c These are pieces of lead metal. Lead is an element.

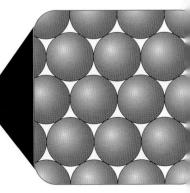

Figure 3a.2 d If we could see the lead atoms, they would look like ▶

SAQ

1 **a** How many elements are there?

   b Give *two* definitions of an *element*.

2 Look at Figures 3a.1 and 3a.2. How do the illustrations show that:

   a silver, copper and lead are elements?

   b silver atoms, copper atom and lead atoms are different?

Compounds

Dalton's atomic theory makes it straightforward to explain what a **compound** is.

- A compound is a substance made of at least two elements chemically combined.
- A compound is a substance made of at least two different types of atom, chemically combined.

The atomic theory explains why a compound has a formula which is always the same. For example, the formula of water is H_2O. We can explain this by saying that water is made of tiny groups of atoms called **molecules**. Every water molecule is made of two hydrogen atoms and one oxygen atom. Therefore the formula of water is H_2O.

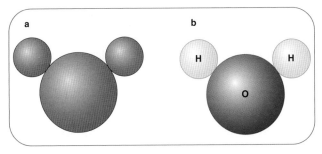

Figure 3a.3 **a** The atoms in a water molecule. **b** The symbols of the elements and the colours are usually added to make the diagrams easier to understand.

Using formulae

The formula of a compound tells us which elements are present in the compound. A formula is written in symbols. The names of the elements can be found on the **Periodic Table** of the elements. There is a Periodic Table on page 345 of this book.

The formula also tells us the number of atoms of each element in one molecule (or **formula unit**) of the compound.

Worked example 1

Which elements are present in sodium hydrogencarbonate $NaHCO_3$?

Answer: the symbol for every element has one capital letter in it. Therefore the elements in sodium hydrogencarbonate are sodium (Na), hydrogen (H), carbon (C) and oxygen (O).

Worked example 2

How many atoms are present in one formula unit of sodium hydrogencarbonate?

Answer: six. They are one sodium atom, one hydrogen atom, one carbon atom and three oxygen atoms.

SAQ

3 a Which elements are present in silver nitrate, $AgNO_3$?

  b How many atoms are present in one formula unit of silver nitrate?

Protons, neutrons and electrons

Even though atoms are amazingly small, they are made of even smaller particles. They have a structure. The smaller particles that atoms are made of are called **protons**, **neutrons** and **electrons**. These particles are not the same as each other. They have different amounts of mass and electric charge.

A proton has the same mass as a neutron. Measured in grams, this mass is 0.000 000 000 000 000 000 000 001 7 g. This number should give you some idea how tiny they are, but it is too awkward a number to use regularly. We use a scale on which a proton and a neutron each has a relative mass of 1. This scale is called the atomic mass scale; it emphasises that protons and neutrons have the same mass as each other. An electron has much less mass than a proton or a neutron. On the same scale, its mass is 0.0005. This is so small that we can call it zero.

A neutron has no electric charge. A proton and an electron both have an electric charge of the same size, but of opposite types. On the scale that is used to describe these charges, the electric charge on a proton is +1; we call this 'plus one'. On the same scale, the electric charge on an electron is −1; we call this 'minus one'. Because a neutron has no charge, it is said to be 'electrically neutral', hence its name.

Atomic structure

The protons and neutrons in each atom are packed together in a tiny lump in the centre of the atom called the **nucleus**. Each proton has a positive electric charge. Each neutron has no electric charge. Because the nucleus of every atom is made of protons and neutrons, the nucleus of every atom has a positive electric charge.

The rest of the atom is space. The electrons move through this space on paths called **shells**. Every electron is constantly moving. Every atom has no overall electric charge. This is because the charge of the protons and the charge of the electrons cancel each other out. Every atom is said to be 'electrically neutral'.

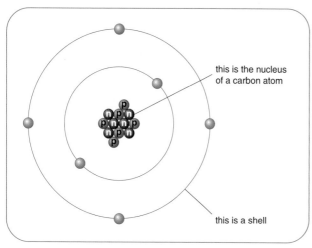

this is the nucleus of a carbon atom

this is a shell

Figure 3a.4 A carbon atom. Each proton is labelled 'p'. Each neutron is labelled 'n'. Each electron is labelled 'e'.

SAQ

4 Look at Figure 3a.4.

   a How many protons are there in a carbon atom?

   b How many neutrons are there in a carbon atom?

   c How many electrons are there in a carbon atom?

   d Where are the protons and neutrons in a carbon atom?

   e Where are the electrons in a carbon atom?

Name of particle	Relative mass	Electric charge	Where it is found in an atom
proton	1	+1	nucleus
neutron	1	0	nucleus
electron	0	−1	shell

Table 3a.1 Protons, neutrons and electrons.

Atomic number

Every element has a different **atomic number**. For example, the atomic number of carbon is six. No other element has an atomic number of six. The atomic number of an element tells us two things:
- the number of protons in the nucleus of each atom of the element
- the position of the element in the Periodic Table. Because carbon has an atomic number of six:
- there are six protons in the nucleus of each carbon atom
- carbon is the sixth element in the Periodic Table.

The elements in the Periodic Table are arranged in order of ascending atomic number. Therefore they are arranged in order of how many protons there are in the nucleus of each atom of the element.

The first element in the Periodic Table is hydrogen. Hydrogen's atomic number is 1. There is one proton in the nucleus of each hydrogen atom. The last natural element in the Periodic Table is plutonium. Plutonium's atomic number is 93. There are 93 protons in the nucleus of each plutonium atom.

SAQ

5 Look at Figure 3a.5.

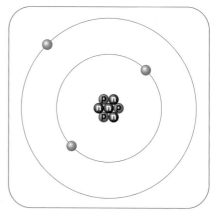

Figure 3a.5
A lithium atom.

Table 3a.1 summarises what you need to know about protons, neutrons and electrons.

a How many protons are there in the nucleus of a lithium atom?

b What is the atomic number of lithium?

c What is lithium's position in the Periodic Table?

d Use the Periodic Table on page 345 to find the symbol used for lithium.

6 Look at Figure 3a.6.

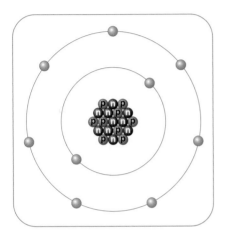

Figure 3a.6 A fluorine atom.

a How many protons are there in the nucleus of a fluorine atom?

b What is the atomic number of fluorine?

c What is fluorine's position in the Periodic Table?

d Use the Periodic Table on page 345 to find the symbol used for fluorine.

Mass number

The total number of protons and neutrons in an atom is called the **mass number**. Look at Figure 3a.4. The carbon atom illustrated has six protons, six neutrons and six electrons. The total number of protons and neutrons in the atom is 12. The mass number of this carbon atom is 12.

SAQ

7 **a** What is the mass number of the lithium atom in Figure 3a.5?

b What is the mass number of the fluorine atom in Figure 3a.6?

Isotopes

Every atom of a particular element has the same number of protons as every other atom of the same element. However, atoms of the same element can vary slightly. They can have different numbers of neutrons from each other. This means they will have different mass numbers. Atoms with the *same* atomic numbers but with *different* mass numbers are called **isotopes**.

Most elements have isotopes. These slightly different atoms are atoms of the same element. Although they have exactly the same chemical properties as each other, they may behave differently in other ways. Figure 3a.7 shows the three different isotopes of hydrogen. The isotope on the right, with mass number three, is radioactive. The other two isotopes are not.

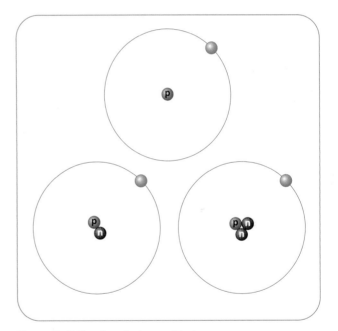

Figure 3a.7 The three isotopes of hydrogen.

SAQ

8 Look at Figure 3a.7.

a What is the atomic number of each atom?

b What is the position in the Periodic Table of each atom?

c What is the mass number of each atom?

Using atomic numbers and mass numbers

You can use the atomic number and mass number of an atom to work out how many protons, neutrons and electrons there are in that atom. You must use the following rules.

- The atomic number is the same as the number of protons in the atom.
- The mass number is the same as the total number of protons and neutrons in the atom.
- The number of electrons in the atom is the same as the number of protons in the atom.

SAQ

9 a Copy and complete Table 3a.2.

   b. Which two atoms in the table are isotopes?

Table 3a.2 explains why an atom is neutral. An atom is neutral because it contains the same number of electrons as protons.

Taking the neon atom (Table 3a.2) as an example:

- the atom has ten protons, total electric charge +10;
- the atom has ten neutrons, total electric charge 0;
- the atom has ten electrons, total electric charge −10;
- the total charge on the neon atom = +10 + 0 − 10 = 0.

The atomic number and mass number of an atom can be included as part of its symbol. For example, the neon atom in Table 3a.2 can be written $^{20}_{10}$Ne. Notice that the atomic number is always written below the mass number.

SAQ

10

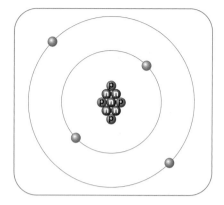

Figure 3a.8 A beryllium atom.

a Write down the symbol of the beryllium atom including its atomic number and mass number.

b Explain why the beryllium atom is electrically neutral.

Electronic structures

Most of the volume of every atom is empty space. The atom's electrons move through this empty space on paths called shells. The number of shells in an atom and number of electrons on each shell of an atom are strictly defined. This is called the **electronic structure**. These rules are followed.

1 Every electron in the atom must be on a shell.
2 The first shell can accommodate a maximum of two electrons.
3 Electrons don't go on the second shell if they can fit on the first shell.
4 The second shell can accommodate a maximum of eight electrons.
5 Electrons don't go on the third shell if they can fit on the first and second shells.

Name of element	Atomic number	Mass number	Number of protons	Number of neutrons	Number of electrons
neon	10	20	10	10	10
chlorine	17	35			
chlorine	17	37			
argon	18	40			
potassium	19	40			
nickel	28	58			
iodine	53	127			

Table 3a.2 Using atomic numbers and mass numbers.

H 6 The third shell can accommodate a maximum of eight electrons.

7 Electrons don't go on the fourth shell if they can fit on the first, second and third shells.

SAQ

11 **a** Use the Periodic Table on page 345 to find the atomic numbers of helium, oxygen and silicon atoms.

H **b** How many electrons are there in one atom of each of these elements?

c Draw a diagram to show the electronic structure of one atom of each of these elements. Use Worked example 3 to help you.

d Write out the electronic structures in numbers.

The biggest scientific instrument in the world

Figure 3a.9 Inside the LEP tunnel. The electrons and positrons travel through the middle of the huge blue magnet seen here.

Figure 3a.10 The LEP tunnel was dug by huge boring machines known as moles.

Of all of the particles described on these pages, electrons are the smallest. In order to find out more about them, and more about matter itself, scientists built a machine over half as wide as Greater London! The machine was called the Large Electron–Positron collider, or LEP for short. The LEP took 6 years and US$800 million to build it. It was sited at a nuclear research establishment in Geneva, Switzerland, called CERN.

The LEP was built in a circular tunnel which goes underneath both Switzerland and France. The inside of the tunnel is 3.8 m wide and the whole tunnel is 27 km in diameter.

The scientists at CERN used the LEP to accelerate electrons until they were travelling almost as fast as the speed of light. The scientists also accelerated similar particles called positrons up to the same speed. The electrons and positrons then did circular laps inside the LEP tunnel – steered by over 4000 magnets. They travelled at such speeds that each 85 km circuit took them less than a thousandth of a second.

The electrons and positrons travelled in opposite directions around the tunnel, so they could be made to smash into each other! The energy of the collisions between the electrons and positrons created particles that do not normally exist. The scientists at CERN were able to observe these particles and make measurements in order to learn more about them. LEP provided scientists with a great deal of evidence between 1989 and 2000, when it was shut down.

If a toddler is given an unfamiliar object they will probably 'investigate' it first by hitting it. In the biggest scientific experiment in the world scientists investigated electrons by hitting them against something else. Rather like the toddler, but it costs more.

H

Worked example 3

The number of electrons on each shell in an atom is called the electronic structure of the atom. The diagram below shows the electronic structure of a chlorine atom. The nucleus is shown as a dot for simplicity.

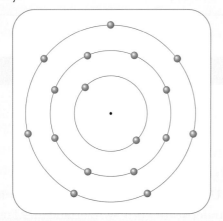

Figure 3a.11 The electronic structure of a chlorine atom. It can be written in numbers as 2.8.7.

Look at the diagram of the electronic structure of a chlorine atom in Figure 3a.11. Notice how it follows the rules listed in the text.

1 You must know how many electrons to put on the diagram. The atomic number of chlorine is 17. Therefore a chlorine atom has 17 protons. Therefore a chlorine atom has 17 electrons. Each of these 17 electrons must be on a shell.
2 Two of these electrons go on the first shell.
3 The first shell is full. There are still 15 electrons left.
4 Eight of these electrons go on the second shell.
5 The second shell is full. There are still seven electrons left
6 These seven electrons go on the third shell.
7 All 17 electrons have been accommodated on the first, second and third shells. There is no need for any electrons to go on the fourth shell.

Summary

You should be able to:

◆ describe an *element* and a *compound*

◆ state how many elements there are in the Periodic Table

◆ use the formula of a compound to work out which elements the compound is made of and state how many of each type of atom are present in one molecule (formula unit) of the compound

◆ state the relative charges and masses of a proton, a neutron and an electron

◆ state that an atom is electrically neutral

H ◆ explain why an atom is electrically neutral

◆ describe an atom, stating where the protons, neutrons and electrons are located

◆ use and define the terms *atomic number*, *mass number* and *isotopes*

◆ state how the elements are arranged in the Periodic Table

H ◆ deduce the number of protons, neutrons and electrons there are in an atom from its atomic number and mass number

◆ recognise which atoms are *isotopes*

◆ deduce the electronic structures of the first 20 elements (hydrogen to calcium)

Questions

1   a   Explain how protons, neutrons and electrons are different from each other.

    b   Where in an atom are protons, neutrons and electrons located?

    c   Are atoms electrically charged?

2   Table 3a.3 describes five atoms. It tells you how many protons, neutrons and electrons there are in each atom. Copy and complete the table by answering these questions.

    a   What is the atomic number of each atom?

    b   What is the mass number of each atom?

    c   What is the position in the Periodic Table of the element that has atoms like this?

    d   What is the name and symbol of the element that has atoms like this?

    e   Which two atoms are isotopes of the same element? Explain your choice.

Atom	Protons	Neutrons	Electrons	Atomic number	Mass number	Position in the Periodic Table	Name of the element	Symbol for the element
i	8	8	8					
ii	11	12	11					
iii	1	0	1					
iv	8	10	8					
v	26	30	26					

Table 3a.3

H 3   'An element is a substance which can't be broken down chemically. Dalton explained this by saying that an element was *a substance made of only one type of atom*. He said that *different atoms were different sizes*.'

Explain why we no longer believe the two statements in italics to be strictly true.

4   A potassium atom has atomic number 19 and mass number 39. A sulfur atom has atomic number 16 and mass number 32.

    a   How many protons, neutrons and electrons will each atom have?

    b   Draw a diagram of the electronic structure of each atom.

    c   Show the electronic structure of each atom with numbers.

    d   Why is each atom electrically neutral?

Ions

A compound is defined as two or more elements chemically bonded together. 'Chemically bonded together' means that the atoms of the different elements remain together in molecules, or formula units. In order to do this, they must be held, or **bonded**, to each other. Sodium chloride doesn't just fall apart into sodium and chlorine, for example. The particles of sodium and chlorine stay bonded together because they have changed from atoms into ions.

Sodium chloride

An **ion** is an electrically charged atom or group of atoms. When an atom gains or loses one, two or three electrons, it becomes an ion. The gain or loss of electrons changes the atom into an electrically charged particle. When sodium and chlorine react together, electrons move between atoms. The sodium atoms change into sodium ions. The chlorine atoms change into chloride ions.

At the instant of reaction, every sodium atom loses an electron. Before this happens, every sodium atom is electrically neutral. When it loses an electron, every sodium atom becomes electrically charged. The sodium atoms are now called ions. Because the sodium atoms have each lost a negative electron, they become **positive ions**. They are called **sodium ions**. Because each atom loses one electron, each ion has a 'one plus' charge. The sodium ion is written Na$^+$.

At the instant of reaction, every chlorine atom gains an electron. Before this happens, every chlorine atom is electrically neutral. When it gains an electron, every chlorine atom becomes electrically charged. The chlorine atoms are now called ions. Because the chlorine atoms have each gained a negative electron, they become **negative ions**. They are called **chloride ions**. Because each atom gains one electron, each ion has a 'one minus' charge. The chloride ion is written Cl$^-$.

The sodium ions are positively charged. The chloride ions are negatively charged. The positive sodium ions and the negative chloride ions attract each other. This attractive force holds them together. The ions are bonded together. The attractive force that bonds them together is called an **ionic bond**. Because of **ionic bonding**, sodium chloride stays together as a compound. It does not separate back into sodium and chlorine.

SAQ

1 a Describe what happens to sodium atoms when they react with chlorine atoms.

   b Do the sodium atoms form positive ions or negative ions in this process?

   c Describe what happens to chlorine atoms when they react with sodium atoms.

   d Do the chlorine atoms form positive ions or negative ions in this process?

Figure 3b.1 a These pieces of sodium consist of electrically neutral sodium atoms.

Figure 3b.1 b The chlorine in this jar consists of electrically neutral chlorine atoms.

Figure 3b.1 c The sodium and the chlorine are reacting. Electrons are moving from the sodium atoms to the chlorine atoms. The atoms are becoming ions.

Figure 3b.1 d The jar now contains sodium chloride. The sodium chloride consists of sodium ions and chloride ions which attract each other and stay together.

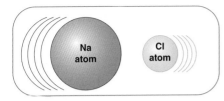

Figure 3b.2 a When they react, a sodium atom collides with a chlorine atom.

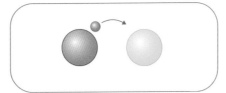

Figure 3b.2 b During the reaction, an electron goes from the sodium atom to the chlorine atom.

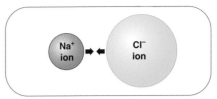

Figure 3b.2 c A sodium ion and a chloride ion have formed. They attract each other and stay together. This is one formula unit of sodium chloride. There are billions of these formula units in every grain of salt.

2 Why do the sodium ions and chloride ions in a grain of salt stay together?

3 A balanced symbol equation for the reaction between sodium and chlorine is

$$2Na + Cl_2 \rightarrow 2Na^+Cl^-$$

The symbols used are Na, Cl_2, Na^+ and Cl^-.

Select from these four symbols:

a the symbol of an ion

b the symbol of an atom

c the symbol of a molecule.

Remember.

● When atoms gain electrons, they gain negative charges to form negatively charged ions. The number of electrons gained equals the negative charge on the ion formed.

● When atoms lose electrons, they lose negative charges and therefore form positively charged ions. The number of electrons lost is equal to the positive charge on the ion formed.

The properties of sodium chloride

Sodium chloride consists of electrically charged particles – sodium ions and chloride ions. This affects the properties of sodium chloride.

● Sodium chloride has a high melting point because the sodium ions and the chloride ions attract each other strongly.

● Sodium chloride dissolves in water. Many (but not all) ionic compounds do this.

● Solid sodium chloride does not conduct electricity. Although the sodium ions and the chloride ions are electrically charged, they cannot move in solid sodium chloride owing to the bonding. Therefore they cannot carry an electric current.

● Sodium chloride solution conducts electricity. When dissolved in water, the sodium ions and chloride ions are free to move. Because the ions are electrically charged and are free to move, the solution conducts electricity.

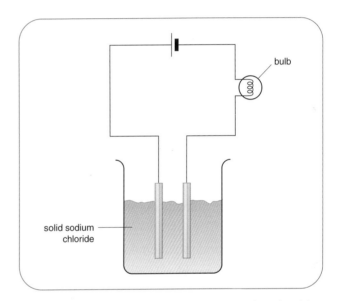

Figure 3b.3 a Solid sodium chloride does not conduct electricity.

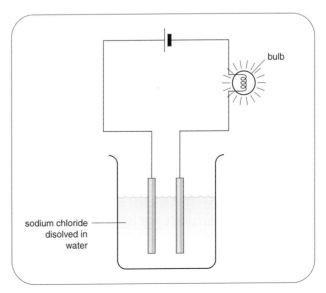

Figure 3b.3 b Sodium chloride solution conducts electricity.

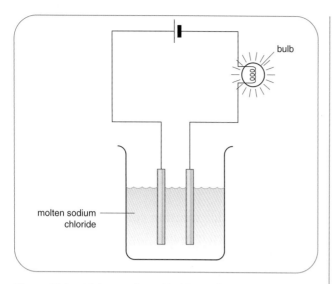

Figure 3b.3 c Molten sodium chloride conducts electricity.

- Molten sodium chloride conducts electricity. When molten, the sodium ions and chloride ions are free to move. Because the ions are electrically charged and are free to move, the molten liquid conducts electricity.

SAQ

4 Explain why:

   a sodium chloride has a high melting point;

   b solid sodium chloride doesn't conduct electricity;

   c sodium chloride solution conducts electricity;

   d molten sodium chloride conducts electricity.

Magnesium oxide

When magnesium reacts with oxygen, magnesium oxide is formed. This reaction is very similar to the reaction of sodium with chlorine, except for one thing. At the instant of reaction, every magnesium atom loses two electrons. Because each atom loses two electrons, each ion has a 'two plus' charge. The magnesium ion is written Mg^{2+}. At the instant of reaction, every oxygen atom gains two electrons. Because each atom gains two electrons, each ion has a 'two minus' charge. The oxide ion is written O^{2-}.

Figure 3b.4 a This piece of magnesium consists of electrically neutral magnesium atoms.

Figure 3b.4 b The oxygen in this gas jar consists of electrically neutral oxygen atoms.

Figure 3b.4 c The magnesium and the oxygen are reacting. Electrons are moving from the magnesium atoms to the oxygen atoms. The atoms are becoming ions.

Figure 3b.4 d White magnesium oxide can now be seen on the tongs. The magnesium oxide consists of magnesium ions and oxide ions, which attract each other and stay together.

SAQ

5 a Describe what happens to magnesium atoms when they react with oxygen atoms.

   b Do the magnesium atoms form one-plus ions or two-plus ions in this process? Explain your answer.

   c Describe what happens to oxygen atoms when they react with magnesium atoms.

   d Do the oxygen atoms form one-minus ions or two-minus ions in this process? Explain your answer.

Figure 3b.5a When they react, a magnesium atom collides with an oxygen atom.

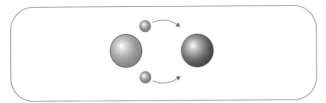

Figure 3b.5b During the reaction, two electrons go from the magnesium atom to the oxygen atom.

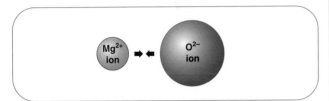

Figure 3b.5c A magnesium ion and an oxide ion have formed. They attract each other and stay together. This is one formula unit of magnesium oxide.

The properties of magnesium oxide

Magnesium oxide consists of electrically charged particles – magnesium ions and oxide ions. This affects the properties of magnesium oxide.

- Magnesium oxide has an extremely high melting point because the two-plus ions and the two-minus ions attract each other very strongly.
- Solid magnesium oxide does not conduct electricity. Although the magnesium ions and the oxide ions are electrically charged, they cannot move in solid magnesium oxide owing to the bonding.
- Molten magnesium oxide conducts electricity. When molten, the magnesium ions and oxide ions are free to move. Because the ions are electrically charged and are free to move, the molten liquid conducts electricity.

SAQ

6 Explain why:

  a magnesium oxide has a higher melting point than sodium chloride;

  b solid magnesium oxide doesn't conduct electricity;

  c molten magnesium oxide conducts electricity.

Ionic compounds

Sodium chloride and magnesium oxide are **ionic compounds**. They consist of positive and negative ions which stay together because they attract each other. The positive ions formed from the metal atoms which lost electrons. The negative ions formed from the non-metal atoms which gained electrons. This is typical of the way metals combine with non-metals to form compounds. When a metal combines with a non-metal, they form an ionic compound.

Dot and cross diagrams

The diagrams in Figures 3b.6 and 3b.7, which show the electronic structures of the atoms and ions involved in these reactions, are called dot and cross diagrams. The electrons from one type of atom are drawn as dots. The electrons from the other type of atom are drawn as crosses. This is not because the electrons are different; it just helps to see where they have come from.

If we draw the electronic structures of the atoms and ions involved in the reactions in this item, a pattern becomes clear.

Sodium chloride

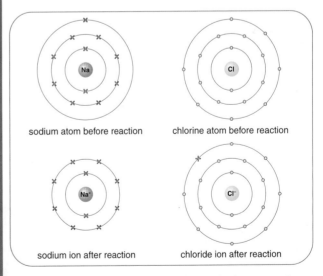

sodium atom before reaction   chlorine atom before reaction

sodium ion after reaction   chloride ion after reaction

Figure 3b.6 Dot and cross diagrams for NaCl. The grey and yellow circles represent the nuclei of the atoms and ions.

The sodium atom loses all the electrons (i.e. one electron) on its outer shell. The chlorine atom fills all the gaps (i.e. one gap) in its outer shell.

	Sodium	Chlorine
Electronic structure of atom	2.8.1	2.8.7
Electrons gained or lost	one lost (the one electron in the third shell)	one gained (filling the one gap in the third shell)
Electronic structure of ion formed	2.8	2.8.8
Ion formed	Na^+	Cl^-
Compound formed	NaCl	

Table 3b.1 Forming sodium chloride.

Magnesium oxide

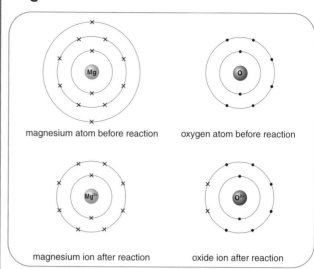

Figure 3b.7 Dot and cross diagrams for MgO.

Magnesite, magnesium and firebricks

One of the main magnesium-containing ores currently being mined is magnesite. Magnesite is magnesium carbonate ($MgCO_3$) and it is mined in many countries including Russia, Korea and Greece. If magnesite is heated, it breaks down into magnesium oxide and carbon dioxide. The magnesium oxide is known as magnesia. The magnesia is used to make firebricks and to line furnaces.

Magnesium oxide is suitable for these applications because of its very high melting point – it melts at 2827 °C. Despite this high melting point, if magnesium oxide firebricks are used to line particularly hot furnaces they do gradually erode away. When the furnace has been running for several months, the firebricks have to be replaced.

Some magnesia is used as a source of magnesium. Demand for this metal is increasing. 400 000 tonnes of magnesium were used worldwide in the year 2000. The amount used per year was predicted to have doubled by 2008. Most magnesium is used for making alloys, because magnesium alloys are strong and resistant to corrosion. These alloys are used to make car body parts and aircraft.

If you have a medicine cupboard at home, it may have a bottle of 'milk of magnesia' in it. This liquid mixture is alkaline, so it relieves the discomfort caused by heartburn or acid indigestion. Because it is an alkali, milk of magnesia neutralises the acid in the stomach that causes acid indigestion. It is called 'milk' because it looks like milk, but it certainly doesn't taste like it!

Figure 3b.8 A piece of magnesite. The blue colour is due to the lighting used.

Figure 3b.9 This pottery kiln is lined with firebricks.

Figure 3b.10 Milk of magnesia.

	Magnesium	Oxygen
Electronic structure of atom	2.8.2	2.6
Electrons gained or lost	two lost (the two electrons in the third shell)	two gained (filling the two gaps in the second shell)
Electronic structure of ion formed	2.8	2.8
Ion formed	Mg^{2+}	O^{2-}
Compound formed	MgO	

Table 3b.2 Forming magnesium oxide.

Full outer shells

In both these compounds (sodium chloride and magnesium oxide), the metal atom loses all the electrons in its outer shell. The outer shells of the ions that form (Na^+ and Mg^{2+}) are their second shells. These shells are full of electrons. In both these compounds, the non-metal atom fills all the gaps in its outer shell. The outer shells of the ions that form (Cl^- and O^{2-}) are full. These non-metal ions also have outer shells that are full of electrons.

When an atom gains or loses electrons and forms an ion, the resulting ion has a full outer shell of electrons.

SAQ

7 The electronic structure of a sodium ion can be written in numbers as 2.8.

   a Write the electronic structures of a chloride ion, a magnesium ion and an oxide ion in numbers.

   b How many protons, neutrons and electrons are there in each of the following particles?

    i $\ _{11}^{23}Na^+$

    ii $\ _{12}^{24}Mg^{2+}$

    iii $\ _{8}^{16}O^{2-}$

    iv $\ _{17}^{35}Cl^-$

Magnesium chloride

Magnesium reacts with chlorine to form a compound called magnesium chloride. The bonding in this compound is ionic. Magnesium atoms have two outer electrons. Chlorine atoms have one gap in their outer shell. In order for all

atoms involved to form ions with full outer shells, one magnesium atom must react with two chlorine atoms.

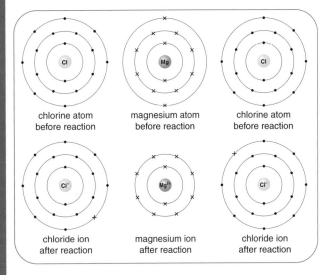

Figure 3b.11 Dot and cross diagrams for $MgCl_2$.

The formula of magnesium chloride is $MgCl_2$. One formula unit of the compound consists of one magnesium ion and two chloride ions. You should notice that, in one formula unit of the compound, the total charge of the ions is zero. This is true of every ionic compound.

$$\left(2+\right) + \left(1-\right) + \left(1-\right) = zero$$

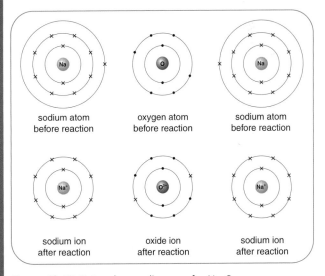

Figure 3b.12 Dot and cross diagrams for Na_2O.

Sodium oxide

Sodium reacts with oxygen to form a compound called sodium oxide. The bonding in this compound is ionic. Sodium atoms have one outer electron. Oxygen atoms have two gaps in their outer shell. In order for all atoms involved to form ions with full outer shells, two sodium atoms must react with one oxygen atom.

The formula of sodium oxide is Na_2O. One formula unit of the compound consists of two sodium ions and one oxide ion. You should notice that, in one formula unit of the compound, the total charge of the ions is zero.

$(1+) + (1+) + (2-)$ = zero

The formulae of ionic compounds

Silver chloride is an ionic compound. The silver ion has a one-plus charge (Ag^+). The chloride ion has a one-minus charge (Cl^-). The formula of silver chloride is $AgCl$ because, in one formula unit of an ionic compound, the total charge of the ions is zero.

$(1+) + (1-)$ = zero

Silver sulfide is an ionic compound. The silver ion has a one-plus charge (Ag^+). The sulfide ion has a two-minus charge (S^{2-}). The formula of silver sulfide is Ag_2S because, in one formula unit of an ionic compound, the total charge of the ions is zero.

$(1+) + (1+) + (2-)$ = zero

SAQ

8 **a** Why don't magnesium and chlorine form a compound with the formula MgCl?

   b Why don't sodium and oxygen form a compound with the formula NaO?

9 Copy and complete Table 3b.3 in order to work out the formulae of the five ionic compounds.

Name of compound	Ions present	Formula of compound
magnesium fluoride	Mg^{2+}, F^-	
sodium sulfide		
magnesium sulfide		
silver oxide		
aluminium oxide	Al^{3+}, O^{2-}	

Table 3b.3 Ionic compounds.

Giant ionic lattices

In a piece of solid sodium chloride or magnesium oxide, the ions are arranged in a highly regular pattern. This pattern is called a **giant ionic lattice**. In this lattice, each positive ion is surrounded by six negative ions and each negative ion is surrounded by six positive ions. The forces of attraction between the positive ions and the negative ions hold the lattice together. These forces are called **electrostatic forces**.

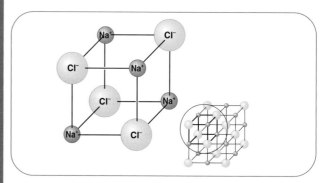

Figure 3b.13 a A tiny part of the sodium chloride lattice is shown, featuring only eight ions.

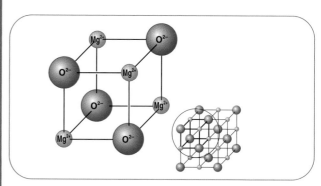

Figure 3b.13 b A tiny part of the magnesium oxide lattice is shown, featuring only eight ions.

If you look at a grain of salt under a hand lens, you will see that it has a cubic shape. The crystal is this shape owing to the pattern of sodium ions and chloride ions in the sodium chloride lattice.

Figure 3b.14 Crystals of sodium chloride.

Summary

You should be able to:

- describe an *ion*

- recognise an ion, an atom and a molecule from given formulae

- describe how positive and negative ions form

- explain how compounds form between metals and non metals

- describe the properties of sodium chloride and magnesium oxide

- describe the ionic bonding of sodium chloride, sodium oxide, magnesium chloride and magnesium oxide using dot and cross diagrams

- explain the significance of a full outer shell

- describe the structures of sodium chloride and magnesium oxide

- explain the physical properties of sodium chloride and magnesium oxide

Questions

1 A balanced symbol equation for the reaction between magnesium and oxygen is

$$2Mg + O_2 \rightarrow 2Mg^{2+}O^{2-}$$

The symbols used are Mg, O_2, Mg^{2+}, O^{2-} and $Mg^{2+}O^{2-}$.

Select from these symbols the symbol of an ion, the symbol of an atom, the formula of a molecule and the formula of an ionic compound.

2 Sodium chloride and magnesium chloride both consist of ions.

   a What is meant by the term 'ion'?

   b Give *one* similarity and *one* difference between a sodium ion and a magnesium ion.

   c Why do the sodium ions and the chloride ions in a salt crystal stay together?

3 List *three* physical properties shown by both sodium chloride and magnesium chloride.

4 Use dot and cross diagrams to show how the ions are formed from the original reacting atoms in sodium oxide, magnesium chloride, magnesium oxide and sodium chloride.

5 Explain the following statements.

   a Magnesium chloride conducts electricity when molten but not when solid.

   b Sodium oxide has a higher melting point than sodium chloride but a lower melting point than magnesium oxide.

Covalent bonding

Two or more non-metal elements can combine together to form a compound. The non-metal atoms bond together to form **molecules**. The atoms in these molecules are held together by **covalent bonds**. **Covalent bonding** is strong and involves the atoms sharing electrons. A covalent bond is a shared pair of electrons.

You have now learned about two types of strong bonding in this module – ionic bonding and covalent bonding.

Water

Water is a covalent compound. The formula of water is H_2O. A water molecule consists of two hydrogen atoms and one oxygen atom. The hydrogen atoms are bonded to the oxygen atom by covalent bonds. Because water molecules are only attracted to neighbouring water molecules by weak forces, water is a liquid. It has a low melting point. Water molecules are electrically neutral, so water does not conduct electricity.

SAQ

1 **a** How many oxygen atoms are there in a water molecule?

   b How many hydrogen atoms are there in a water molecule?

   c How many atoms are there in total in a water molecule?

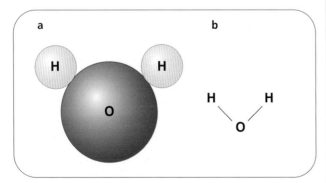

Figure 3c.1 a A water molecule consists of two hydrogen atoms and one oxygen atom. **b** This representation of a water molecule is called a displayed formula. The lines between the O and H symbols represent the covalent bonds.

d How are the atoms in a water molecule bonded to each other?

e Why does water have a low melting point?

f Why doesn't water conduct electricity?

Carbon dioxide

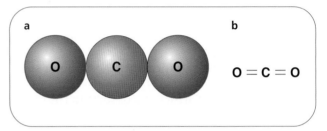

Figure 3c.2 a A carbon dioxide molecule consists of two oxygen atoms and one carbon atom. **b** This representation of a carbon dioxide molecule is called a displayed formula. The lines between the O and C symbols represent the double covalent bonds.

Carbon dioxide is a covalent compound. The formula of carbon dioxide is CO_2. A carbon dioxide molecule consists of two oxygen atoms and one carbon atom. The oxygen atoms are bonded to the carbon atom by covalent bonds. The bonding between the carbon atom and each oxygen atom consists of four electrons and is called a double covalent bond.

Because carbon dioxide molecules are only attracted to neighbouring carbon dioxide molecules by very weak forces, carbon dioxide is a gas. It has a low melting point. Carbon dioxide molecules are electrically neutral, so carbon dioxide does not conduct electricity.

Atoms, ions, molecules and formula units

An **atom** is:
● a single particle consisting of a nucleus surrounded by electrons moving on shells;
● electrically neutral.

An **ion** is:
● an atom, or small group of atoms, that has gained or lost one or more electrons;
● electrically charged.

A **molecule** is:

- two or more atoms covalently bonded together – the atoms in the molecule may be the same as each other, in which case it is a molecule of an element, or the atoms in the molecule may be different from each other, in which case it is a molecule of a compound;
- electrically neutral.

A **formula unit** is:

- all of the particles specified in the formula of an ionic compound – there is at least one positive ion and one negative ion in the formula unit; the total amount of positive charge is equal to the total amount of negative charge;
- electrically neutral.

SAQ

2 This question is about the following eight symbols:

H_2O Na NaCl CO_2 Na^+ MgO Mg O^{2-}

In this list which of the symbols represent:

a two atoms?

b two ions?

c two molecules?

d two formula units?

Dot and cross diagrams

When atoms make covalent bonds, they share enough electrons to gain a full outer shell of electrons. That is, the total of the electrons in the atom's outer shell plus the electrons it shares from other atoms gives the atom a full outer shell. This can be shown using dot and cross diagrams.

Hydrogen

The atoms in hydrogen gas are bonded in pairs. The formula of hydrogen is therefore H_2. A pair of hydrogen atoms bonded together is called a hydrogen molecule. The two atoms bond by sharing a pair of electrons (Figure 3c.3).

- The atomic number of hydrogen is one. Each hydrogen atom therefore has one electron. On the diagrams of the atoms before bonding, each hydrogen atom has one electron. This electron is on the first shell, which is also the outer shell.

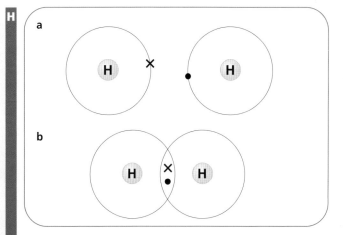

Figure 3c.3 a Two unbonded hydrogen atoms. **b** A hydrogen molecule.

- When the atoms are drawn bonded together, their outer shells are drawn overlapping. Each atom shares one electron with the other atom. These electrons are drawn on the shell overlap. This shared pair of electrons is the covalent bond.
- If the electrons on the shell overlap are counted, each atom now has two electrons on its outer shell. Each atom now has a full outer shell.

Chlorine

The atoms in chlorine gas are bonded in pairs. The formula of chlorine is therefore Cl_2. A pair of chlorine atoms bonded together is called a chlorine molecule. The two atoms bond by sharing a pair of electrons. The dot and cross diagram is shown in Figure 3c.4.

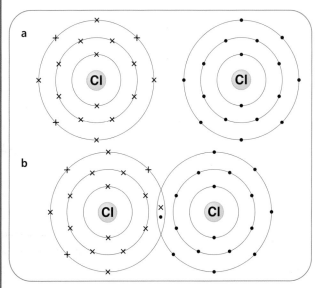

Figure 3c.4 a Two unbonded chlorine atoms. **b** A chlorine molecule.

H

- The atomic number of chlorine is 17. Each chlorine atom therefore has 17 electrons: two on the first shell, eight on the second shell and seven on the third. On the diagrams of the atoms before bonding, each chlorine atom has 17 electrons.
- When the atoms are drawn bonded together, their outer shells are drawn overlapping. Each atom shares one electron with the other atom. These electrons are drawn on the shell overlap. This shared pair of electrons is the covalent bond.
- If the electrons on each outer shell including the shell overlap are counted, each atom now has eight electrons on its outer shell. Each atom now has a full outer shell.
- The electrons on the inner shells do not take part in bonding. They can be omitted from the dot and cross diagram. Figure 3c.5 is a dot and cross diagram of chlorine, omitting the inner shells.

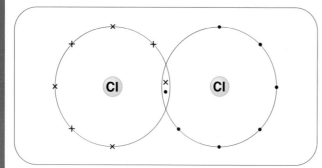

Figure 3c.5 The inner shells can be omitted from a dot and cross diagram.

SAQ

3 **a** How many electrons does an independent chlorine atom have on its outer shell?

   b When a chlorine atom bonds to another chlorine atom how many electrons does the other chlorine atom share with it?

   c How many electrons does each chlorine atom have on its outer shell now?

   d Why can the first and second shells on the chlorine atom be omitted from the dot and cross diagram?

H **Methane**

The atoms in methane gas are bonded together by covalent bonds. The formula of methane is CH_4. One carbon atom is bonded to four hydrogen atoms. The atoms bond by sharing pairs of electrons. The dot and cross diagram looks like this.

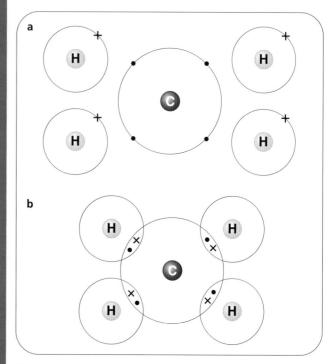

Figure 3c.6 a The five atoms before bonding. **b** A methane molecule.

- The atomic number of carbon is six. Each carbon atom therefore has six electrons, two on the first shell and four on the second shell. On these diagrams, the carbon atom's first shell has been omitted.
- If the electrons on its outer shell including the shell overlaps are counted, the carbon atom now has eight electrons on its outer shell. Each hydrogen atom has two electrons on its outer shell. Each atom now has a full outer shell.

SAQ

4 **a** How many electrons does an independent carbon atom have on its outer shell?

   b When a carbon atom bonds to a hydrogen atom, how many electrons does the hydrogen atom share with it?

   c Why does a carbon atom bond to four hydrogen atoms?

H Water

The atoms in water are bonded together by covalent bonds. The formula of water is H_2O. One oxygen atom is bonded to two hydrogen atoms. The atoms bond by sharing pairs of electrons. The dot and cross diagram looks like this.

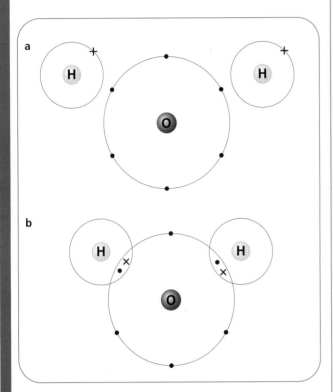

Figure 3c.7 a The three atoms before bonding. **b** A water molecule.

- The atomic number of oxygen is eight. Each oxygen atom therefore has eight electrons, two on the first shell and six on the second shell. On these diagrams, the oxygen atom's first shell has been omitted.
- If the electrons on its outer shell including the shell overlaps are counted, the oxygen atom now has eight electrons on its outer shell. Each hydrogen atom has two electrons on its outer shell. Each atom now has a full outer shell.

SAQ

5 a How many electrons does an independent oxygen atom have on its outer shell?

b When an oxygen atom bonds to a hydrogen atom, how many electrons do they share?

c Why does an oxygen atom bond to two hydrogen atoms?

H Carbon dioxide

The atoms in carbon dioxide are bonded together by covalent bonds. The formula of carbon dioxide is CO_2. One carbon atom is bonded to two oxygen atoms. The atoms bond by sharing pairs of electrons. The dot and cross diagram looks like this.

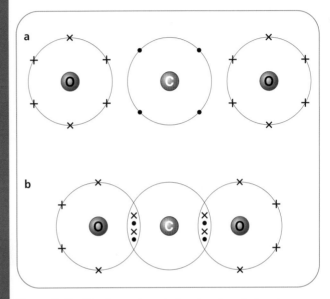

Figure 3c.8 a The three atoms before bonding. **b** A carbon dioxide molecule.

- The atomic number of carbon is six. Each carbon atom therefore has six electrons, two on the first shell and four on the second shell. On these diagrams, the carbon atom's first shell has been omitted.
- The atomic number of oxygen is eight. Each oxygen atom therefore has eight electrons, two on the first shell and six on the second shell. On these diagrams, the oxygen atom's first shell has been omitted.
- The carbon atom shares two electrons with each oxygen atom. Each oxygen atom shares two electrons with the carbon atom. These electrons are drawn on the shell overlap. The two shared pairs of electrons are called a double covalent bond.
- If the electrons on its outer shell including the shell overlaps are counted, the carbon atom now has eight electrons on its outer shell. Each oxygen atom has eight electrons on its outer shell. Each atom now has a full outer shell.

H *SAQ*

6 Why does the bonding between the carbon atom and each oxygen atom in carbon dioxide involve four electrons?

The properties of carbon dioxide and water

The bonding that holds the three atoms together in a water molecule is strong, but the bonds between neighbouring water molecules are weak. The bonds between neighbouring water molecules are called **intermolecular forces**. The intermolecular forces in water are weak. This is why water is a liquid with a low melting point. Molecules like water molecules that have weak intermolecular forces are known as **simple molecules**.

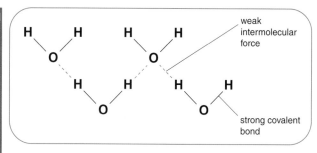

Figure 3c.9 Covalent bonding and intermolecular forces in water.

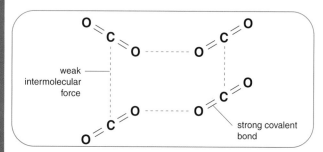

Figure 3c.10 Covalent bonding and intermolecular forces in carbon dioxide.

Dry ice

Carbon dioxide has one very unusual property. It doesn't form a liquid under normal conditions! Below −78 °C, carbon dioxide exists as a solid. If a piece of solid carbon dioxide is allowed to warm up, it changes directly into a gas at −78 °C. It doesn't melt first. This is why it's called *dry* ice. When water ice warms up, it melts and forms liquid water, which is wet. Dry ice doesn't do this: it goes straight from being a solid to being a gas.

Figure 3c.11 The 'smoke' coming out of this container has been produced using dry ice.

You have probably seen a rock band live on television with clouds of 'dry ice smoke' swirling around their ankles. However, what you're seeing is not carbon dioxide and it's not smoke. It's actually clouds of tiny water

droplets. The effect is produced with dry ice though, hence the name.

The 'smoke' is produced by putting blocks of dry ice into water. The dry ice changes into carbon dioxide gas, which bubbles through the water. The bubbles of carbon dioxide gas carry water vapour with them as they rise. The bubbles of gas are very cold. When they reach the air, the water vapour condenses back into liquid water. This forms a cold, dense white fog which slowly billows over the stage.

You can even hire a 'dry ice smoke' generator for your wedding. Then your first dance really will be in the clouds!

Apparently, mosquitoes are attracted to the carbon dioxide we breathe out. They home in on it and then they bite us. If you are having a party outside and being bothered by biting insects, a block of dry ice might help. Place it well away from you and your guests. The dry ice will turn into carbon dioxide, which will attract the mosquitoes. The insects will be very puzzled and disappointed when they find that their favourite smell has not led them to a meal. For you and your friends, this will be good news.

H

Carbon dioxide molecules are also simple molecules. The bonds between neighbouring carbon dioxide molecules are weak intermolecular forces. This is why carbon dioxide is a gas.

Water molecules and carbon dioxide molecules are electrically neutral. Water and carbon dioxide do not have free electrons. The electrons cannot move from one molecule to another molecule. This is why water and carbon dioxide do not conduct electricity.

The Periodic Table

The **Periodic Table** arranges the elements in order of increasing atomic number. The first element is therefore hydrogen, with atomic number one. The elements are arranged in rows and columns. The rows are called **periods**, the columns are called **groups**.

SAQ

7 Use the Periodic Table on page 345 to:

a name two elements in the same group as sulfur (S);

b name two elements in the same period as sulfur.

Periodic Table groups

Elements in the same group have the same number of electrons in their outer shell.

Because of this the elements in a group have similar chemical properties to each other. The number of electrons in the outer shell is the same as the group number, for example the elements in Group 5 have five outer electrons.

Group 1

The elements in Group 1 all have one electron in their outer shells. They all have similar chemical properties. They are all reactive metals called the **alkali metals**. You will learn more about the alkali metals in Item C3d.

Figure 3c.13 a Lithium. **b** Sodium. **c** Potassium.

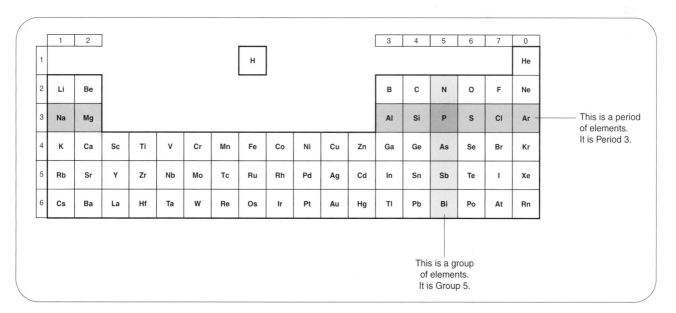

Figure 3c.12 A reduced version of the Periodic Table. More information about each element is in the large Periodic Table on page 345. (You may find Group 0 called Group 8 on some versions of the Periodic Table.)

Figure 3c.14 a Chlorine is a yellow–green gas. **b** Bromine is a dark orange liquid. **c** Iodine is a dark grey solid.

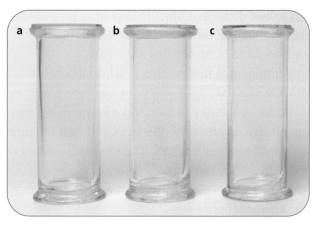

Figure 3c.15 The noble gases are all colourless. **a** Helium. **b** Neon. **c** Argon.

Group 7

The elements in Group 7 (Figure 3c.14) all have seven electrons in their outer shells. They all have similar chemical properties. They are all reactive non-metals called the **halogens**. You will learn more about the halogens in Item C3e.

Group 0 (zero)

The elements in Group 0 (Figure 3c.15) all have full outer shells. They all have similar chemical properties. They are all unreactive gases called the **noble gases**. This group is sometimes called Group 8.

SAQ

8 a Xenon is in Group 0. Which period is it in?

　b How many gaps does a xenon atom have in its outer shell?

　c Predict one property of xenon.

Periodic Table periods

Elements in the same period of the Periodic Table have the same number of occupied shells. This is the same as the period number.

For example, boron and neon are both in Period 2. Both elements have electrons in their first shell and their second shell but not in any other shells. Both elements have two occupied shells.

SAQ

9 How many occupied shells are there in:

　a a carbon atom?　**b** a chlorine atom?

　c a helium atom?　**d** a calcium atom?

10 Identify the group and period of the element whose electronic structure is:

　a 2.5;　**b** 2.8.8.2;

　c 2.8.1;　**d** 2.3.

Summary

You should be able to:

◆ describe a *molecule*

◆ state the number and type of atoms in a molecule given its formula

◆ name two types of strong bonding

◆ describe covalent bonding between non-metal atoms

◆ describe some properties of water and carbon dioxide

H ◆ use dot and cross diagrams to describe the bonding in hydrogen, chlorine, water, methane and carbon dioxide

continued on next page

Summary – *continued*

- ◆ relate the properties of carbon dioxide and water to their structure

- ◆ use the terms *group* and *period* correctly to describe elements in the Periodic Table

- ◆ recognise that elements in the same group have similar properties and the same number of electrons in their outer shell

- ◆ recognise that elements in the same period have the same number of occupied shells

- ◆ deduce the group and period to which an element belongs from its electronic structure

Questions

1 a How many oxygen atoms are there in a carbon dioxide molecule?

   b How many carbon atoms are there in a carbon dioxide molecule?

   c How many atoms are there in total in a carbon dioxide molecule?

   d How are the atoms in a carbon dioxide molecule bonded to each other?

   e Why doesn't carbon dioxide conduct electricity?

2 Use the Periodic Table on page 345 to answer this question.

   a Name two elements in Group 5.

   b Name two elements in Period 3.

   c Name the element which is in Group 5 and Period 3.

   d i How many electrons does the element from part **c** have in its outer shell?

      ii How many occupied shells does it have?

H 3 Draw diagrams to show the electronic structures of:

   a a hydrogen atom (1 electron);   b a carbon atom (6 electrons);

   c an oxygen atom (8 electrons);   d a chlorine atom (17 electrons).

4 Draw dot and cross diagrams to show the bonding in:

   a hydrogen (H_2);   b water (H_2O);   c methane (CH_4);

   d carbon dioxide (CO_2);   e chlorine (Cl_2).

5 a Why does an oxygen atom bond to two hydrogen atoms but a carbon atom bond to four hydrogen atoms?

   b Why does water not conduct electricity?

   c Why does carbon dioxide have a low boiling point?

The alkali metals

The elements in Group 1 of the Periodic Table are lithium, sodium, potassium, rubidium, caesium and francium. They are all metals. They have a bright, shiny grey appearance when clean and they all conduct electricity. They also have some more unusual properties. They all react vigorously with water to produce an alkaline solution. The elements in this group are known as the **alkali metals**.

The alkali metals react with the oxygen in the air as well as with water. Because of this, they are stored in bottles of oil. The oil prevents the metals coming into contact with oxygen or water vapour from the air. The alkali metals do not react with the oil.

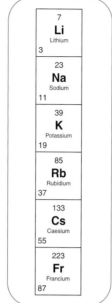

Figure 3d.1 Group 1 of the Periodic Table.

Figure 3d.2 The alkali metals are stored under oil because they react with oxygen and water.

SAQ

1 Give two properties of the alkali metals that are typical of metals.

The reaction of lithium with water

If a piece of lithium is put into a container of water, it floats on the water. The lithium reacts with the water immediately. Two products are formed. Bubbles of gas are produced; this gas is hydrogen. The other product is lithium

hydroxide, which dissolves in the water. Lithium hydroxide is an **alkali**. The word equation for this reaction is:

lithium + water → lithium hydroxide + hydrogen

The balanced symbol equation for this reaction is

$$2Li + 2H_2O \rightarrow 2LiOH + H_2$$

You should remember from Item C3c that, in the element hydrogen, the atoms form H_2 molecules. Each pair of hydrogen atoms is held together by a covalent bond.

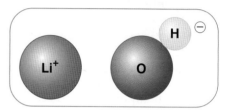

Figure 3d.3 One formula unit of lithium hydroxide. The lithium ion has a 1+ charge. All of the Group 1 metals are present in their compounds as 1+ ions. The hydroxide ion has a 1− charge. The oxygen atom and the hydrogen atom are held together by a covalent bond.

Figure 3d.4 Lithium reacting with water. Hydrogen bubbles can be seen.

The reaction of sodium with water

Sodium reacts with water in a similar way. The products of the reaction are hydrogen and sodium hydroxide. Sodium hydroxide is an alkali. The reaction of sodium with water is more vigorous than the reaction of lithium with water.

If the two reactions are compared, this difference in reactivity is noticed in two ways.

● If similar sized pieces are used, the sodium will finish reacting before the lithium.
● The sodium gets hot enough to melt during the reaction, the lithium does not.

Figure 3d.5 Sodium reacting with water. Hydrogen bubbles can be seen.

SAQ

2 a Write a word equation for the reaction of sodium with water.

b Write a balanced symbol equation for the reaction of sodium with water.

The reaction of potassium with water

Potassium reacts with water in a similar way, but it reacts more vigorously than either lithium or sodium. The products of the reaction are hydrogen and potassium hydroxide. Potassium hydroxide is an alkali.

Figure 3d.6 Potassium reacting with water. The lilac flame can be seen.

● As soon as the potassium touches the water, a lilac-coloured flame is seen on the piece of metal. There is a flame because the reaction is hot enough to set light to the hydrogen that forms. The lilac colour is due to the potassium.
● The reaction is very rapid and usually ends with a small explosion.

SAQ

3 a Write a word equation for the reaction of potassium with water.

b Write a balanced symbol equation for the reaction of potassium with water.

4 What evidence tells us that, with water:

a sodium is more reactive than lithium?

b potassium is more reactive than sodium?

Rubidium

The alkali metals get more reactive as we go down Group 1. This change in behaviour is called a **trend**. The metals show a trend in hardness too. Lithium can be cut with a sharp knife if you press hard. Sodium is easier to cut with a knife than lithium is. Potassium is easier to cut with a knife than sodium; in fact, potassium is quite soft.

In other ways lithium, sodium and potassium are the same as each other. They are all metals, they are shiny grey when clean, they conduct electricity and they float on water. These similarities and trends make it possible to predict the properties of rubidium. Rubidium is the element below potassium in Group 1. We expect it to have the following properties:

● it is a metal
● it is shiny grey when clean
● it conducts electricity
● it floats on water
● it is softer than potassium
● it has to be stored under oil
● it reacts with water, giving the products rubidium hydroxide and hydrogen
● it reacts with water more vigorously than potassium.

All of these predictions about rubidium are true. This is an illustration of how useful the Periodic Table is. Known facts about three elements have been used to make accurate predictions about a fourth element.

SAQ

5 The two elements below potassium in Group 1 of the Periodic Table are rubidium and caesium.

   a Write a balanced symbol equation for the reaction of rubidium with water.

   b Predict *one* property of rubidium hydroxide.

   c Write a balanced symbol equation for the reaction of caesium with water.

   d Make *six* predictions about properties you would expect caesium to have.

Electronic structure

Every Group 1 metal consists of atoms with one electron in the outer shell. This is why the Group 1 metals have similar properties. They all consist of atoms with one electron in the outer shell.

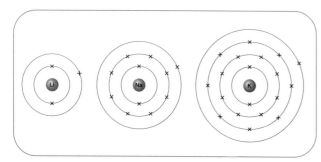

Figure 3d.7 Electronic structures of lithium, sodium and potassium. Each atom has one electron in the outer shell.

Lithium atoms lose one electron each

When lithium reacts with water, it forms lithium hydroxide. Lithium hydroxide is an ionic compound. Its formula is Li^+OH^-. This means that lithium is present in lithium hydroxide as lithium ions, Li^+. In this reaction, lithium changes from lithium atoms to lithium ions. Each lithium atom loses the one electron on its outer shell when it reacts. This can be described with an equation like this:

$$Li \rightarrow Li^+ + e^-$$

This is called an **ionic half-equation**. The 'e⁻' here represents the electron the lithium atom has lost. The lithium ion that forms is stable because it has a full outer shell.

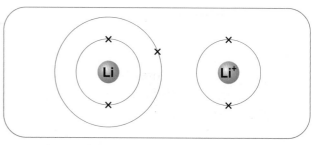

Figure 3d.8 A lithium atom (left) loses the electron from its outer shell when it reacts and becomes a lithium ion (right).

Lithium atoms are reactive. This is because lithium atoms don't have a full outer shell and are unstable. Lithium ions are unreactive. This is because lithium ions have a full outer shell and are stable.

Sodium and potassium

Sodium atoms also lose one electron each when they react. So do potassium atoms and the atoms of other Group 1 metals. This is why they have similar properties. The elements become more reactive as we go down the group.

They become more reactive because it is easier for the larger atoms (e.g. caesium) to lose their outer shell electron. The outer shell electron is lost more easily because it is further from the nucleus. It therefore experiences less attraction from the nucleus. It is harder for the smaller atoms (e.g. lithium) to lose the outer shell electron. Their outer shell electrons are nearer the nucleus and therefore experience a greater attractive force.

SAQ

6 a Write an ionic half-equation to show a sodium atom losing an electron to form a sodium ion.

   b Use electronic structure diagrams as in Figure 3d.8 to show a sodium atom losing an electron to form a sodium ion.

   c Why is potassium more reactive with water than sodium is?

H Oxidation

When an atom loses one or more electrons, we say that it has been oxidised. **Oxidation** is loss of electrons. Group 1 metal atoms each lose one electron when they react. This can be seen from the ionic half-equation:

$$K \rightarrow K^+ + e^-$$

The potassium atom loses an electron as it forms the potassium ion. The potassium atom is oxidised in this reaction.

Flame tests

If a lithium, sodium or potassium compound is put into a blue Bunsen flame, the flame changes colour. The colour seen depends on the alkali metal in the compound. These tests are called **flame tests**.

A flame test can tell you whether or not lithium, sodium or potassium is present in an unknown compound. To do a flame test, you need a clean piece of nichrome wire. Moisten the wire with some uncontaminated dilute hydrochloric acid, then touch the wire on the compound being tested. Put the wire in a blue Bunsen flame. Look to see if the flame changes colour.

SAQ

7 Which alkali metal is present in the compound if the flame goes:

   a lilac?

   b golden yellow?

   c crimson?

Figure 3d.9 A lithium compound on the wire loop has made the flame turn crimson.

Figure 3d.10 A sodium compound on the wire loop has made the flame turn golden yellow.

Figure 3d.11 A potassium compound on the wire loop has made the flame turn lilac.

	Lithium	Sodium	Potassium	Rubidium
Metal/non-metal	metal	metal	metal	metal
Appearance	shiny grey	shiny grey	shiny grey	shiny grey
Storage	stored in oil	stored in oil	stored in oil	stored in oil
Electrical conductivity	very good	very good	very good	very good
Hardness	quite hard	soft	very soft	extremely soft
Reactivity with water	fairly reactive	reactive	very reactive	extremely reactive
Products of reaction with water	hydrogen gas and an alkali	hydrogen gas and an alkali	hydrogen gas and an alkali	hydrogen gas and an alkali
Flame test colour	crimson	golden yellow	lilac	deep red

Table 3d.1 Summary of alkali metal behaviour.

The alkali metals and us

We have very different amounts of lithium, sodium and potassium in our bodies. You probably contain less than one-hundredth of a gram of lithium but around one hundred grams of both sodium and potassium.

Compounds of all of the alkali metals have become important in medicine. Lithium carbonate was the first medication that was found to help mentally ill patients suffering from manic depression. This illness can be so severe that sufferers used to be locked in mental hospitals. When the drug was first introduced,

Figure 3d.12 This child is being given a rehydrating drink made by dissolving glucose, salt and small amounts of other chemicals in water.

it helped some patients so much they were able to return home.

Sodium chloride is just common salt, but it can be a life-saver. Young children can get very severe diarrhoea. This can cause their bodies to lose so much water that they die of dehydration. Water alone will not rehydrate them. However, if they drink water with glucose and salt dissolved in it, this does the trick.

The use of potassium you are about to find here is not exactly a medical use. In the USA, convicted murderers are executed in some states. One of the means of execution is called 'lethal injection'. An anaesthetic is injected first, followed by a large amount of potassium chloride solution. Death is caused immediately by heart attack.

The death sentence can also be carried out by electrocution or gassing, but these forms of execution damage the organs of the body. Some condemned prisoners agree to donate their organs for transplants. No damage is caused to the body's organs by potassium chloride. The organs can be used for transplants if this method of execution is used.

Summary

You should be able to:

◆ state the names of the first three metals in Group 1

◆ state the 'family name' of the Group 1 metals

◆ explain why the Group 1 metals are stored in oil

◆ state how the Group 1 metals react with water and recall the order of reactivity of the first three metals in the group

◆ use a word equation to describe the reaction of a Group 1 metal with water

H ◆ use a balanced symbol equation to describe the reaction of a Group 1 metal with water

◆ predict the properties of rubidium and caesium

continued on next page

Summary – *continued*

- ◆ explain why Group 1 metals have similar properties

H ◆ use electronic structures to explain why Group 1 metals have similar properties

- ◆ explain why the Group 1 metals get more reactive going down the group

- ◆ use ionic half-equations to explain the process of oxidation

- ◆ describe how to perform a flame test

- ◆ state the flame test colours for lithium, sodium and potassium compounds

- ◆ interpret information about flame tests

Questions

1   a   Describe how you would do a flame test on an unidentified white powder.

    b   What colour change will you see in a blue Bunsen flame if you do a flame test with the following compounds?

        i   Sodium chloride

        ii   Potassium sulfate

        iii   Lithium carbonate

2   You are given a piece of sodium metal, a piece of lithium metal and a piece of potassium metal. You are not told which is which.

    a   How could you find out which is which by putting the pieces of metal into water?

    b   What safety precautions would you have to take in order to do this experiment?

3   a   Describe what you see when potassium reacts with water.

    b   Name the gaseous product of this reaction.

    c   Name the alkaline product of this reaction.

    d   Write a word equation for the reaction of potassium with water.

H   e   Write a balanced symbol equation for the reaction of potassium with water.

4   a   Why are alkali metals stored under oil?

    b   Why do the Group 1 metals have similar properties?

H   c   Why is sodium more reactive than lithium?

    d   Lithium is oxidised when it reacts with water. Explain this statement with the aid of an ionic half-equation.

Halogens

The elements in Group 7 of the Periodic Table are called the **halogens**. The halogens are a family of elements with similar properties. They are all non-metals. The first element in the group is fluorine. Fluorine is followed by chlorine, bromine, iodine and astatine. These elements are coloured. Chlorine is a yellow–green gas. Bromine is a dark, orange–brown liquid. Iodine is a dark grey solid.

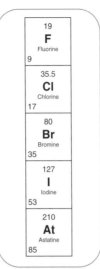

19	
F	
Fluorine	
9	
35.5	
Cl	
Chlorine	
17	
80	
Br	
Bromine	
35	
127	
I	
Iodine	
53	
210	
At	
Astatine	
85	

Figure 3e.1 Group 7 of the Periodic Table.

Uses of the halogens

The halogens have many uses. Some of these uses are illustrated in Figures 3e.2–3e.5

Figure 3e.2 Chlorine has been used to kill micro-organisms in this drinking water.

Figure 3e.3 These carrots are being sprayed with a pesticide which kill pests that would damage the crop. The pesticide is a chlorine compound.

Figure 3e.4 The insulation on this electric cable is made of polyvinylchloride (PVC). PVC is a chlorine compound.

Figure 3e.5 Iodine is an antiseptic – it kills micro-organisms. Here it is being used to sterilise a wound before an operation.

Uses of sodium chloride

The chemical compound sodium chloride is commonly known as salt. Some of the uses of salt are illustrated in Figures 3e.6–3e.8.

Figure 3e.6 This fish has been salted. The salt will preserve the fish and stop it decaying.

Figure 3e.7 Salt acts as a flavour enhancer. It improves the flavour of savoury food, such as these chips.

Figure 3e.8 Chlorine is made from sodium chloride in these mercury cells.

SAQ

1 **a** Give *three* uses of chlorine.

   b Give *three* uses of sodium chloride.

Reaction with alkali metals

The halogens have similar chemical properties to each other. The halogens all react with the elements in Group 1 to form compounds called **salts** or **metal halides**. For example, when lithium is reacted with fluorine, lithium fluoride is formed. Lithium fluoride can be classified as a salt or as a metal halide. The word equation for this reaction is:

lithium + fluorine → lithium fluoride

Lithium fluoride is an ionic compound. Lithium atoms and fluorine atoms are reacting together to make a compound consisting of lithium ions and fluoride ions. Each lithium atom has one electron in its outer shell. The lithium atom loses this electron in the reaction and becomes an Li^+ ion. Each fluorine atom has seven electrons in its outer shell. The fluorine atom gains an electron in the reaction and becomes an F^- ion.

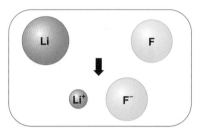

Figure 3e.9 Lithium reacts with fluorine to give lithium fluoride.

All the halogens react similarly with lithium and with the other alkali metals. For example, a word equation for the reaction of potassium with bromine is:

potassium + bromine → potassium bromide

The halogens all react similarly with Group 1 metals. The halogens have similar properties to each other because they all have atoms with seven electrons in the outer shell.

SAQ

2   Write word equations for the following reactions:

   a   caesium with bromine

   b   potassium with fluorine

   c   lithium with chlorine

   d   rubidium with iodine.

3   Why do all the halogens have similar properties?

H Balanced symbol equations

When you write a balanced symbol equation for the reaction between an alkali metal and a halogen, you must remember these rules.

● The halogens consist of atoms covalently bonded in pairs to form molecules. The formulae of these molecules are F_2, Cl_2, Br_2 and I_2.

● The alkali metal halides are ionic. The alkali metal ions all have a 1+ charge. The halide ions all have a 1– charge. One formula unit of an alkali metal halide contains one alkali metal ion and one halide ion, e.g. NaCl or LiF.

H Worked example 1

Write a balanced symbol equation for the reaction between lithium and fluorine.

Step 1: start with a word equation.

Lithium + fluorine → lithium fluoride

Step 2: put in the correct formulae.

$Li + F_2 \rightarrow LiF$

Step 3: balance the equation.

$2Li + F_2 \rightarrow 2LiF$ ✓

Because each LiF formula unit contains one Li^+ ion and one F^- ion, each F_2 molecule reacts with two Li atoms. This produces two LiF formula units.

SAQ

4   Write balanced symbol equations for:

   a   rubidium reacting with fluorine

   b   potassium reacting with iodine

   c   lithium reacting with bromine

   d   caesium reacting with chlorine.

5   When lithium reacts with fluorine, the lithium is oxidised. Explain this statement.

Reactivity

The halogens show a clear trend of reactivity. Fluorine is more reactive than chlorine, which is more reactive than bromine, which is more reactive than iodine. The halogens get less reactive as we go down the group. This can be seen in their reactions with sodium (Figure 3e.10).

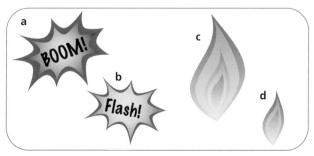

Figure 3e.10 a Fluorine reacts explosively with sodium. **b** Chlorine reacts quickly with sodium. **c** Bromine reacts steadily with sodium. **d** Iodine reacts slowly with sodium.

SAQ

6 **a** Write word equations for the reactions in Figure 3e.10.

b Write balanced symbol equations for the reactions in Figure 3e.10.

Displacement reactions

If chlorine solution is added to a solution of sodium bromide, a **displacement reaction** occurs. The products are sodium chloride and bromine. A word equation for this reaction is:

sodium bromide + chlorine →

sodium chloride + bromine

Look at the word equation. The chlorine has taken the place of the bromine in the sodium halide compound. We say that the chlorine has **displaced** the bromine.

This reaction has an observable result. The bromine that forms makes the solution go orange.

Figure 3e.11 Chlorine displaces bromine from sodium bromide.

An element that displaces another element from one of its compounds is more reactive than the element it displaces. The reaction shown in Figure 3e.11 proves that chlorine is more reactive than bromine. The displacement reaction can be tried in reverse. Bromine solution can be added to a solution of sodium chloride. Nothing happens. Bromine cannot displace chlorine. This proves that bromine is less reactive than chlorine.

Bromine displaces iodine because bromine is more reactive than iodine.

Sodium iodide + bromine →

sodium bromide + iodine

Iodine can't displace bromine because iodine is less reactive than bromine.

Sodium bromide + iodine → no reaction

SAQ

7 Look at Figure 3e.12.

Figure 3e.12 The reaction between chlorine solution and sodium iodide solution.

a Write a word equation for the reaction taking place.

b What observable evidence is there that this reaction occurs?

c What does this reaction tell us about chlorine and iodine?

d What happens if iodine solution is added to sodium chloride solution? Explain your answer.

8 Write balanced symbol equations for the following reactions:

a chlorine and sodium bromide

b chlorine and sodium iodide

c bromine and sodium iodide.

Gaining electrons

When chlorine reacts with sodium, it forms sodium chloride. Sodium chloride is an ionic compound. Its formula is Na^+Cl^-. This means that chlorine is present in sodium chloride as chloride ions, Cl^-. In this reaction, chlorine changes from chlorine atoms to chloride ions. Each chlorine atom gains one electron and fills the gap in its outer shell when it reacts. This can be described with an equation like this:

$$Cl + e^- \rightarrow Cl^-$$

H This is called an **ionic half-equation**. The 'e⁻' here represents the electron the chlorine atom has gained. The chloride ion that forms is stable, because it has a full outer shell.

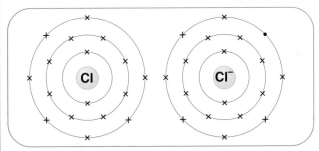

Figure 3e.13 A chlorine atom gains an electron and fills the gap in its outer shell when it reacts and becomes a chloride ion.

Chlorine atoms are reactive. This is because chlorine atoms don't have a full outer shell and are relatively unstable. Chloride ions are unreactive. This is because chloride ions have a full outer shell and are relatively stable.

Bromine and iodine

Bromine atoms also gain one electron each when they react. So do iodine atoms and the atoms of other halogens. Each atom has seven electrons in its outer shell, so each atom has one gap in its outer shell. This is why the halogens have similar properties.

The elements become more reactive as we go up the group. They become more reactive because it is easier for the smaller atoms (e.g. chlorine) to gain an electron and fill the gap in the outer shell. It is harder for the larger atoms (e.g. iodine) to gain an electron and fill the gap in the outer shell.

The smaller atoms are more reactive because the gap in their outer shell is nearer the nucleus. The nucleus therefore exerts a stronger attractive force on an electron which approaches the gap. Smaller atoms gain electrons more easily than larger atoms.

H **SAQ**

9 a Write an equation to show a fluorine atom gaining an electron to form a fluoride ion.

 b Use electronic structure diagrams as in Figure 3e.13 to show a fluorine atom gaining an electron to form a fluoride ion.

 c Why is fluorine more reactive than chlorine?

Reduction

When an atom gains one or more electrons, we say that it has been reduced. **Reduction** is gain of electrons. In Item C3d, you learned that oxidation is loss of electrons. There is an easy way to remember this:

Oxidation Is Loss, **R**eduction Is **G**ain – OIL RIG.

Halogen atoms each gain one electron when they react. This can be seen from the ionic half-equation:

$$Cl + e^- \rightarrow Cl^-$$

The chlorine atom gains an electron as it forms the chloride ion. The chlorine atom is reduced in this reaction.

Physical properties

The boiling points of chlorine, bromine, and iodine increase steadily down the group.

- Chlorine: −35 °C.
- Bromine: +59 °C.
- Iodine: +170 °C.

SAQ

10 a Predict the colours of fluorine and astatine. (Hint – look at the start of this item.)

 b Predict the boiling points of fluorine and astatine.

	Chlorine	Bromine	Iodine
Metal/non-metal	non-metal	non-metal	non-metal
State	gas	liquid	solid
Appearance	yellow–green	dark orange–brown	dark grey
Reactivity	very reactive	reactive	fairly reactive
Reaction with sodium	vigorous, white solid product	steady, white solid product	slow, white solid product
Boiling point	−35 °C	+59 °C	+170 °C

Table 3e.1 Summary of the properties of the halogens.

Good news or bad news?

A typical adult body contains about 5 grams of fluorine. The element fluorine is highly reactive and very dangerous, but all of the fluorine in our bodies is in the form of compounds. Fluorine compounds are not reactive. Most of the fluorine compounds in our bodies are found in our teeth and bones. Teeth and bones contain a hard compound called fluoroapatite.

Fluoroapatite in our tooth enamel makes the teeth resistant to corrosion by the acids that are formed in our mouths by plaque bacteria. For this reason, toothpaste manufacturers add fluoride compounds to toothpaste. In the UK, and in many other parts of the world, water companies add small amounts of fluoride compounds to drinking water. The toothpaste manufacturers and water companies believe they are helping public health by adding fluoride.

Is this true? Is fluoride good for us? A 10 g dose of sodium fluoride can be fatal, but the amounts in toothpaste and drinking water are very low. Even so, the National Pure Water Association (NPWA) was set up in England in 1960 to campaign against the fluoridation of drinking water. The NPWA has even gone so far as to suggest that the UK water industry might be poisoning children.

In this sort of situation, we need scientists to collect evidence that shows whether or not fluoride in drinking water is a danger to children. We need scientists to collect evidence that shows how much benefit the fluoride gives to the teeth and bones. When science has provided better evidence, we can make better decisions.

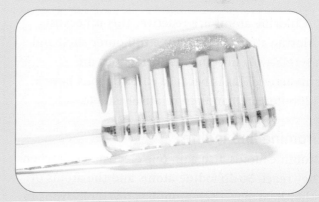

Figure 3e.14 This toothpaste contains the fluoride compound sodium fluorophosphate.

Summary

You should be able to:

- state the 'family name' of the elements in Group 7

- name the elements in Group 7

- state the colours of chlorine, bromine and iodine

- state some uses of chlorine, iodine and sodium chloride

- describe how the halogens react with alkali metals, including word equations

(H) - write balanced symbol equations for the reactions of the halogens with alkali metals

- recall the order of reactivity of the halogens

- describe the displacement reaction between a halogen and a solution of a metal halide, including a word equation

continued on next page

Summary - *continued*

- ◆ use electronic structures to explain why the halogens have similar properties

H ◆ write a balanced symbol equation for the displacement reaction between a halogen and a solution of a metal halide

- ◆ explain why the halogens get more reactive going up Group 7

- ◆ use ionic half-equations to explain the process of reduction

- ◆ predict some physical properties of fluorine and astatine

Questions

1   a   Which group of elements in the Periodic Table is known as the halogens?

    b   Name *three* of the halogens and state what colour they are.

    c   Would you expect a piece of iodine to conduct electricity? Explain your answer.

    d   What happens to the reactivity of the halogens as we go down the group?

    e   You are given a solution of a halogen and a solution of the sodium compound of another halogen. What experiment could you do to find out which of the two halogens is most reactive? Explain your answer.

2   Give *one* use of iodine, *two* uses of chlorine and *three* uses of sodium chloride.

3   Fluorine reacts with caesium to form a white crystalline solid.

    a   Name the white crystalline solid.

    b   Would you expect this reaction to be fast or slow? Explain your answer.

    c   Write a word equation for the reaction.

H 4   Potassium reacts with chlorine.

    a   Write a balanced symbol equation for this reaction.

    b   Write an ionic half-equation to show what happens to each potassium atom in this reaction.

    c   Write an ionic half-equation to show what happens to each chlorine atom in this reaction.

    d   Which substance is oxidised in this reaction and which substance is reduced? Explain your answer.

    e   What type of bonding holds the potassium and chlorine particles together in potassium chloride?

    f   What is the structure of potassium chloride?

Changes made by electricity

When a direct electric current (DC) flows through a metal wire under normal conditions, the wire is not changed by the current. This is not the case when a direct electric current flows through a melted or dissolved ionic compound. The electricity causes permanent changes in the compound. When a compound undergoes permanent change caused by electric current, the process is called **electrolysis**.

Electrolysis of molten aluminium oxide

The apparatus in Figure 3f.1 shows what happens when a direct electric current is passed through molten aluminium oxide.

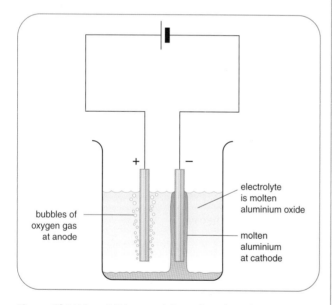

Figure 3f.1 When DC is passed through molten aluminium oxide, oxygen gas is formed at the anode and aluminium metal is formed at the cathode.

Electrolysis vocabulary

The vocabulary words that follow are important when discussing electrolysis.

Anode

The **anode** is the positive electrode. It attracts negative ions.

Cathode

The **cathode** is the negative electrode. It attracts positive ions.

Anion

An **anion** is a negative ion. It is attracted to the positive electrode – the anode. In molten aluminium oxide the anions are the oxide ions, O^{2-}. The oxide ions are attracted to the anode.

Cation

A **cation** is a positive ion. It is attracted to the negative electrode – the cathode. In molten aluminium oxide the cations are the aluminium ions, Al^{3+}. The aluminium ions are attracted to the cathode.

Electrolyte

An **electrolyte** is a liquid that conducts electricity and is permanently changed by a direct electric current. Ionic compounds, either when molten or when dissolved in water, act as electrolytes. In Figure 3f.1 the electrolyte is molten aluminium oxide.

SAQ

1 a Does the passing of DC change a molten ionic compound?

   b Does the passing of DC change a metal wire?

2 Give a definition for each of the following terms.

   a Anion   d Electrolyte

   b Anode   e Cathode

   c Cation   f Electrolysis

Making aluminium

The aluminium we use is made by the electrolysis of molten aluminium oxide. The molten aluminium oxide is decomposed by a direct electric current. The source of the aluminium oxide is an ore called **bauxite**. Bauxite is impure aluminium oxide; it looks like red earth.

Figure 3f.2 A bauxite mine in Jamaica.

The DC makes the molten aluminium oxide decompose into aluminium and oxygen. The word equation for this is:

aluminium oxide → aluminium + oxygen

- The aluminium oxide has to be molten or it will not conduct electric current.
- The aluminium ions are positive. They are cations. They are attracted to the negative electrode. Aluminium metal forms at the cathode. The cathode is made of graphite (carbon).
- The oxide ions are negative. They are anions. They are attracted to the positive electrode. Oxygen gas forms at the anode. The anode is made of graphite.
- The graphite anodes react with the oxygen and burn away. The anodes need to be replaced regularly.
- Producing aluminium by electrolysis requires a lot of electrical energy. This makes aluminium expensive.
- Aluminium oxide has a melting point of over 2000 °C. Running the process at this temperature would make it even more expensive. In order to lower the melting point of the aluminium oxide, a mineral called cryolite is added. The mixture of aluminium oxide and cryolite melts at around 850 °C.

Figure 3f.3 a This dam and hydroelectric power station provide electrical energy for aluminium smelting plant like this one (**b**).

Figure 3f.4 These cables are made of aluminium.

SAQ

3 **a** Name the ore that is a common source of aluminium oxide.

  b Write a word equation for the decomposition of aluminium oxide.

  c Why is aluminium expensive?

4 When molten aluminium oxide is electrolysed:

  a why does the aluminium oxide have to be molten?

  b what are the anode and cathode made of?

  c why do the anodes need replacing regularly?

Electrolysis of molten aluminium oxide

The reaction at the anode

The oxide ions are negative – they are O^{2-} ions. They are attracted to the anode, which is positive. At the anode, each oxide ion loses two electrons and becomes an oxygen atom. An ionic half-equation for the reaction at the electrode is:

$$O^{2-} \rightarrow O + 2e^-$$

Oxygen atoms bond together covalently to form oxygen molecules, O_2, so the equation can be written:

$$2O^{2-} \rightarrow O_2 + 4e^-$$

Therefore oxygen gas is formed at the anode.

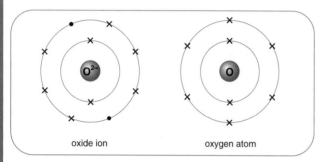

Figure 3f.5 An oxide ion losing two electrons to form an oxygen atom.

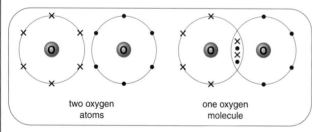

Figure 3f.6 Two oxygen atoms forming an oxygen molecule.

SAQ

5 a Explain why oxygen is said to be oxidised in this reaction.

b How many oxygen atoms are there in an oxygen molecule?

The reaction at the cathode

The aluminium ions are positive – they are Al^{3+} ions. They are attracted to the cathode, which is negative. At the cathode, each aluminium ion gains three electrons and becomes an aluminium atom. An ionic half-equation for the reaction at the electrode is:

$$Al^{3+} + 3e^- \rightarrow Al$$

Therefore liquid aluminium metal is formed at the cathode.

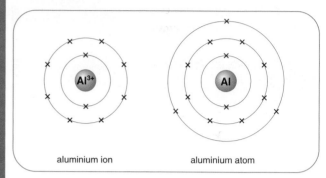

Figure 3f.7 An aluminium ion gaining three electrons to form an aluminium atom.

SAQ

6 Explain why aluminium is said to be *reduced* in this reaction.

Electrolysis of dilute sulfuric acid

When a direct electric current is passed through dilute sulfuric acid, oxygen and hydrogen gases are formed. The apparatus in Figure 3f.8 shows how this can be done.

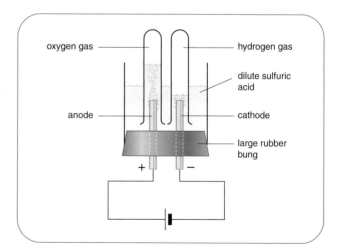

Figure 3f.8 When DC is passed through dilute sulfuric acid, oxygen gas is collected at the anode and hydrogen gas is collected at the cathode.

Identifying oxygen

When DC is passed through dilute sulfuric acid, oxygen gas forms at the anode. The oxygen can be collected in test tubes using the apparatus in Figure 3f.8. The oxygen gas can be identified using a glowing splint. If a glowing splint is put into oxygen gas, the glowing splint relights. Only oxygen does this. When the splint relights, this identifies the gas as oxygen.

Figure 3f.9 The glowing splint test proves that an unknown gas is oxygen if the splint relights.

Oxygen and hydrogen

Early scientists did not suspect the existence of many of the gases we now know about. They thought air was a single pure substance. They even believed the air to be an element – one of the fundamental substances that made up everything else.

The first scientist to make oxygen was Carl Scheele, who made oxygen in his laboratory in Sweden in 1772. The news of his discovery did not spread far, however. When the English scientist Joseph Priestley made oxygen in 1774, he believed he had made a new discovery. In these days of modern telecommunications it is incredible to think of such an important scientific discovery going unnoticed for two years!

Oxygen is now an important industrial chemical. Over 100 million tonnes of pure oxygen are obtained from the air and used around the world every day. Oxygen is used by hospitals to help people with breathing problems, in steel making and by the chemical industry.

Liquid oxygen is a blue liquid with some amazing properties. It is magnetic. If liquid oxygen is poured between the poles of a powerful magnet, a pulsating blue blob of liquid oxygen will hang in the air between the poles.

Henry Cavendish discovered hydrogen in 1766 in Clapham, England. He reacted dilute sulfuric acid with iron filings and collected the bubbles of gas that were given off. This gas burnt very easily and was much lighter than air. When hydrogen burns, the only product is water, and this is how the gas got its name. *Hydro* means *water* and *gen* means *maker*.

Hydrogen molecules are the lightest of any substance. This causes the pitch of sounds produced in hydrogen to become higher, like the high-pitched squeaky voice people have if they breathe helium.

Figure 3f.10 Carl Scheele.

Figure 3f.11 Liquid oxygen suspended between the poles of a strong magnet.

Figure 3f.12 Henry Cavendish

Identifying hydrogen

When a direct electric current is passed through dilute sulfuric acid, hydrogen gas forms at the cathode. The hydrogen can be collected in test tubes using the apparatus in Figure 3f.8. The hydrogen gas can be identified using a lighted splint. If a lighted splint is put into hydrogen gas you will hear a squeaky 'pop' noise. Only hydrogen does this. When a lighted splint causes a pop, this identifies the gas as hydrogen.

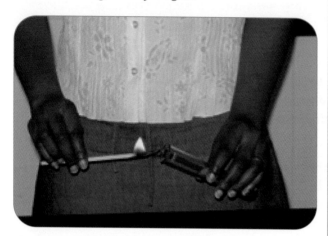

Figure 3f.13 The lighted splint test proves that an unknown gas is hydrogen if there is a squeaky popping noise.

SAQ

7 When DC is passed through dilute sulfuric acid:

   a where does oxygen form?

   b where does hydrogen form?

   c how can each gas be identified?

   d what substance acts as an electrolyte?

The ions involved in the electrolysis of dilute sulfuric acid

The formula of sulfuric acid is H_2SO_4. Dilute sulfuric acid contains hydrogen ions, H^+, and sulfate ions, SO_4^{2-}, owing to the sulfuric acid present. There are also ions present owing to the water: whenever water is present, there are hydrogen ions, H^+, and hydroxide ions, OH^-.

Dilute sulfuric acid therefore contains one type of cation, H^+, and two types of anion, SO_4^{2-} and OH^-.

The reaction at the cathode

The hydrogen ions are positive – they are H^+ ions. They are attracted to the cathode, which is negative. At the cathode, each hydrogen ion gains one electron and becomes a hydrogen atom. An ionic half-equation for this reaction is:

$$H^+ + e^- \rightarrow H$$

Hydrogen atoms bond together covalently to form hydrogen molecules, H_2, so the equation can be written:

$$2H^+ + 2e^- \rightarrow H_2$$

Therefore hydrogen gas forms at the cathode.

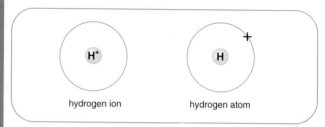

hydrogen ion   hydrogen atom

Figure 3f.14 A hydrogen ion gaining one electron to form a hydrogen atom.

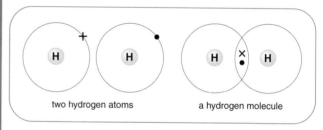

two hydrogen atoms   a hydrogen molecule

Figure 3f.15 Two hydrogen atoms forming a hydrogen molecule.

The reaction at the anode

Two types of anion are present, the sulfate ions, SO_4^{2-}, and the hydroxide ions, OH^-. They are attracted to the anode, which is positive. The sulfate ions do not change, because any change that can occur requires too much energy. The only ions that are changed at the anode are the hydroxide ions. At the anode, hydroxide ions lose electrons. Water and oxygen are formed. An ionic half-equation for this reaction is:

$$4OH^- \rightarrow 2H_2O + O_2 + 4e^-$$

Therefore oxygen gas is formed at the anode. Water forms at the anode, too, but it is not noticed because it simply mixes in with the rest of the solution.

Summary

You should be able to:

♦ describe electrolysis and use the terms *anode*, *cathode*, *anion*, *cation* and *electrolyte* correctly in the context of electrolysis

♦ describe how aluminium is extracted from bauxite by the electrolysis of molten aluminium oxide

♦ write a word equation for the decomposition of aluminium oxide

H ♦ write an ionic half-equation for the reaction occurring at each electrode during the electrolysis of molten aluminium oxide

♦ explain why this electrolysis process is expensive and explain the role of cryolite in this process

♦ state that dilute sulfuric acid can be broken down by electrolysis to give oxygen and hydrogen, knowing at which electrode each gas forms

♦ describe identifying tests for oxygen and hydrogen

H ♦ write an ionic half-equation for the reaction occurring at each electrode during the electrolysis of dilute sulfuric acid

Questions

1 In the electrolysis of molten aluminium oxide:

   a Which product forms at the cathode?

   b Why does this product form at the cathode?

   c Which product forms at the anode?

   d Why does this product form at the anode?

2 Air is 21% oxygen. Pure oxygen is 100% oxygen. Explain why a splint that is glowing in the air relights in pure oxygen, You should use terms such as 'concentration' and 'collisions' in your answer.

3 Identify the three anions and the three cations in the list that follows:

   K^+, Cl^-, Ca^{2+}, Na^+, F^-, OH^-.

H 4 a How many electrons are shared between the oxygen atoms in an oxygen molecule?

   b Name this type of covalent bond.

   c Name another substance with atoms that are held together by this type of covalent bond.

5 Explain why hydrogen ions can be said to be reduced when DC is passed through dilute sulfuric acid.

Typical metals

The central block of the Periodic Table consists of elements that are all typical metals. These elements are called the **transition elements**, or the transition metals. Many familiar metals are found here, like iron, copper, silver and gold. Although there are some exceptions, the transition metals have typical metal properties.

- They are good conductors of heat and electricity.
- They have a bright, shiny appearance.
- They are malleable and ductile.

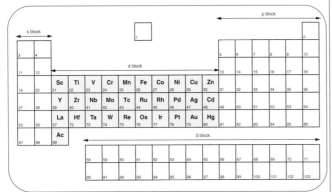

Figure 3g.1 Transition elements are found in the centre of the Periodic Table. The block of elements at the bottom of the table are also transition elements but they are not studied in this Item; see also page 345.

SAQ

1 Which of the following atomic numbers refer to transition elements? 12, 26, 34, 56, 79.

2 Element 74 is a transition element. Give its name and symbol, and suggest *three* properties it is likely to have.

Coloured compounds

Transition elements have coloured compounds. For example:

- copper(II) compounds (which contain the Cu^{2+} ion) are usually blue or greeny-blue;
- iron(II) compounds (which contain the Fe^{2+} ion) are usually light green;
- iron(III) compounds (which contain the Fe^{3+} ion) are usually orange or brown.

This property is typical of the transition elements but is not typical of all metals. Sodium and aluminium are metals but they are not transition elements. Most sodium and aluminium compounds are colourless or white.

Figure 3g.2 a Copper(II) nitrate. **b** Iron(II) sulfate. **c** Iron(III) chloride.

Catalysts

A **catalyst** is a substance that speeds up the rate of a chemical reaction while remaining unchanged itself. Many transition elements and compounds of transition elements can act as catalysts.

- Iron is used as a catalyst in the Haber process. This process is essential in the manufacture of many fertilisers. You will learn more about it in Item C4d.
- Nickel is used as a catalyst when margarine is made. The nickel catalyst speeds up the reaction between hydrogen and liquid vegetable oils that changes the liquid oils into spreadable solids.

Having catalytic properties is typical of the transition elements and their compounds but is not typical of all metals.

SAQ

3 Predict the colours of the following compounds.
  a Iron(II) chloride.
  b Iron(III) nitrate.
  c Copper(II) sulfate.

4 What is meant by a 'catalyst'?

5 State *two* properties that are typical of transition elements but are not typical of all metals.

Transition element carbonates

The transition elements form carbonate compounds. Four transition-metal carbonates are shown in Figure 3g.3.

Figure 3g.3 a Iron(II) carbonate, $FeCO_3$. **b** Copper(II) carbonate, $CuCO_3$. **c** Manganese(II) carbonate, $MnCO_3$. **d** Zinc carbonate, $ZnCO_3$.

The compounds in Figure 3g.3 are ionic. In each of these compounds, the transition element is present as a 2+ ion. The formula of the carbonate ion is CO_3^{2-}, so the carbonate ion is a 2− ion. If the formula of iron(II) carbonate is written with the electric charges included it looks like this: $Fe^{2+}CO_3^{2-}$.

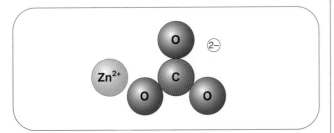

Figure 3g.4 One formula unit of zinc carbonate. The zinc ion has a 2+ charge. The carbonate ion has a 2− charge. The carbon atom and the oxygen atoms are held together by covalent bonds.

SAQ

6 Which of the ions present in zinc carbonate is an anion and which is a cation?

Thermal decomposition reactions

The transition element carbonates undergo thermal decomposition reactions. This means they break down when they are heated. They do not break down to the elements they are made of. They break down to two compounds. These compounds are the transition element's oxide and carbon dioxide. For example:

copper(II) carbonate →
copper(II) oxide + carbon dioxide

This thermal decomposition reaction can be investigated using the apparatus in Figure 3g.5.

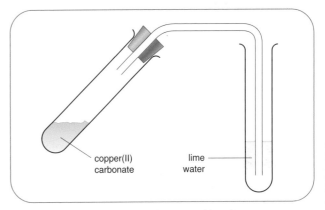

Figure 3g.5 a Before heating the copper(II) carbonate.

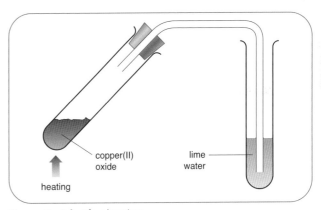

Figure 3g.5 b After heating.

Copper(II) carbonate is green, copper(II) oxide is black. A colour change occurs. This happens when all of these carbonate compounds decompose, although the actual colours are different in different cases. Because all of these carbonate compounds give off carbon dioxide when they decompose, all of them will make the lime water go cloudy.

7 Write word equations for the thermal decomposition of:

a manganese(II) carbonate

b zinc carbonate.

Balanced symbol equations

The balanced symbol equation for each thermal decomposition reaction is straightforward. The metal ions involved all have a 2+ charge. An oxide ion has a 2– charge. The metal oxides that form therefore have one metal ion and one oxide ion in each formula unit of the compound. Remember, the total electric charge in one formula unit of an ionic compound must add up to zero.

$(2+) + (2-) =$ zero

The formula of copper(II) oxide is CuO. The formula of manganese(II) oxide is MnO. The formula of zinc oxide is ZnO. The formula of iron(II) oxide is FeO. The formula of carbon dioxide is CO_2. The balanced symbol equation for the thermal decomposition of copper(II) carbonate is:

$$CuCO_3 \rightarrow CuO + CO_2$$

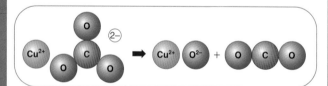

Figure 3g.6 One copper(II) carbonate formula unit decomposes when heated to form one copper(II) oxide formula unit and one carbon dioxide molecule.

8 Write balanced symbol equations for the thermal decomposition of:

a manganese(II) carbonate

b zinc(II) carbonate.

Precipitation reactions

If a solution of sodium hydroxide is added to a solution containing Cu^{2+} ions, Fe^{2+} ions or Fe^{3+} ions, a solid product forms. A reaction like this in which two solutions are mixed and an insoluble solid product forms is called a **precipitation reaction**. The solid product is called a **precipitate**. The solid products formed are different colours.

● When sodium hydroxide solution is added to a solution containing Cu^{2+} ions, a pale blue precipitate forms.

● When sodium hydroxide solution is added to a solution containing Fe^{2+} ions, a dirty green precipitate forms.

● When sodium hydroxide solution is added to a solution containing Fe^{3+} ions, an orangey-brown precipitate forms.

Figure 3g.7 a Sodium hydroxide solution from the pipette is about to be added to the solution containing Cu^{2+} ions. **b** A pale blue precipitate has formed.

Figure 3g.8 a Sodium hydroxide solution from the pipette is about to be added to the solution containing Fe^{2+} ions. **b** A dirty green precipitate has formed.

Figure 3g.9 a Sodium hydroxide solution from the pipette is about to be added to the solution containing Fe^{3+} ions. **b** An orangey-brown precipitate has formed.

These precipitation reactions can be used to identify a transition element ion in a solution. If sodium hydroxide solution is added to an unknown solution and a pale blue precipitate is seen then the unknown solution contains Cu^{2+} ions.

SAQ

9 What transition element ion is present in an unknown solution if:

a sodium hydroxide solution is added and an orangey-brown precipitate is seen?

b sodium hydroxide solution is added and a green precipitate is seen?

Transition elements around us

Some of the transition elements, such as copper and iron, are found in common uses all around us. Others have less familiar names but are still quite commonplace. Ordinary light bulbs have filaments that are made of tungsten and held up by support wires containing molybdenum.

Figure 3g.10 The filament in this light bulb is made of tungsten (element 74). The support wires are made of an alloy containing molybdenum (element 42).

Niobium has a bluish colour. This colour, and the fact that it doesn't irritate the skin, has made niobium popular as a jewellery metal. The modern trend for navel piercings has introduced many people to niobium for the first time.

Palladium is used to make the catalytic converters that help to reduce the amounts of pollutants in car exhausts.

Iridium is used to make the tips of the spark plugs in petrol engines.

Tantalum is a very expensive metal. This has limited the amount of it we use, even though this metal is amazingly strong. Many of the parts of modern fighter–bomber aircraft contain tantalum alloys.

We are even finding new uses for some of the best-known metals. Silver has been used in jewellery for over 5000 years, but silver has recently been incorporated into the dressings used to protect wounds. Silver has very powerful antibacterial properties, so the new dressings stop the wounds from going septic.

Figure 3g.11 These burn gloves are silver coated. The silver helps to prevent the infections that can cause major problems in burnt flesh.

But is this really a new use, or is it just a new spin on an old idea? For centuries, it was the custom in some parts of the world to throw a silver coin into a well as soon as it was dug. Doubtless this started out as a superstition. Perhaps it was thought of as a way of pleasing the 'water spirits' so the spirits would keep the well water clean and fresh. We cannot be sure why the coin was thrown in the well, but we know it worked. The silver from the coin would have greatly lowered the number of bacteria that could survive in the well water. The coin really did help keep the water clean and the people healthy.

H Metal hydroxides

The precipitates that form when sodium hydroxide solution is added to a solution containing a transition element ion are **metal hydroxides**. Metal hydroxides are ionic compounds in which the positive ion is a metal and the negative ion is the hydroxide ion, OH^-.

The formation of the metal hydroxide can be shown as a symbol equation. Taking the formation of copper(II) hydroxide as an example:

$$Cu^{2+} + 2OH^- \rightarrow Cu(OH)_2$$

You should notice the following points.

● This equation does not tell the whole story. The solution containing Cu^{2+} must also contain negative ions, which are not included in this equation. The solution containing OH^- also contains sodium ions, which are not included in this equation. The equation focuses on the ions that produce the precipitate.

● The formula of copper(II) hydroxide is $Cu(OH)_2$. The copper(II) ion has a 2+ charge. The hydroxide ion has a 1– charge. Each formula unit of copper(II) hydroxide contains one copper(II) ion and two hydroxide ions. This makes the total electric charge in one formula unit add up to zero.

$$\boxed{2+} + \boxed{1-} + \boxed{1-} = zero$$

SAQ

10 a Write a symbol equation for the formation of iron(II) hydroxide.

b Write a symbol equation for the formation of iron(III) hydroxide.

Transition metal ion	Formula of ion	Colour of solution	Colour of precipitate with NaOH	Formula of precipitate
copper(II)	Cu^{2+}	blue	pale blue	$Cu(OH)_2$
iron(II)	Fe^{2+}	pale green	green	$Fe(OH)_2$
iron(III)	Fe^{3+}	pale orange–yellow	brown	$Fe(OH)_3$

Table 3g.1 Summary of transition metal ions and transition metal hydroxides.

Summary

You should be able to:

◆ identify a transition element, giving its name and symbol, using the Periodic Table

◆ state that iron and copper are transition elements

◆ describe the typical properties of transition elements, including the formation of coloured compounds and their ability to act as catalysts

◆ define *thermal decomposition*

◆ describe the thermal decomposition of a transition element carbonate compound, including a word equation and reference to colour change and the identification of the gas given off

H ◆ construct balanced symbol equations for the thermal decomposition of four transition element carbonates

continued on next page

Summary - *continued*

◆ define *precipitation reaction* and *precipitate*

◆ describe what happens when sodium hydroxide solution is added to separate solutions containing Cu^{2+}, Fe^{2+} and Fe^{3+} ions

H ◆ construct symbol equations for the reactions between sodium hydroxide solution and separate solutions containing Cu^{2+}, Fe^{2+} and Fe^{3+} ions

Questions

1 Iron(II) carbonate undergoes thermal decomposition when heated.

   a Draw and label an apparatus that could be used to investigate this in a laboratory.

   b State *two* visible changes that would show that the decomposition had taken place.

   c Write a word equation for the thermal decomposition of iron(II) carbonate.

H d Write a balanced symbol equation for the thermal decomposition of iron(II) carbonate.

2 You are given two unidentified solutions. You suspect that one of them contains Cu^{2+} ions and the other one contains Fe^{2+} ions.

   a Describe how you could use sodium hydroxide solution to see whether you are right.

   b Describe the results you will get if you are right.

3 A, B, C and D are four unknown gases. They all look alike. Each gas is tested with lime water, a lighted splint and a glowing splint. The results of the tests are shown in Table 3g.2.

	A	B	C	D
Lime water test	no change	lime water goes cloudy	no change	no change
Lighted splint test	no change	no change	squeaky pop is heard	no change
Glowing splint test	glowing splint relights	no change	no change	no change

Table 3g.2

   a Which gas cannot be identified?

   b Suggest the identity of each of the other three gases.

H 4 a Explain why $FeCO_3$ is the correct formula of iron(II) carbonate and writing Fe_2CO_3 or $Fe(CO_3)_2$ would be wrong.

   b Explain why $Fe(OH)_3$ is the correct formula of iron(III) hydroxide and writing $FeOH$ or $Fe(OH)_2$ would be wrong.

Metal properties

It is no coincidence that the two periods of human history immediately following the Stone Age are named after metals. The discovery of new materials, or the ability to make new materials, has always propelled technology forward. The Bronze Age replaced the Stone Age because bronze gave people superior tools and weapons. The Iron Age replaced the Bronze Age because iron gave people superior tools and weapons.

Metals are so important because of their unique combination of properties. The properties illustrated in the following photographs are called physical properties. Physical properties do not involve chemical reactions.

SAQ

1 a Give *six* typical physical properties of metals.

b For each property that you chose, give *one* use that depends on that property.

Properties and uses of metals

Figures 3h.1–3h.6 show the typical properties of metals. Different metals show these properties to different extents, so different metals are used in different situations. The properties, and cost, of a metal have to match the purpose.

Iron is used to make steel. Steel is one of the strongest of the metals that are cheap to produce. Steel is the metal of choice where large amounts of a strong metal are required. Steel is used to make car bodies and bridges, and to reinforce concrete.

Figure 3h.1 Metals are **hard** and have **high densities**. These properties make steel suitable for making anvils.

Figure 3h.2 Metals are **lustrous** – bright and shiny – when they are clean. This property makes silver suitable for making ear-rings.

Figure 3h.3 Metals have **high tensile strength**. This means they are strong when pulled. This property makes steel suitable for making the cables in this suspension bridge.

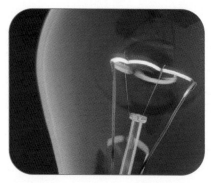

Figure 3h.4 Metals have **high melting points** and **high boiling points**. This property makes tungsten suitable for making the filament in this light bulb.

Figure 3h.5 Metals are **good conductors of heat**. This property makes copper suitable for making the base of this saucepan. Heat from a gas flame or hot plate can flow easily through the copper and reach the food in the pan quickly.

Figure 3h.6 Metals are **good conductors of electricity**. This property makes gold suitable for plating the contacts on this circuit board, which you can see at the bottom.

Figure 3h.7 The steel used to make this car body is cheap and strong.

Copper is one of the best electrical conductors so it is used for domestic electric wiring. It is also mixed with zinc to make the alloy called brass.

SAQ

2 **a** Which metals have to be mixed to make brass?

b Why is steel used to make the body panels of most cars?

c Why is most electric wiring made of copper?

Metallic bonds

Metals are crystalline, because the atoms in a piece of metal are arranged in a regular pattern.

The atoms in a piece of metal are held closely together by **metallic bonds**. Metallic bonds are strong. Typical metals have high melting and boiling points because the atoms in the metal are held together by strong metallic bonds.

SAQ

3 There are three types of strong bonding: ionic bonding, metallic bonding and covalent bonding. Which of these types of bond holds the particles together in:

a a water molecule?

b a sodium chloride crystal?

c a piece of nickel?

Ⓗ Delocalised electrons

In a piece of metal, the electrons on the outer shell of every atom can move freely from atom to atom throughout the piece of metal. Because these electrons are not confined to any one atom, they are said to be **delocalised**. The electrons are called a 'sea of delocalised electrons'. Because they have contributed their outer electrons to the 'sea', all of the metal atoms in the piece of metal have become positive ions.

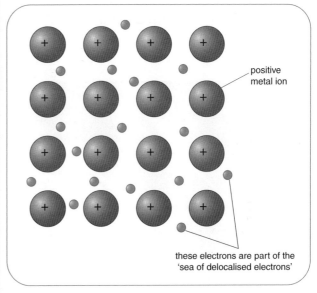

positive metal ion

these electrons are part of the 'sea of delocalised electrons'

Figure 3h.8 The positive ions and the sea of delocalised electrons attract each other strongly. This strong attraction is called metallic bonding. These metal particles are bonded in a square pattern.

Because the metal ions are positively charged and the sea of delocalised electrons is negatively charged, there is a strong force of attraction between them. This strong attractive force is what we call metallic bonding. When a metal is melted or boiled, this strong attractive force has to be overcome. This is why metals have high melting and boiling points.

SAQ

4 **a** What does 'delocalised' mean?

b Which electrons does each atom contribute to the sea of delocalised electrons?

c Why do the metal ions and the sea of delocalised electrons attract each other?

d Why do metals have high melting and boiling points?

Conduction and resistance

An electric current can flow through metal. The electric current consists of electrons flowing through the piece of metal. The electrons move from metal atom to metal atom around a circuit.

The electrons that move are the delocalised electrons. Metals conduct electricity because these delocalised electrons can move easily.

Under normal conditions, a piece of metal has what is known as 'electrical resistance'. This means that, although an electric current can flow through the metal, the electrical energy is transferred to heat energy within the metal. Unless the electric current is continually supplied with energy, either from the mains or from a battery, it will stop flowing.

Superconductivity

Some materials lose their electrical resistance if the metal is made cold enough. If an electric current is put into a loop of lead wire at a temperature of −267 °C, this current will continue to flow for years − even though the wire is not connected to the mains or a battery.

Figure 3h.9 Because of electrical resistance in the wires of this toaster, electrical energy is being transferred to heat energy.

Figure 3h.10 These overhead electric cables have resistance. As electric current flows through them, some electrical energy is transferred to heat energy. If they were superconducting, no energy would be lost as heat.

Medical scans

In 1911, Heike Onnes, a Dutch scientist, discovered that mercury metal loses all its electrical resistance if it is cooled to −272 °C. Superconductivity had been discovered.

Since Heike Onnes made his discovery, superconductivity has come a long way. Have you ever had an MRI scan at a hospital? MRI stands for 'magnetic resonance imaging'. MRI scanning allows doctors to see inside your body without cutting you open. If you have damaged your knee, for example, an MRI scanner can build up a very detailed picture of the inside of the joint, and you won't feel a thing.

The MRI scanner needs extremely powerful electromagnets. These have coils of wire that are cooled to superconducting temperatures by liquid helium at −270 °C. The superconducting coils allow us to produce the strong magnets that the scanner needs.

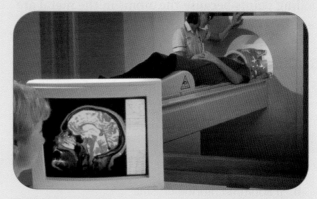

Figure 3h.11 A patient about to go inside an MRI scanner. Part of the scan of their brain can be seen on the monitor. It is interesting to think how near they are to liquid helium at −270 °C!

This property of losing electrical resistance at very low temperatures is called **superconductivity**. A superconductor is a material that conducts electricity with little or no resistance. Superconductivity brings with it many new possibilities.

Figure 3h.13 This superconducting magnet is part of the PETRA particle accelerator in Hamburg, Germany. The magnet contains coils made of superconducting niobium wire. The very high currents in the superconducting coils help make this such a strong electromagnet.

Figure 3h.12 The electronic circuits in a modern computer are fast, but they would be even faster if they were superconducting.

SAQ

5 **a** What has a piece of metal lost if it has become a superconductor?

b What conditions are necessary for a metal to become a superconductor?

c Use the photographs in Figures 3h.10–3h.13 to describe three possible uses of superconducting metals.

The big problem with superconductivity is that metals and other materials only lose their resistance at extremely low temperatures. It is difficult to imagine how the power lines in Figure 3h.10 could be kept cold all year round. In order to reap the benefits of superconductors, we need them to work at normal temperatures, for example around 20 °C.

Summary

You should be able to:

◆ describe the typical physical properties of metals and explain why a particular metal is suited to a given use

◆ describe the arrangement of particles in a solid metal

◆ recognise that the particles in a metal are held together by metallic bonds

H ◆ describe metallic bonding and use it to explain why metals have high melting and boiling points and can conduct electricity

◆ describe an electric current in a metal

◆ describe superconductors and some of their potential benefits

H ◆ explain the drawbacks of superconductors

Questions

1 Table 3h.1 lists the melting points, electrical conductivities and densities of five metals. The higher the electrical conductivity value the better the conductor. The higher the density value, the greater the mass in kg of one cubic metre of the metal.

Metal	Melting point	Relative electrical conductivity	Density
copper	1083 °C	593	8 920 kg per m^3
iron	1535 °C	100	7 860 kg per m^3
sodium	98 °C	218	970 kg per m^3
tungsten	3410 °C	180	19 400 kg per m^3
nickel	1453 °C	145	8 900 kg per m^3

Table 3h.1

  a Which metal has the highest melting point?

  b Which metal has the lowest electrical conductivity?

  c If you had five blocks of identical volume, one made of each metal, which would be heaviest?

  d State *two* things about sodium that appear to be unusual.

  e Why is tungsten used to make light bulb filaments?

2 You attach one end of a length of wire to one terminal of a battery and the other end of the wire to the other terminal of the battery.

  a Why does an electric current flow in the wire?

  b Why does the wire get hot after a short time?

  c Why does the current decrease as time passes and eventually stop flowing?

3 Explain the meaning of the following terms.

  a Electrical resistance

  b Superconductor

H 4 a Explain what is meant by the term *metallic bonding*.

  b How does the theory of metallic bonding explain why metals have high melting and boiling points?

  c How does the theory of metallic bonding explain why metals conduct electricity?

Acids and bases

Universal indicator

Universal indicator is a mixture of coloured dyes. Most of these dyes are obtained from plants. When universal indicator is put into water, the water turns green.

Figure 4a.1 When universal indicator is added to water, the colour seen is green.

When universal indicator is put into solutions of certain compounds, different colours are seen.

Figure 4a.2 a When universal indicator is added to carbon dioxide solution, the colour seen is yellow.

Figure 4a.2 b When universal indicator is added to lemonade, the colour seen is orange.

Figure 4a.2 c When universal indicator is added to a solution of hydrogen chloride, the colour seen is red.

One family of compounds that changes the colour of universal indicator is the **acids**. Figure 4a.2 shows universal indicator being added to three acidic solutions. These acids have changed the colour of the universal indicator to yellow, orange and red.

Figure 4a.3 a When universal indicator is added to ammonia solution, the colour seen is blue.

Figure 4a.3 b When universal indicator is added to sodium hydroxide solution, the colour seen is purple.

Another family of compounds that change the colour of universal indicator is the **alkalis**. Figure 4a.3 shows universal indicator being added to solutions of two alkalis. These alkalis have changed the colour of the universal indicator to blue and purple.

SAQ

1 What colour or colours might you see when universal indicator is added to:

 a a solution of an acid?

 b a solution of an alkali?

 c water?

The pH scale

Universal indicator can tell us whether a solution contains an acid or an alkali. It can also tell us how acidic or how alkaline a solution is. In Figure 4a.2, for example, the most acidic solution is the hydrogen chloride solution. Strongly acidic solutions turn universal indicator red. In Figure 4a.3, the more alkaline solution is the sodium hydroxide solution. Strongly alkaline solutions turn universal indicator purple.

We represent the amount of acidity or alkalinity in a solution using the **pH scale**. This scale goes from 1 to 14. A substance like water that is neither acidic nor alkaline has a pH number of 7. Substances like this are called **neutral** substances. A solution of a neutral substance dissolved in water has a pH of 7. Acidic solutions have pH numbers below 7. The lower the number, the more acidic the solution. Alkaline solutions have pH numbers above 7. The higher the number, the more alkaline the solution.

SAQ

2 Use Figures 4a.1–4a.4 to find the pH number of:

 a water

 b carbon dioxide solution

 c lemonade

 d hydrochloric acid solution

 e ammonia solution

 f sodium hydroxide solution.

3 Which solution in Figure 4a.2 is the stronger acid, the carbon dioxide solution or the lemonade?

Neutralisation

When equal amounts of an acid and an alkali are mixed together, they react to produce a neutral solution. This is called **neutralisation**. Neutralisation is a chemical reaction; one product of a neutralisation reaction is always a chemical called a **salt**. Salts are neutral compounds. Neutralisation is illustrated in Figures 4a.5–4a.7.

SAQ

4 a Use Figure 4a.4 to give the pH number of each solution shown in Figure 4a.5.

 b What happens to the pH number as each portion of alkali is added to the acid?

5 a Use Figure 4a.4 to give the pH number of each solution shown in Figure 4a.6.

 b What happens to the pH number as each portion of acid is added to the alkali?

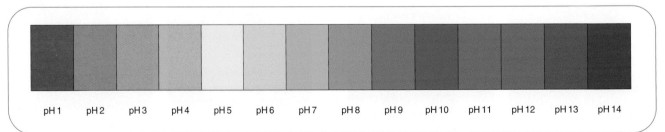

Figure 4a.4 This chart is used to find the pH number of a solution after you have added universal indicator to it.

Figure 4a.5 The first picture shows a small beaker of acid with universal indicator added to it. The following pictures show how the indicator colour changes as alkali is added bit by bit.

Figure 4a.6 The first picture shows a small beaker of alkali with universal indicator added to it. The following pictures show how the indicator colour changes as acid is added bit by bit.

Figure 4a.7 The first picture shows a small beaker of nitric acid with universal indicator added to it. The following pictures show how the indicator colour changes when powdered calcium carbonate is added.

If alkali is added to an acid, the pH increases. If acid is added to an alkali, the pH decreases.

Bases

A substance that can neutralise an acid is called a **base**. Some bases are metal oxide compounds, such as copper oxide and magnesium oxide. These two substances will both neutralise an acid if enough of the base is added to the acid. When an acid is neutralised by a metal oxide, the products are water and a salt.

Acid + metal oxide → water + salt

Some bases are metal carbonate compounds, such as sodium carbonate and calcium carbonate. These two substances will both neutralise an acid if enough of the base is added to the acid. When an acid is neutralised by a metal carbonate, the products are carbon dioxide gas, water and a salt.

Acid + metal carbonate
→ carbon dioxide + water + salt

SAQ

6 Which gas is causing the frothing effect in Figure 4a.7?

The term **alkali** is a specific term for a base that is soluble in water. Many alkalis are metal hydroxide compounds, such as sodium hydroxide and potassium hydroxide. These two substances will both neutralise an acid, if enough of the alkali is added to the acid. When an acid is neutralised by a metal hydroxide, the products are water and a salt.

Acid + metal hydroxide → water + salt

One common alkali that is not a metal hydroxide compound is ammonia solution. Ammonia is a strong-smelling gas which dissolves in water to produce an alkaline solution.

SAQ

7 Copy the passage that follows, filling in the gaps.

An acid can be neutralised by adding a ____ or an ____ to it. An ____ is a soluble ____. An alkali can be neutralised by adding an ____ to it.

Naming the salt

Salts are ionic compounds with names consisting of two words. For example **sodium chloride** and **copper(II) sulfate** are both salts. When a salt is made by a neutralisation reaction, it is possible to name the salt using the following rules.

● The first word in the salt's name comes from the base that is involved in the reaction.
● If the base contains a metal, the name of the metal is the *first* word in the salt's name.
● If the base is ammonia, ammonium is the *first* word in the salt's name.
● The second word in the salt's name comes from the acid that is involved in the reaction.
● If the acid is hydrochloric acid, chloride is the *second* word in the salt's name.
● If the acid is nitric acid, nitrate is the *second* word in the salt's name.
● If the acid is sulfuric acid, sulfate is the *second* word in the salt's name.

Worked example 1

Sulfuric acid solution is neutralised by adding copper(II) oxide to it. What is the name of the salt that forms?

Step 1: the first word in the salt's name comes from the base that is involved in the reaction. The base involved is copper(II) oxide. The base contains copper(II) so copper(II) is the first word in the name of the salt.

Step 2: the second word in the salt's name comes from the acid that is involved in the reaction. The acid involved is sulfuric acid, so sulfate is the second word in the name of the salt.

Answer: the name of the salt formed when sulfuric acid solution is neutralised by adding copper(II) oxide is therefore *copper(II) sulfate*.

Figure 4a.8 a Solid copper(II) oxide is added to sulfuric acid in a beaker. Universal indicator has *not* been added. **b** The sulfuric acid has been neutralised by the copper(II) oxide. A salt called copper(II) sulfate has been made. Copper(II) sulfate is blue.

SAQ

8 **a** Write a word equation for the reaction between sulfuric acid and copper(II) oxide.

   b When sulfuric acid is neutralised by copper(II) oxide, why does the solution turn blue?

9 Name the salts produced in the following neutralisation reactions:

   a hydrochloric acid + potassium hydroxide

   b sulfuric acid + sodium hydroxide

   c the reaction shown in Figure 4a.7.

Hydrogen ions and hydroxide ions

When an acid dissolves in water, the solution produced contains hydrogen ions – H^+ ions. The higher the concentration of hydrogen ions in a solution, the lower the pH. For example when hydrogen chloride molecules dissolve in water the solution produced contains H^+ ions and Cl^- ions. The H^+ ions make the solution acidic.

When an alkali dissolves in water, the solution produced contains hydroxide ions, OH^-. The higher the concentration of hydroxide ions in a solution, the higher the pH. For example, when sodium hydroxide dissolves in water, the solution produced contains Na^+ ions and OH^- ions. The OH^- ions make the solution alkaline.

When an acid reacts with an alkali, the H^+ ions in the acid combine with the OH^- ions in the alkali to make water. This is why they neutralise each other.

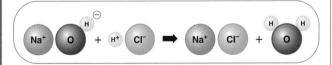

Figure 4a.9 Sodium hydroxide and hydrochloric acid react to give sodium chloride and water.

The ionic equation for the combination of hydrogen ions and hydroxide ions to give water is

$$H^+ + OH^- \rightarrow H_2O$$

This ionic equation focuses our attention on the part of the reaction resulting in neutralisation of the acid and alkali.

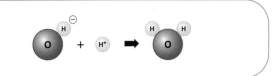

Figure 4a.10 Hydrogen ions and hydroxide ions combine to give water molecules.

The ions involved in the SAQs in this Item, including the hydrogen ion and the hydroxide ion, are summarised in Table 4a.1.

Positive ions		Negative ions	
Name	Formula	Name	Formula
hydrogen	H^+	hydroxide	OH^-
sodium	Na^+	chloride	Cl^-
potassium	K^+	nitrate	NO_3^-
ammonium	NH_4^+	oxide	O^{2-}
calcium	Ca^{2+}	carbonate	CO_3^{2-}
copper(II)	Cu^{2+}	sulfate	SO_4^{2-}

Table 4a.1

Formulae of the acids and bases

Neutralisation reactions can be described using balanced symbol equations. First you must learn the formulae of the substances involved.

Acids

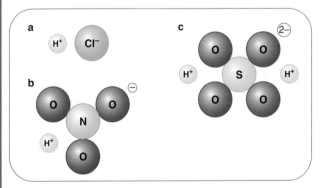

Figure 4a.11 a Hydrochloric acid is HCl. The hydrogen ion has a 1+ charge. The chloride ion has a 1– charge. **b** Nitric acid is HNO_3. The NO_3^- ion is called the nitrate ion. The hydrogen ion has a 1+ charge. The nitrate ion has a 1– charge. The nitrogen atom and the oxygen atoms are held together by covalent bonds. **c** Sulfuric acid is H_2SO_4. The SO_4^{2-} ion is called the sulfate ion. The hydrogen ion has a 1+ charge. The sulfate ion has a 2– charge. The sulfur atom and the oxygen atoms are held together by covalent bonds.

H **Bases**

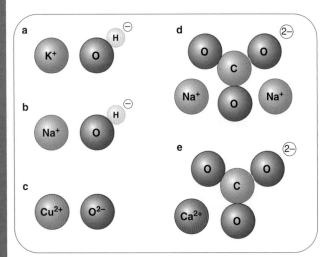

Figure 4a.12 a Potassium hydroxide is KOH. The OH^- ion is called the hydroxide ion. The potassium ion has a 1+ charge. The hydroxide ion has a 1– charge. The oxygen atom and the hydrogen atom are held together by a covalent bond. **b** Sodium hydroxide is NaOH. **c** Copper(II) oxide is CuO. **d** Sodium carbonate is Na_2CO_3. The CO_3^{2-} ion is called the carbonate ion. The sodium ion has a 1+ charge. The carbonate ion has a 2– charge. The carbon atom and the oxygen atoms are held together by covalent bonds. **e** Calcium carbonate is $CaCO_3$.

The formula of ammonia is NH_3, but when ammonia dissolves in water it forms ammonium hydroxide, NH_4OH.

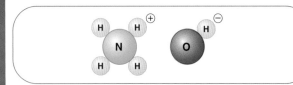

Figure 4a.13 Ammonium hydroxide is NH_4OH. The NH_4^+ ion is called the ammonium ion. The ammonium ion has a 1+ charge. The nitrogen atom and the hydrogen atoms are held together by covalent bonds. The hydroxide ion has a 1– charge. The oxygen atom and the hydrogen atom are held together by a covalent bond.

SAQ

10 What makes ammonia solution alkaline?

11 One formula unit of sodium carbonate contains two sodium ions, but one formula unit of calcium carbonate only contains one calcium ion. Explain why this is.

H **Balanced symbol equations**

Worked example 2

Write a balanced symbol equation for the neutralisation reaction between nitric acid and sodium hydroxide.

Step 1: write a word equation. You will need to use the section titled *Bases* to find the general equation:

acid + metal hydroxide → water + salt

Then fill in the names of the acid and the metal hydroxide on the left of the arrow:

nitric acid + sodium hydroxide → water + salt

Then use the section titled *Naming the salt* to fill in the correct name of the salt:

nitric acid + sodium hydroxide
→ water + sodium nitrate

Step 2: rewrite the equation using the correct formula of each substance:

$HNO_3 + NaOH → H_2O + NaNO_3$

The formula of sodium nitrate is $NaNO_3$ because the charge on the sodium ion is 1+ and the charge on the nitrate ion is 1–. In one formula unit of the compound, the total charge of the ions must be zero, so the formula is $NaNO_3$.

Step 3: balance the equation. There must be the same number of each type of atom on each side of the arrow. On the left of the arrow there are 2H, 1N, 4O and 1Na. On the right of the arrow there are 2H, 1N, 4O and 1Na. This equation is already balanced. The answer is:

$HNO_3 + NaOH → H_2O + NaNO_3$ ✓

H

Worked example 3

Write a balanced symbol equation for the neutralisation reaction between hydrochloric acid and calcium carbonate.

Step 1: write a word equation. You will need to use the section titled *Bases* to find the general equation:

acid + metal carbonate →
        carbon dioxide + water + salt

Fill in the names of the acid and the metal carbonate on the left of the arrow:

hydrochloric acid + calcium carbonate →
        carbon dioxide + water + salt

Use the section titled *Naming the salt* to fill in the correct name of the salt:

hydrochloric acid + calcium carbonate →
     carbon dioxide + water + calcium chloride

Step 2: rewrite the equation using the correct formula of each substance:

$HCl + CaCO_3 \rightarrow CO_2 + H_2O + CaCl_2$

The formula of calcium chloride is $CaCl_2$ because the charge on the calcium ion is 2+ and the charge on the chloride ion is 1–. In one formula unit of the compound the total charge of the ions must be zero, so the formula is $CaCl_2$.

Step 3: balance the equation. There must be the same number of each type of atom on each side of the arrow. On the left of the arrow there are 1H, 1Cl, 1Ca, 1C and 3O. On the right of the arrow there are 2H, 2Cl, 1Ca, 1C and 3O. This equation is not balanced. To be balanced, there has to be a second HCl involved on the left of the arrow. The answer is:

$2HCl + CaCO_3 \rightarrow CO_2 + H_2O + CaCl_2$ ✓

H

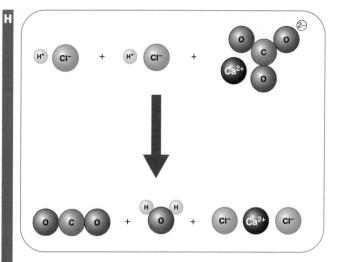

Figure 4a.14 Two hydrochloric acid formula units react with one calcium carbonate formula unit to give one carbon dioxide molecule, one water molecule and one calcium chloride formula unit.

SAQ

12 Write word equations and balanced symbol equations for the following neutralisation reactions:

  a hydrochloric acid + sodium carbonate

  b sulfuric acid + sodium hydroxide

  c nitric acid + copper oxide

  d nitric acid + ammonia (hint: ammonia solution contains ammonium hydroxide; use the acid + metal hydroxide general equation, even though the ammonium ion is not a metal ion).

Sulfuric acid

Sulfuric acid is one of the most important industrial chemicals there is. You can judge how industrialised a nation is by how much sulfuric acid it uses every year. The more sulfuric acid used, the more industrialised or developed a country is.

The manufacture of sulfuric acid uses some quite simple ingredients. Sulfur is found as the pure element in several countries. First of all the sulfur is extracted from the ground. Next, the sulfur is burnt, combining with oxygen from the air to form sulfur dioxide. After a second burning stage, sulfur trioxide is formed. The sulfur trioxide then reacts with water to make sulfuric acid.

Sulfuric acid has an enormous number of uses. It is used to make a fertiliser called ammonium sulfate. This fertiliser is actually a salt and is made by neutralising sulfuric acid with ammonia. Plants need sulfur, which can be provided by ammonium sulfate fertiliser to increase crop yield. (See Figure 4c.7 on page 180 and Item B4d.)

Before a steel car body is painted, it has to be made completely clean. This is done by immersing the car body in sulfuric acid. This acid has other uses in the automobile industry, for example, car batteries contain sulfuric acid.

Some uses of sulfuric acid are quite surprising. It is sprayed on fields of potatoes a few days before the potatoes are harvested. This kills off the potato plants above ground and makes the potatoes themselves easier to lift from the soil mechanically. It also causes the plant to put all of its resources into growing the potatoes below the soil rather than the plants above the soil. This results in bigger potatoes.

Figure 4a.15 These car bodies were cleaned by immersion. They will now be painted.

Summary

You should be able to:

◆ describe how universal indicator is used

◆ use the pH scale.

◆ describe some uses of sulfuric acid in agriculture and in the automobile industry

◆ recall the names of some substances that will neutralise acids, using the terms *base* and *alkali*

◆ describe and explain the pH changes during neutralisation

◆ predict the name of the salt formed when an acid is neutralised by a base

H ◆ describe neutralisation using an ionic equation

◆ construct word equations and balanced symbol equations for the neutralisation of acids by bases

Questions

1 Solution A has a pH of 4. Solution B has a pH of 10. Solution C has a pH of 7.

   a Classify solutions A, B and C as neutral, acidic or alkaline.

   b What colour would you see if universal indicator was added separately to each of solutions A, B and C?

   c i Which *two* out of solutions A, B and C could mix to give a neutral solution?

      ii What type of reaction is taking place between these two solutions?

2 An underdeveloped country cannot afford to make or import sulfuric acid. How might this affect the following industries in that country?

   a Agriculture.

   b Car manufacture.

3 For each of the following neutralisation reactions, name the salt produced and write a word equation.

   a Hydrochloric acid + copper(II) oxide

   b Nitric acid + potassium hydroxide

   c Sulfuric acid + sodium carbonate

   d Hydrochloric acid + sodium hydroxide.

4 Hydrochloric acid solution can be neutralised by adding potassium hydroxide solution.

   a What ion makes hydrochloric acid solution acidic?

   b What ion makes potassium hydroxide solution alkaline?

   c Explain with the aid of an ionic equation why these two solutions will neutralise each other.

5 Write a balanced symbol equation for each of the following reactions:

   a hydrochloric acid + copper(II) oxide

   b nitric acid + potassium hydroxide

   c sulfuric acid + sodium carbonate

   d hydrochloric acid + sodium hydroxide.

   e hydrochloric acid + ammonia solution

   f sulfuric acid + ammonia solution.

Relative atomic mass

The Periodic Table consists of the symbols of all the known elements. In the box with the symbol of each element there are two numbers. The smaller number is the atomic number – the number of protons in the nucleus of one atom of that element. The larger number is the **relative atomic mass** of the element. This number tells you how heavy the atoms of the element are, relative to atoms of other elements.

For example, carbon atoms have a relative atomic mass of 12. Magnesium atoms have a relative atomic mass of 24. This means magnesium atoms are twice as heavy as carbon atoms.

In this Item, the value of the relative atomic mass of copper should be rounded to 64 and the value for chlorine to 35.

SAQ

1 Use the Periodic Table on page 345 to find the atomic numbers of:

    a oxygen   **b** sulfur   **c** bromine.

2 Use the Periodic Table on page 345 to find the relative atomic masses of:

    a oxygen   **b** sulfur   **c** bromine.

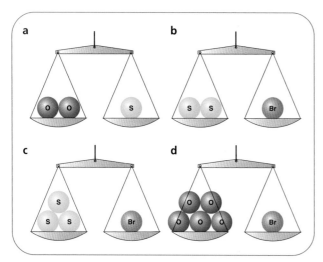

Figure 4b.1 The circles represent atoms of oxygen, sulfur and bromine.

3 Look at Figure 4b.1.

    a On which two of the four scales would be the masses be balanced? Explain your choice.

b Look at the two scales you decided would not be balanced. Which is the heavier side in each case?

Relative formula mass

If you know the formula of a substance you can find the relative mass of one molecule or formula unit of the substance. This is called the **relative formula mass** of the substance. The relative formula mass is calculated by adding the relative masses of the atoms in the formula.

Worked example 1

What is the relative formula mass of a carbon dioxide (CO_2) molecule?

Step 1: a carbon dioxide molecule consists of one carbon atom (relative atomic mass = 12) and two oxygen atoms (relative atomic mass = 16 each).

The relative formula mass of a carbon dioxide molecule is:

$12 + 16 + 16 = 44$

The answer is 44.

Worked example 2

What is the relative formula mass of a glucose ($C_6H_{12}O_6$) molecule?

Step 1: a glucose molecule consists of six carbon atoms (relative atomic mass = 12 each), twelve hydrogen atoms (relative atomic mass = 1 each) and six oxygen atoms (relative atomic mass = 16 each).

The relative formula mass of a carbon dioxide molecule is:

$(12 \times 6) + (1 \times 12) + (16 \times 6)$

$72 + 12 + 96 = 180$

The answer is 180.

Use the Periodic Table on page 345 to find the relative atomic masses of the atoms needed to answer SAQs **4** and **5**.

4   What is the relative formula mass of

   a   a water (H_2O) molecule?

   b   a copper(II) sulfate ($CuSO_4$) formula unit? (Remember that the value of the relative atomic mass of copper should be rounded to 64.)

   c   a calcium carbonate ($CaCO_3$) formula unit?

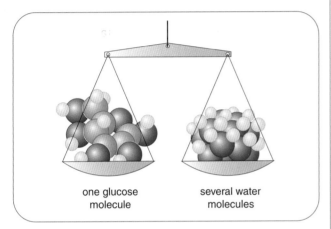

Figure 4b.2

5   The scales in Figure 4b.2 have one glucose molecule on one side and several water molecules on the other side. The scales are balanced. How many water molecules must there be on the right hand pan?

Formulae with brackets

Some ionic compounds have formulae which include brackets. Examples include copper(II) nitrate, $Cu(NO_3)_2$, and iron(III) hydroxide, $Fe(OH)_3$.

In the formula of copper(II) nitrate, the little '2' means that there are two NO_3^- ions in each formula unit.

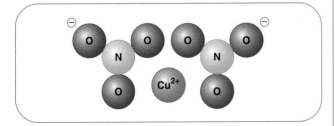

Figure 4b.3 Copper(II) nitrate.

6   The charge on a copper(II) ion is 2+. The charge on a nitrate ion is 1−. Explain why the formula of copper(II) nitrate is $Cu(NO_3)_2$.

7   What is the relative formula mass of copper(II) nitrate? (Remember that the value of the relative atomic mass of copper should be rounded to 64.)

In the formula of iron(III) hydroxide, the little '3' means that there are three OH^- ions in each formula unit.

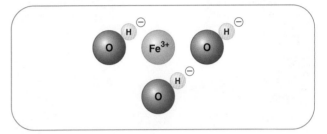

Figure 4b.4 Iron(III) hydroxide.

8   The charge on a iron(III) ion is 3+. The charge on a hydroxide ion is 1−. Explain why the formula of iron(III) hydroxide is $Fe(OH)_3$.

9   What is the relative formula mass of iron(III) hydroxide?

10   What is the relative formula mass of

   a   calcium hydroxide, $Ca(OH)_2$?

   b   ammonium sulfate, $(NH_4)_2SO_4$?

Reacting masses

In any chemical reaction, the total mass of products formed is equal to the total mass of reactants used up. If more products are needed, more reactants have to be used. The amount of products formed will increase if the amount of reactants at the start is increased.

SAQ

11 24 g of magnesium reacts with 16 g of oxygen to make magnesium oxide. What mass of magnesium oxide is formed?

12 Look at Figure 4b.5. If 48 g of magnesium reacts with enough oxygen to make 80 g of magnesium oxide, what mass of oxygen is needed?

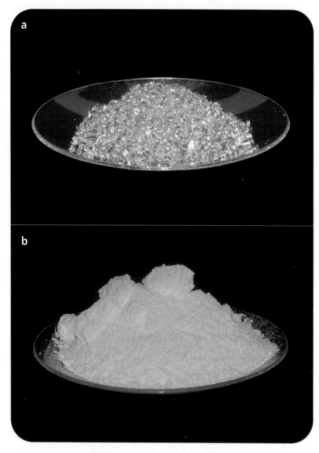

Figure 4b.5 48 g of magnesium (**a**) will react with oxygen to make 80 g of magnesium oxide (**b**).

Using ratios

The ratio of the amount of reactant used up to the amount of product formed is a constant. For example, consider the reaction:

copper(II) oxide + sulfuric acid →

copper(II) sulfate + water

The mass of copper(II) sulfate that forms is always double the mass of copper(II) oxide that reacts. So, if 10 g of copper(II) oxide reacts, 20 g of copper(II) sulfate are formed. If 2.5 g of copper(II) oxide reacts, 5 g of copper(II) sulfate are formed.

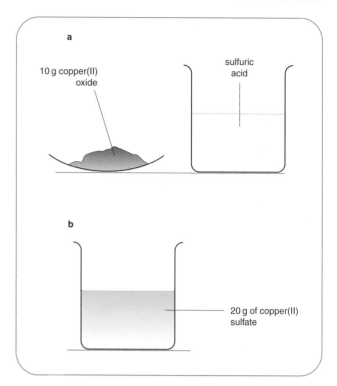

Figure 4b.6 a The 10 g of copper(II) oxide will react with the sulfuric acid in the beaker. **b** 20 g of copper(II) sulfate have been formed and are dissolved in solution.

SAQ

13 What mass of copper(II) sulfate forms if:

    a 6 g of copper(II) oxide reacts with sulfuric acid?

    b 70 kg of copper(II) oxide reacts with sulfuric acid?

14 What mass of copper(II) oxide reacts with sulfuric acid if:

    a 30 g of copper(II) sulfate are formed?

    b 500 g of copper(II) sulfate are formed?

15 In the reaction:

copper(II) oxide + sulfuric acid →

copper(II) sulfate + water

    a which substances are the products?

    b which substances are the reactants?

Why mass is conserved

Mass is always conserved in chemical reactions. The mass of the reactants used up is always equal to the mass of the products formed. This is because all of the atoms in the reactants are also present in the products. The balanced symbol equation for a reaction tells us this.

Worked example 3

Copper(II) oxide + sulfuric acid →

copper(II) sulfate + water

$CuO + H_2SO_4 \rightarrow CuSO_4 + H_2O$

The balanced symbol equation says that, when a copper(II) oxide formula unit reacts with a sulfuric acid formula unit, a copper(II) sulfate formula unit and a water molecule are formed. The atoms that make up the reactants are exactly the same as the atoms that make up the products.

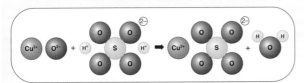

Figure 4b.7 The balanced symbol equation says that each time two reactant formula units collide and react, they consist of these nine atoms and they make products which consist of the same nine atoms.

The equation can be written with the relative formula mass of each substance underneath it.

Remember that the value of the relative atomic mass of copper should be rounded to 64.

$CuO + H_2SO_4 \rightarrow CuSO_4 + H_2O$

80 + 98 → 160 + 18

This explains why the mass of copper(II) sulfate that forms is twice the mass of copper(II) oxide that reacts. Copper(II) sulfate formula units are twice the mass of copper(II) oxide formula units.

SAQ

Remember that the value of the relative atomic mass of copper should be rounded to 64.

16 If 80 g of copper(II) oxide reacts with 98 g of sulfuric acid:

   a what mass of copper(II) sulfate will be formed?

   b what mass of water will be formed?

17 If 160 g of copper(II) oxide reacts with sulfuric acid and makes 320 g of copper(II) sulfate:

   a what mass of sulfuric acid must be reacting?

   b what mass of water will be formed?

Worked example 4

Magnesium + oxygen → magnesium oxide

$2Mg + O_2 \rightarrow 2MgO$

The balanced symbol equation says that when two magnesium atoms react with an oxygen molecule two magnesium oxide formula units are formed. The atoms that make up the reactants are exactly the same as the atoms that make up the products.

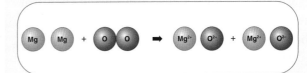

Figure 4b.8 The balanced symbol equation says that each time two magnesium atoms and one oxygen molecule collide and react, they consist of these four atoms and they make products which consist of the same four atoms.

The equation can be written with the masses of the two magnesium atoms, the oxygen molecule, and the two magnesium oxide formula units, underneath it.

$2Mg + O_2 \rightarrow 2MgO$

48 + 32 → 80

SAQ

18 If 480 g of magnesium reacts with 320 g of oxygen, what mass of magnesium oxide will be formed?

19 If 4.8 g of magnesium reacts with oxygen and makes magnesium oxide:

   a what mass of oxygen must be reacting?

   b what mass of magnesium oxide will be formed?

20 a What mass of oxygen would react with 3.00 g of magnesium?

   b What mass of magnesium oxide would form in this case?

   c What mass of oxygen would react with 0.63 g of magnesium?

   d What mass of magnesium oxide would form in this case?

Yield

When chemicals react, it is possible to calculate the mass of products that you expect to get. For example, Figure 4b.6 shows that, when 10 g of copper(II) oxide reacts with sulfuric acid, you expect to get 20 g of copper(II) sulfate. The amount of product you get is called the **yield**.

If you end up with the amount of product that you expected to get, this is called getting a **100% yield**. If the experiment fails completely and you get no product, this is called getting a **0% yield**. Normally when we make things using chemical reactions, the actual yield we get is somewhere between 0% and 100%.

The percentage yield is calculated using this formula:

$$percentage\ yield = \frac{actual\ yield}{expected\ yield} \times 100\%$$

Worked example 5

If the expected yield of copper(II) sulfate is 20 g but the actual yield is 15 g, what is the percentage yield?

Figure 4b.9 You hoped to make 20 g of copper(II) sulfate (**a**) but at the end of the experiment (**b**) you only got 15g.

$$percentage\ yield = \frac{actual\ yield}{expected\ yield} \times 100\%$$

$$percentage\ yield = (15\ g \div 20\ g) \times 100\%$$

$$percentage\ yield = 0.75 \times 100\%$$

$$percentage\ yield = 75\%$$

The percentage yield was 75%.

SAQ

21 Five laboratories make paracetamol. Calculate the percentage yield at each factory.

   a Laboratory A expected to make 900 g but actually got 450 g.

   b Laboratory B expected to make 1200 g but actually got 960 g.

   c Laboratory C expected to make 400 g but actually got 280 g.

   d Laboratory D expected to make 600 g but actually got 240 g.

   e Laboratory E expected to make 300 g but actually got 255 g.

22 Look at your answers to SAQ **21**. Put them into rank order according to how successful they were, starting with the least successful.

Where the product is lost

Chemical processes very rarely result in 100% yield. It is almost impossible to avoid losing some of the product. Figures 4b.10–4b.12 illustrate some of the ways in which product can be lost.

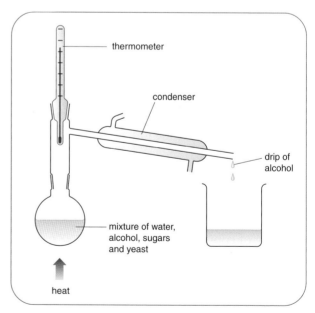

Figure 4b.10 This apparatus has been used to make alcohol. The alcohol is being collected as it drips out of the right-hand end of the condenser. However, some of the alcohol will be lost into the air by evaporation.

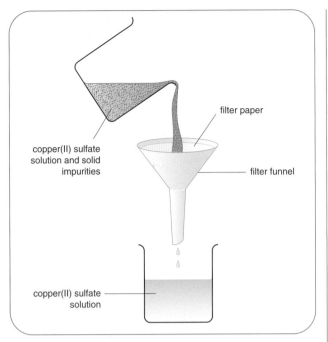

Figure 4b.11 Copper(II) sulfate solution has been made and is being separated from solid impurities by filtering it. Some drips of the copper(II) sulfate solution will stay in the upper beaker – product is lost when transferring liquids. Some of the copper(II) sulfate solution will be left soaked into the filter paper – product can be lost during filtration.

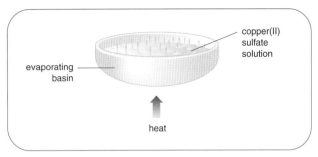

Figure 4b.12 The copper(II) sulfate solution that was filtered in Figure 4b.11 is now being evaporated in order to obtain the crystals. If it is heated too strongly, some of the copper(II) sulfate may spit out of the evaporating basin or may get so hot that it decomposes. Product can be lost during heating.

Measuring atomic mass

How can you measure the mass of something as small as an atom? This was a real challenge to early chemists. The English chemist John Dalton put forward his atomic theory in 1803. It was over half a century later, in 1860, when Stanislao Cannizzaro, who came from Sicily, calculated a reliable set of relative atomic masses from experimental results. Cannizzaro worked out his atomic masses relative to the mass of an oxygen atom, which he gave a mass of 16.

Modern atomic mass data is given relative to carbon atoms that have six protons and six neutrons. These atoms are given a mass of 12 units and are known as 'carbon-12' atoms. A powerful tool called the mass spectrometer is used to find the masses of other atoms relative to a carbon-12 atom.

A mass spectrometer finds the relative mass of atoms of different elements by first of all changing the atoms into ions. These ions are then fired through a magnetic field. Lighter ions are deflected a lot from their original path by the magnetic field. Heavier ions are deflected by a smaller amount. If scientists measure how much the ions are deflected, they can work out the mass of the ions.

The world's most accurate mass spectrometer is at the Massachusetts Institute of Technology. It can measure the relative mass of different atoms to a staggering accuracy of ten decimal places! As we make progress in science and technology, we need more and more accurate measuring instruments.

Figure 4b.13 A mass spectrometer.

Summary

You should be able to:

- look up relative atomic masses from the Periodic Table

- calculate the formula mass of a substance

- recognise that the total mass of reactants used up is equal to the total mass of product formed

- use ratios to calculate the masses of reactants and products

- **H** explain why mass is conserved in a chemical reaction

- use balanced symbol equations to calculate the masses of reactants and products

- use percentage yield to describe the amount of product obtained from a chemical reaction

- recognise possible reasons why the percentage yield of a reaction is less than 100%

- calculate the percentage yield of a reaction

Questions

1 a What are the relative atomic masses of zinc, carbon and oxygen?

 b What is the relative formula mass of zinc carbonate ($ZnCO_3$)?

2 When calcium carbonate is heated, it undergoes thermal decomposition to form calcium oxide and carbon dioxide. When 1 kg of calcium carbonate decomposes, 560 g of calcium oxide are formed.

 a How much calcium oxide will form when the following masses of calcium carbonate decompose?

  i 3 kg

  ii 100 g

  iii 1 g

 b When 1 kg of calcium carbonate decomposes, the expected yield of calcium oxide is 560 g. If 1 kg of calcium carbonate is decomposed in a lab and the actual yield of calcium oxide is 532 g, what is the percentage yield?

continued on next page

Questions - *continued*

3   Sodium hydroxide and hydrochloric acid react together to make sodium chloride and water. The balanced formula equation for this reaction is

$$NaOH + HCl \rightarrow NaCl + H_2O$$

  a   What type of reaction is this?

  b   Use the Periodic Table on page 345 to find the relative atomic masses of sodium, oxygen, hydrogen and chlorine.

  c   Work out the relative formula masses of sodium hydroxide, hydrochloric acid, sodium chloride and water.

In this question, remember to round the value of the relative atomic mass of chlorine to 35.

4   a   Describe *four* reasons why the actual yield of a chemical reaction might be less than 100%.

  b   Calculate the percentage yield for each of the following situations.

      i   You expected to make 700 g of product but only made 140 g.

      ii You expected to make 0.60 g of product but only made 0.15 g.

      iii You expected to make 2.00 g of product but only made 1.36 g.

H 5   a   Write a balanced symbol equation for the thermal decomposition of zinc carbonate.

  b   Write the relative formula mass of each substance underneath its formula in the equation.

  c   What mass of carbon dioxide is formed when 125 g of zinc carbonate decomposes?

  d   What mass of carbon dioxide is formed when 25 g of zinc carbonate decomposes?

  e   What mass of carbon dioxide is formed if 162 g of zinc oxide is formed?

6   Ammonia and nitric acid react to make the fertiliser ammonium nitrate. A balanced symbol equation for this reaction is:

$$NH_3 + HNO_3 \rightarrow NH_4NO_3$$

  a   Use the Periodic Table on page 345 to find the relative atomic masses of nitrogen, hydrogen and oxygen.

  b   Calculate the relative formula masses of ammonia, nitric acid and ammonium nitrate.

  c   What mass of ammonium nitrate forms when 3.40 g of ammonia reacts with excess nitric acid?

  d   What mass of ammonium nitrate forms when 630 tonnes of nitric acid reacts with excess ammonia?

  e   What mass of ammonia and nitric acid must react together to form 400 tonnes of ammonium nitrate?

Plants need certain elements

Plants absorb water from the soil through their roots. Dissolved in the soil water are many chemical compounds. Some of these chemical compounds are compounds of elements that are essential to the plant. For example, three elements that are essential to plants are nitrogen, phosphorus and potassium. Plants need these elements in order to grow healthily. Plants get these elements, combined with other elements, from compounds dissolved in the soil water.

If the soil does not contain compounds of these essential elements, the soil will be infertile. Soil like this is said to give a low 'crop yield'. Infertile soil can be improved by adding **fertilisers** that contain compounds of the elements lacking in the soil. Such fertilisers will increase crop yield by causing the crops to grow bigger and faster.

Figure 4c.1 The soil at the top of this photograph contains all the nutrients the crop needs. The soil at the bottom does not.

SAQ

1 The following chemical compounds can be used as fertilisers. Use the formula of the substance and the Periodic Table on page 345 to name the element or elements in each compound that is essential for crop growth.

   a Ammonium nitrate, NH_4NO_3

   b Ammonium phosphate, $(NH_4)_3PO_4$

   c Ammonium sulfate, $(NH_4)_2SO_4$

   d Potassium nitrate, KNO_3

2 Use the relative atomic masses from the Periodic Table on page 345 to calculate the relative formula masses of the compounds in SAQ **1**.

Because nitrogen, phosphorus and potassium are so important to soils, fertilisers often have 'NPK numbers' printed on their bags. These numbers are the percentages by mass of the three elements nitrogen (N), phosphorus (P) and potassium (K).

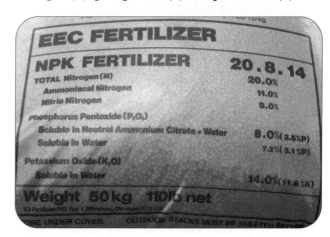

Figure 4c.2 This fertiliser contains 20% nitrogen, 8% phosphorus and 14% potassium.

The effect of fertilisers

As plant roots absorb water from the soil they absorb dissolved compounds of essential elements too. The rate at which these compounds are absorbed can be greater than the rate at which they are naturally replaced. In this case, the amount of compounds of essential elements in the soil decreases and the soil becomes infertile. Using chemical fertilisers can replace these compounds or add similar ones to the soil, making the soil fertile again. (See Item B4d *Plants need minerals too*.)

Nitrogen is a particularly important essential element for plants. Plants use nitrogen compounds in order to make proteins. Proteins are essential in any animal or plant for health and growth. Natural processes replace the nitrogen compounds in soil that plants remove. However, if additional fertilisers are added, it is possible to boost the level of nitrogen compounds in the soil, causing greater crop growth than is usual. Fertilisers that add nitrogen compounds to the soil are called nitrogenous fertilisers.

SAQ

3 Intensive farming involves growing the greatest density of crops the soil can support. Explain why intensive farming may result in a farmer needing to use nitrogenous fertiliser.

Figure 4c.3 The soil in the land at the top and bottom of this photograph lacks nitrogen compounds. The soil in the fields in the centre has been treated with nitrogenous fertiliser.

Eutrophication

Fertilisers must be soluble in water. If they are not, they cannot be taken up by plant roots. This means that heavy rain will dissolve the fertiliser and may cause it to wash into rivers. This is known as run-off. Unfortunately, run-off can have a serious effect on the life in the river. A sequence of steps occurs.

● Heavy rain causes fertiliser run-off into a river.
● Algae are plants that live in rivers. The fertiliser will cause them to grow very quickly. The algae can then cover the entire surface of the river. This is known as an algal bloom. It is possible for an algal bloom to happen overnight.
● When the algae are covering the surface of the river, light will not be able to reach the plants growing on the river bed. These plants will die.
● The dead plants will rot. This involves aerobic bacteria. The aerobic bacteria use up the oxygen in the river water as they decompose the dead plants.
● The fish, amphibian and other animal life in the river all need oxygen. When the aerobic bacteria have used up the oxygen in the water, these animals will die.

This progressive process is called **eutrophication**.

Figure 4c.4 This river has undergone eutrophication.

SAQ

4 Describe eutrophication in your own words. Each of the five bullet points above should be reduced to as few words as possible. Can you do it in fewer than 20 words?

Many fertilisers are salts

Ammonia is a base. If ammonia is used to neutralise an acid, the product of the reaction is a salt. Some of the salts that can be made like this are fertilisers. This means they contain one or more of the essential elements that plants need.

● Ammonium nitrate is made by neutralising ammonia with nitric acid. More ammonium nitrate is used in the world each year than any other artificial fertiliser.
● Ammonium phosphate is made by neutralising ammonia with phosphoric acid.
● Ammonium sulfate is made by neutralising ammonia with sulfuric acid.

Another fertiliser that is made using ammonia as a starting material is urea. Potassium nitrate is a fertiliser which can be made by neutralising an acid with a base: if nitric acid is neutralised with potassium hydroxide, the salt produced is potassium nitrate.

SAQ

5 Look at the artificial fertilisers in Figure 4c.7.
   a What reactants could have been used to make each of these fertilisers?
   b What essential element (or elements) does each fertiliser contain?

Getting it right

The farming areas on the banks of the Mississippi river have been using huge amounts of fertiliser every year for a great many years. The fertiliser run-off causes a problem called eutrophication in the Mississippi. Eutrophication causes the level of oxygen dissolved in the water to fall.

Because of the problem caused by the fertiliser, a huge volume of oxygen-depleted water travels south into the Gulf of Mexico every minute. This has caused a 'dead zone' – 18 000 square kilometres of ocean with little or no life in it. This is equivalent to a circle of ocean 150 km in diameter.

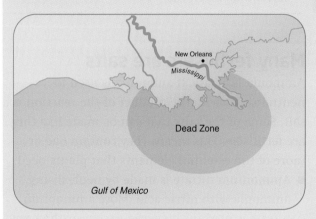

Figure 4c.5 The Mississippi delta.

One of the causes of the eutrophication is the use of excessive amounts of nitrogenous fertilisers. Often, the soil doesn't actually need the amount of fertiliser being applied, but the farmers add it anyway 'to be on the safe side'. The fertilisers are cheap and the farmers are not prepared to risk the disaster of a failed crop.

The US Department of Agriculture thinks that it may have a solution to the problem. Special infra-red light detectors are used to measure the colour of the leaves growing on all areas of a farm. The results are used to gauge the health of the leaves growing in each area. These results in turn can be used to get an accurate measure of exactly how much nitrogenous fertiliser the soil in each area actually needs.

The farmer then applies the correct amount of fertiliser to each area of the farm. In this way the farmer avoids adding too little fertiliser, which would risk low crop yields. The farmer also avoids adding too much, which would waste money and cause fertiliser run-off and eutrophication. The US Department of Agriculture hopes that this 'win–win' situation can solve a problem which they currently say is 'bigger than we'd even like to talk about.'

Figure 4c.6 The farming areas near the Mississippi river are famous for growing cotton. The use of nitrogenous fertilisers prevents the soil from becoming infertile.

 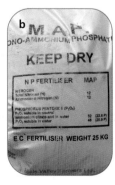 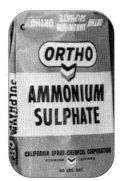

Figure 4c.7 a Ammonium nitrate fertiliser.
b Ammonium phosphate fertiliser.
c Ammonium sulfate fertiliser.

Making fertilisers by neutralisation

A salt can be made by neutralising an acid with a base. When this is done in a laboratory, the base is dissolved in water, giving an alkali. The technique used is called **titration**. This technique involves measuring exactly what volume of the acid will neutralise a particular volume of the alkali. In the photo sequence that follows, the acid used is sulfuric acid and the alkali used is ammonia solution. The salt being made is therefore ammonium sulfate.

● First, the acid is put into an accurate measuring tube called a **burette**.

Figure 4c.8 A funnel is used to fill the burette with sulfuric acid.

● A measured amount of alkali is put into a conical flask. This can be done using a measuring cylinder but a **graduated pipette** is a more accurate way.

Figure 4c.9 A graduated pipette is used to measure 25 cm³ of ammonia solution into a conical flask.

● Indicator is added to the alkali in the conical flask.

Figure 4c.10 An indicator called litmus solution is added to the ammonia solution, which turns blue.

● The acid from the burette is added to the alkali in the conical flask until the indicator changes colour. The acid and alkali have now neutralised each other and made a salt. The salt is not pure at this stage, however – it is contaminated by the indicator. The volume of acid needed to neutralise the 25 cm³ of alkali is noted down.

Figure 4c.11 12.5 cm³ of sulfuric acid from the burette have been added to the 25 cm³ of alkali in the conical flask. The litmus has gone purple, showing that this volume of acid was just enough to neutralise the alkali.

- The experiment is repeated without using any indicator. Exactly the same volumes of acid and alkali are used. We know that these volumes are exactly enough to neutralise each other. The conical flask now contains a pure solution of the salt we want. It is not contaminated by extra acid, extra alkali or indicator. This is why an accurate technique like titration has been used.

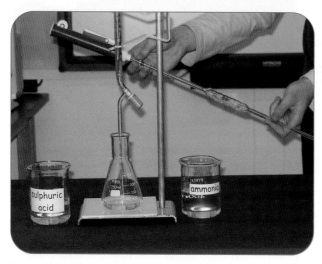

Figure 4c.12 The conical flask contains ammonium sulfate solution.

- If the water in the salt solution is carefully evaporated off, the solid salt (the fertiliser) is obtained.

Figure 4c.13 When the water is evaporated off, solid ammonium sulfate is obtained.

SAQ

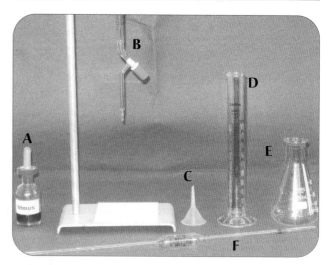

Figure 4c.14

6 Look at Figure 4c.14. Give the correct names for the pieces of apparatus labeled A–F.

Percentages of essential elements

It is possible to use relative atomic masses and relative formula masses to calculate the percentage of each essential element in a fertiliser.

Worked example 1

What is the percentage of nitrogen in ammonium nitrate?

Step 1: the formula of ammonium nitrate is NH_4NO_3. The formula mass of ammonium nitrate is $14+1+1+1+1+14+16+16+16 = 80$.

Step 2: there are two nitrogen atoms in one formula unit of ammonium nitrate. These two nitrogen atoms have a total mass of 28.

Step 3: the percentage nitrogen in each ammonium nitrate formula unit is:

$(28 \div 80) \times 100\% = 35\%$

Ammonium nitrate is 35% nitrogen.

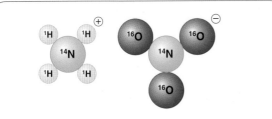

Figure 4c.15 One formula unit of ammonium nitrate consists of these nine atoms. The relative atomic mass of each atom is written with its symbol. The relative formula mass is 80. The two nitrogen atoms contribute 28 of this. The percentage of nitrogen is (28/80) × 100% = 35%.

SAQ

7 Use the Periodic Table on page 345 to find the relative atomic masses needed to answer the following questions.

  a Calculate the percentage of phosphorus in ammonium phosphate.

  b Calculate the percentage of nitrogen in ammonium phosphate.

  c Calculate the percentage of nitrogen in potassium nitrate.

  d Calculate the percentage of potassium in potassium nitrate.

8 A bag of ammonium nitrate may be labelled with the NPK numbers '35-0-0'. Explain what this means.

Summary

You should be able to:

♦ explain how plants absorb fertilisers

♦ describe the advantages of using fertilisers

♦ explain how fertilisers increase crop yield

♦ describe the process of eutrophication

♦ recall the names of *three* essential elements needed for plant growth and recognise their presence in a fertiliser from its formula

♦ calculate the relative formula mass of a fertiliser

♦ give the names of nitrogenous fertilisers manufactured from ammonia

♦ name the acid and alkali which would react together to make certain fertilisers

♦ describe how a fertiliser can be made by titration in a laboratory

♦ calculate the percentage by mass of each essential element in a fertiliser

Questions

1 a Why do farmers use fertilisers?

   b Why must an effective fertiliser:

      i be soluble in water?

      ii be a compound of nitrogen, phosphorus or potassium?

2 A sample of dry ammonium nitrate powder can be made in a school lab.

   a Which reactants would you use to make ammonium nitrate?

   b What is the name of the technique you would use to make ammonium nitrate?

   c Show how you would make ammonium nitrate by sketching each step in the procedure. Label each piece of apparatus the first time you draw it.

3 Name the salt that would be made by each of the following neutralisation reactions.

   a Phosphoric acid reacting with ammonia.

   b Nitric acid reacting with potassium hydroxide.

   c Ammonia reacting with sulfuric acid.

H 4 The formula of ammonium sulfate is $(NH_4)_2SO_4$.

   a Write a balanced symbol equation for the reaction between ammonia and sulfuric acid to give ammonium sulfate.

   b Use the Periodic Table on page 345 to calculate the relative formula mass of ammonium sulfate.

   c What is the percentage of nitrogen in ammonium sulfate?

   d What is the percentage of sulfur in ammonium sulfate?

5 Read the context box *Getting it right* on page 180. Turn back to Item B4d (page 78) and read the section on GPS technology. Suggest possible advantages and disadvantages of using infra-red light detectors or GPS technology to measure how much fertiliser is needed.

Industry is expensive to run

Many of the substances that we use are manufactured by the chemical industry. The cost of making a substance depends on many factors.

- The cost of the materials that are used to make the new substance. These starting materials are also known as 'raw materials'.
- The cost of the equipment used. This equipment is also called the '**plant**'.
- The cost of the energy needed to power the plant. Electricity and gas are commonly used.
- The amount of money paid to the workforce.
- The rate at which the product is produced.

SAQ

1 Each of the following cartoons illustrates one factor affecting the cost of making a new substance. Describe the factor shown in each cartoon.

Figure 4d.1

Uses of ammonia

Ammonia is a compound that is made in large amounts by the chemical industry. It is a compound of nitrogen and hydrogen. The formula of ammonia is NH_3. Ammonia has many uses including:

- making fertilisers
- making nitric acid
- making cleaning agents.

The world's growing population causes an ever-increasing demand for food. Fertilisers made from ammonia are used widely in order to grow larger crops and meet this demand. Ammonia has become an extremely important industrial chemical.

Figure 4d.2 This spray-on oven cleaner contains ammonia.

Ammonia is made from nitrogen and hydrogen

We make ammonia by reacting nitrogen and hydrogen. The nitrogen is obtained from the air. The hydrogen is obtained either from natural gas or as a byproduct of the cracking of heavy fractions of crude oil.

Ammonia is then made in a highly automated industrial process called the **Haber process**. 'Automated' means that much of the work is done automatically, under computer control rather than human control. The industrial installation where this happens is called a Haber process plant. Because the process is automated, the plant can save a lot of money by employing fewer people.

Figure 4d.3 Ammonia is made by the Haber process at this plant in Mississippi, USA.

SAQ

2 **a** What is the formula of ammonia?

   b What is the name of the process by which ammonia is made industrially?

   c Which *two* substances are reacted together to make ammonia?

   d What raw materials are the sources of these two substances?

   e Give *three* uses of ammonia.

The word equation for the formation of ammonia is:

nitrogen + hydrogen ⇌ ammonia

This reaction is **reversible**. Nitrogen and hydrogen react together to make ammonia. Ammonia can also decompose to give nitrogen and hydrogen. This is what the ⇌ arrow means. It means that the reaction that makes ammonia is a **reversible reaction**. It can proceed in a forward direction, with nitrogen and hydrogen reacting to make ammonia. It can proceed in a backward direction, with ammonia reacting to make nitrogen and hydrogen.

The reaction has to be speeded up

The reaction of nitrogen with hydrogen to make ammonia is naturally very slow. Unless we do something about this, the slow reaction will result in very little ammonia being made. This would result in very low profits. The reaction is speeded up in several ways.

Iron catalyst

A catalyst speeds up a reaction without being used up itself. The catalyst used in the Haber process is iron. The iron catalyst increases the profitability of the plant because it increases the rate of the reaction.

High pressure

The mixture of nitrogen and hydrogen is pressurised using a compressor. When the gases are at high pressure, their molecules are closer together. The gas molecules collide more often, so increasing the rate of the reaction.

The process cannot be run at too high a pressure, however. A plant that runs at very high pressure is very expensive to build owing to the high cost of the compressors. The plant becomes very expensive to maintain because the high pressure inside it causes joints to split and leak.

Raised temperature

The gases are heated to 450 °C before they react on the surface of the catalyst. Raising the temperature increases the rate of the reaction because the molecules move faster. When the molecules collide, the collisions have more energy and are more likely to be successful. In theory, even higher temperatures could be used but this would mean that more energy would have to be used. Energy costs money!

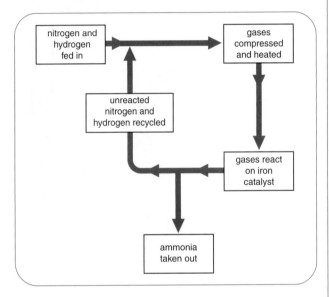

Figure 4d.4 Flow diagram of the Haber process.

Recycling starting materials

The nitrogen and hydrogen gases are heated, pressurised, and passed over an iron catalyst. Ammonia is made. Some of the reactants do not become ammonia but stay as nitrogen and hydrogen. These gases are not wasted; they are recycled. They are mixed with more nitrogen and hydrogen, heated and pressurised again, and passed over the catalyst again. The nitrogen and hydrogen have cost money so they must not be wasted.

SAQ

3 Explain:

   a the meaning of the $\rightleftharpoons$ arrow

   b why high pressures and high temperatures are expensive

   c how automation can save money.

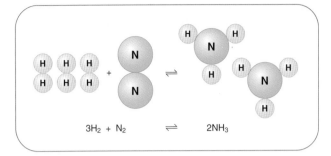

$$3H_2 + N_2 \rightleftharpoons 2NH_3$$

Figure 4d.5 Three hydrogen molecules react with one nitrogen molecule to make two molecules of ammonia.

Making a profit

The conditions used to manufacture ammonia are dictated by the economics of the process. The plant is set up to make the maximum profit. Three main factors are involved.

Yield

The plant has to make as much ammonia as it can from each tonne of nitrogen and hydrogen mixture. The percentage of each tonne of mixture that is converted into ammonia is called the percentage yield. A high percentage yield is very desirable.

Rate

The plant has to make the yield as quickly as possible. The rate of the reaction is therefore extremely important. A high rate is very desirable.

Plant costs

It may be that an excellent yield and an excellent rate are achievable, but only by having high plant costs. The costs of the plant, energy requirements and workforce must be kept to a minimum.

SAQ

4 Write a balanced symbol equation for the reaction of nitrogen with hydrogen making ammonia.

Developing the Haber process

In AD 850, a Chinese writer called Cheng Yin wrote about gunpowder. He warned people of the dangers of making it, writing that it had caused the loss of many people and their houses. Gunpowder contains sodium nitrate as one of its key ingredients.

During the First World War (1914–1918), all the explosives being used were nitrogen compounds. The need for nitrogen compounds to make explosives did not put pressure on the British and French. They imported compounds such as sodium nitrate by sea from South America. The sodium nitrate was then used to make explosives.

It was a different story in Germany. The British navy stopped ships from elsewhere in the world reaching German ports. Germany couldn't import sodium nitrate. This caused two serious problems. Firstly, Germany needed the sodium nitrate to make explosives. Secondly, Germany needed the sodium nitrate to make fertilisers. As the war went on, food became in very short supply in Germany. Germany needed another source of nitrogen compounds urgently.

Back in 1908, Fritz Haber had developed a process for making ammonia from nitrogen and hydrogen. Although this reaction had been known about for a long time, it was not thought useful. There were two problems. The

reaction was very slow and the yield of ammonia was very low. Haber solved these problems by using a catalyst, high pressure and a raised temperature. The Haber process was developed further in 1909 by Karl Bosch.

Figure 4d.6 a Fritz Haber. **b** Karl Bosch.

The Haber process was developed before the start of the First World War. However, it was the needs of the war that made it important. Suddenly an industrial process of comparatively minor importance had to be developed into a full-scale industry. By the end of the war, the future of the Haber process was secure.

H Compromise

The iron catalyst is cheap and causes a faster reaction without affecting the percentage yield. Unfortunately, the other conditions used do not have such a straightforward and positive effect. It is not possible to choose conditions for the Haber process to give all of highest yield, highest rate and lowest cost.

- Very high temperatures give a very high rate, but high temperatures decrease the yield. Using

very high temperatures also increases energy costs. 450 °C is chosen as a temperature that gives an acceptable rate without causing too great a decrease in yield or increase in energy costs. 450 °C is known as the 'optimum' temperature.

- Very high pressures give a very high rate and they also increase the yield. Unfortunately, it is extremely costly to build, maintain and run a plant operating at very high pressures.

The conditions used are those that give the highest final profit. This means making ammonia at the lowest cost per tonne. The conditions used are said to be a compromise. Many Haber process plants are run under conditions that give a low percentage yield each time the reactants pass over the catalyst. This is compensated for by recycling the unreacted gases. Recycling ensures that all the nitrogen and hydrogen react eventually.

SAQ

5 Describe the advantages and disadvantages of running a Haber process plant at:

a high pressure

b high temperature.

6 Explain why the use of an iron catalyst can be described as having no disadvantages.

Summary

You should be able to:

♦ describe some uses of ammonia, recognising its importance in relation to world food production

♦ state the raw materials used to manufacture ammonia

♦ write a word equation for the formation of ammonia from nitrogen and hydrogen

♦ describe how ammonia is made in the Haber process

♦ describe the factors that affect the cost of making a new substance

♦ describe how these factors affect the cost

♦ understand the meaning of the term *reversible reaction* and recognise the $\rightleftharpoons$ double arrow symbol

♦ write a balanced symbol equation for the formation of ammonia from nitrogen and hydrogen

♦ explain how the conditions used in the Haber process affect the rate of the reaction and the percentage yield

♦ explain how these conditions are determined by economic considerations

Questions

1 a What is made in the Haber process?

b Why is the manufacture of this chemical important?

c Write a word equation for the chemical reaction involved in the Haber process.

d Name the *two* reactants involved in the Haber process.

e What raw material is used as the source of each reactant?

continued on next page

Questions - *continued*

2   Table 4d.1 shows the percentage yield of ammonia when a mixture of nitrogen and hydrogen reacts together at different pressures. The temperature in each case is 450 °C.

a   Plot a graph of this data, with pressure on the horizontal axis and yield on the vertical axis. Join the points with a smooth curve.

b   What is the percentage yield at a pressure of 400 atmospheres?

c   What pressure would give a percentage yield of 20%?

d   What happens to the yield as pressure increases?

e   Estimate the pressure that would be necessary to give a percentage yield of 60%.

Pressure (atmospheres)	Percentage yield of ammonia
100	17%
200	28%
300	37%
400	44%
500	50%

Table 4d.1

3   Sulfuric acid is manufactured in the contact process. The most important step in this process is the reaction of sulfur dioxide and oxygen to make sulfur trioxide:

$$2SO_2 + O_2 \rightleftharpoons 2SO_3$$

This is a reversible reaction. A divanadium pentoxide catalyst is used. The temperature used is 400 °C. The pressure used is normal atmospheric pressure (1 atmosphere).

Table 4d.2 shows the percentage yield of sulfur trioxide when a mixture of sulfur dioxide and oxygen reacts together at different temperatures. The pressure in each case is 1 atmosphere.

Temperature (°C)	Percentage yield of sulfur trioxide
350	98%
400	93%
450	89%
500	86%
550	84%

Table 4d.2

a   Explain the meaning of the term 'reversible reaction'

b   Explain why a catalyst is used in the contact process.

c   Suggest why the contact process is run at 400 °C even though a higher yield is obtained at 350 °C.

d   Give *one* advantage and *two* disadvantages of running the contact process at a temperature of 550 °C rather than 400 °C.

e   If the pressure is increased, the yield of sulfur trioxide increases. Why is the contact process run at a pressure of 1 atmosphere despite this?

Detergents

The ingredients in a washing powder

When we put dirty clothes into an automatic washing machine, we also put in **washing powder**. This can also be added in liquid form. Washing powders and washing liquids are made by blending several ingredients.

- The active **detergent** removes the dirt from the clothes. Many detergents are salts. They are made by a neutralisation reaction between an acid and an alkali.

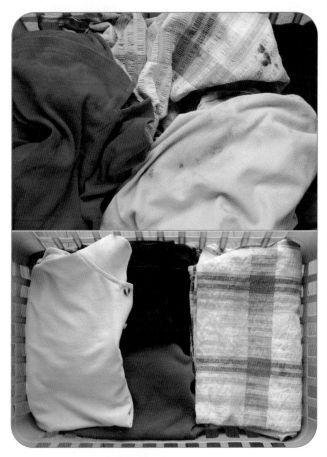

Figure 4e.1 The dirt on these clothes was removed in the washing machine by the detergent.

- Most water from domestic water supplies contains dissolved chemicals that make it difficult for detergents to clean clothes properly. Water like this is called **hard water**. As well as affecting the detergent's cleaning power, hard water produces **limescale**. Limescale is a hard, insoluble substance that forms on the washing machine's heating element. The limescale makes the machine less efficient.

To solve these problems, a **water softener** is added to the washing powder. This removes the chemicals that cause hardness. This enables the detergent to do its job and stops the build-up of limescale.

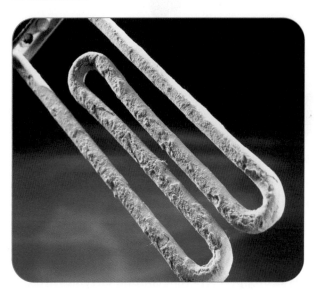

Figure 4e.2 This heating element is covered in scale and will do its job inefficiently. Using a washing powder containing water softener would have helped to prevent this.

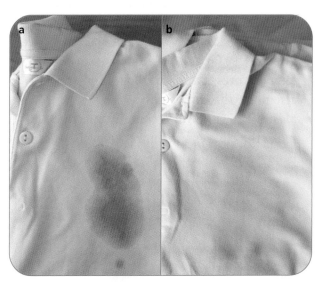

Figure 4e.3 a This red wine stain is unlikely to be washed out by the detergent. Bleach does remove the colour of the stain, however (**b**).

- Some deeply coloured stains will not be completely washed off by detergents. Washing powders may contain **bleaches** to remove such coloured stains.
- White clothes will appear to be an even brighter white if an **optical brightener** is added to the washing powder.

Figure 4e.4 Washing powder containing an optical brightener has made the shirt on the right appear 'whiter than white'.

- Many foods are difficult to wash off clothes. **Enzymes** are added to the washing powder to digest the food. This enables the detergent to remove the food stain from the clothes. Washing powders that contain enzymes are known as biological washing powders. The enzymes work best at lower washing temperatures – around 40 °C. This is much cheaper than using a hotter wash because it saves energy. Lower washing temperatures are often better for the clothes too. High washing temperatures can make colours run and can also make the clothes shrink.

- Most washing powders contain a perfume to make the clothes smell nice when they have been washed and dried.

Figure 4e.5 The enzymes in a biological washing powder will digest the food stains and turn this back into a white T-shirt! The enzymes include proteases, which digest protein, and lipases, which digest fats.

SAQ

1 Explain why washing powders contain each of the following:

   a enzymes   **b** detergents

   c optical brighteners   **d** bleaches

   e water softeners.

Washing labels

Different clothes are made of different fabrics. They have to be cleaned in different ways, otherwise they may get damaged. The washing label on each garment tells you how to clean it.

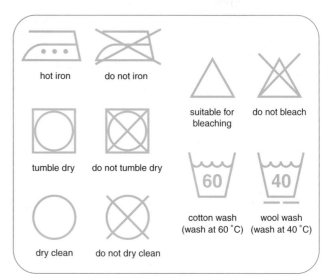

Figure 4e.6 These symbols are found on washing labels. They tell you how to treat the garment in order to keep it in good condition.

SAQ

2 Look at the washing label. How should you treat this garment?

            Figure 4e.7

3 Look at the washing label. What sort of treatment would damage this garment?

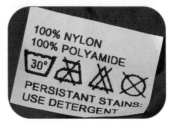

            Figure 4e.8

Controversial laundry?

Washing your clothes sounds like an unlikely area for controversy. However, as washing powders have come to contain more and more artificial and unusual ingredients, some concerns have arisen for the consumer and the environment.

Many of the detergents used in washing powders are artificial and non-biodegradable. They go from our washing machines into the sewers and eventually into rivers. They don't decompose, so they stay there, causing unsightly foaming and possibly having toxic effects on water life. Biodegradable detergents have been developed to help minimise these problems.

Figure 4e.9 This photograph of the Saale River in Germany was taken in the 1990s. It shows the effect of detergent pollution – froth and scum on the water.

Optical brighteners work by absorbing ultraviolet and violet light, and re-emitting blue light. This counteracts the tendency of white objects to look slightly yellow. Optical brighteners are also used to whiten paper and disposable nappies. Is this necessary? The brighteners make these products *look* cleaner without them actually *being* cleaner. Should the chemical industry be using up resources and should consumers be spending money on these products?

Some people are allergic to the enzymes used in biological washing powders. The enzymes give them a skin rash. Because of this, most manufacturers make two versions of their product, one with enzymes and one without. The washing powder without enzymes is known as a 'non-bio'.

Figure 4e.10 The powder on the left contains enzymes, the powder on the right doesn't.

Washing powders also contain chemicals called phosphates. They act as water softners. Phosphates also help to stop dirt that has been washed off one part of a garment from sticking back onto it, somewhere else. Unfortunately, phosphates help to cause eutrophication when they eventually drain into rivers.

When washing powders were first developed as an alternative to laundry soap, none of these issues would have been imagined. There are now so many humans washing so many dirty clothes every day that they cannot be ignored.

How detergents work

A lot of the dirt that gets onto clothes is of an oily or greasy nature. Dirt like this won't dissolve in water.

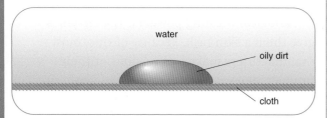

Figure 4e.11 The oily dirt does not dissolve in water. It sticks to the clothes.

Detergent molecules enable the dirt to dissolve in water and be washed away. This is possible because the two ends of each detergent molecule are different.

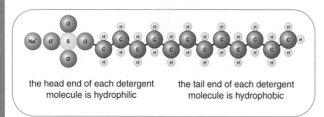

the head end of each detergent molecule is hydrophilic

the tail end of each detergent molecule is hydrophobic

Figure 4e.12 A detergent molecule.

One end of each detergent molecule, called the tail, is **hydrophobic**. This means water-hating. Although this end of the molecule hates water, it likes grease. This end of each detergent molecule mixes with greasy, oily dirt.

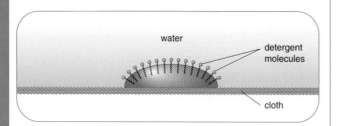

Figure 4e.13 The hydrophobic tails have mixed with the oily dirt.

The other end of each detergent molecule, called the head, is **hydrophilic**. This means water-loving. The blob of oily dirt now has water-loving detergent heads sticking out of it. This enables it to dissolve in water. The blob of dirt comes off the clothes into the water and is washed away.

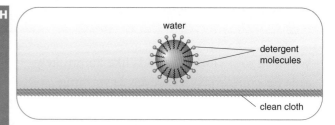

Figure 4e.14 The hydrophilic heads enable the oily dirt to dissolve in the water.

SAQ

4 a Explain the meaning of the terms *hydrophobic* and *hydrophilic*.

 b Why won't oily dirt wash off clothes into pure water?

 c Explain how detergents enable oily dirt to wash off clothes.

Solvents and solutes

When dirt washes off clothes in a washing machine, the dirt dissolves in the water. The water is acting as a **solvent**. The dirt is acting as a **solute**. The mixture of dirt dissolved in water is a **solution**.

Some dirt dissolves easily in water. Such dirt is **soluble** in water. Some dirt doesn't dissolve easily in water. Such dirt is **insoluble** in water. Oily dirt is insoluble in water. The detergent does its job by helping insoluble dirt to dissolve in water. Enzymes in biological washing powders do their job by helping insoluble 'food dirt' to dissolve in water.

Dry cleaning

Having clothes dry cleaned still involves a solvent, but that solvent isn't water. A solvent is used in which oily dirt is soluble. This solvent, called a 'dry cleaning solvent', does not need to have detergents added to it. Some clothes lose their shape if they are put into water. Dry cleaning these clothes is less likely to damage them.

SAQ

5 Explain why it may be better to have clothes dry cleaned than to put them in the washing machine.

Intermolecular forces

Intermolecular forces cause the molecules in solvents to attract each other. There are two types of **intermolecular force** – polar forces and non-polar forces. Polar forces cause water molecules to attract each other.

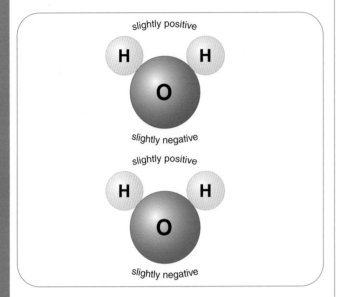

Figure 4e.15 The oxygen side of a water molecule has a slightly negative electric charge. The hydrogen side of a water molecule has a slightly positive electric charge. This makes water molecules attract each other. These intermolecular forces are called 'polar forces'.

Non-polar forces cause dry cleaning solvent molecules to attract each other. Molecules of oily dirt are also attracted to each other by non-polar forces.

If a solute and a solvent have similar types of intermolecular forces, the solute is liable to be soluble in the solvent. If they have different types of intermolecular forces, the solute is liable to be insoluble in the solvent. This is why oily dirt is soluble in dry cleaning solvent. The molecules of the two substances attract each other by non-polar forces. Oily dirt is not soluble in water. Water molecules attract each other by polar forces. Oil molecules attract each other by non-polar forces.

SAQ

6 **a** Explain why oily dirt is insoluble in water
   b Explain why oily dirt is soluble in dry cleaning solvent.

Washing-up liquid

The liquids that we use to wash up plates and cutlery are a mixture of several ingredients.

● The active detergent cleans the food off the washing up. Most food dirt is oily. The detergent helps it to dissolve in the washing-up water.

● A rinsing agent is added so that the washing-up water drains off the plates and cutlery quickly. If the washing-up water fails to do this, it might dry on the plates. The detergent and dissolved food would then be deposited on the plates.

Figure 4e.16 These plates and glasses have been washed and they are now draining. The rinse agent in the washing-up liquid helps the washing-up water drain away. If the washing-up water dries on the plates, detergent and dissolved food will get deposited on them.

● A water softener is added to soften the water.
● Colour and perfume are added to make the washing-up liquid more attractive.
● A thinning agent is added so that the washing-up liquid squirts out of its bottle easily.

SAQ

7 What would be wrong with a washing-up liquid that did not contain:

   a a rinse agent?
   b detergent?
   c a thinning agent?
   d added fragrance (perfume)?

Summary

You should be able to:

♦ recall that many detergents are salts and that they can be made by neutralising an acid with an alkali

♦ describe the function of each ingredient in a washing powder

♦ use washing labels to decide on the ideal washing conditions for particular garments

♦ explain the advantages of low temperature washes

[H] ♦ explain how detergents work

♦ use the terms *solute*, *solvent*, *solution*, *insoluble* and *soluble*

♦ recognise that different solutes dissolve in different solvents

♦ describe dry cleaning

[H] ♦ use intermolecular forces to explain how a dry cleaning solvent works

♦ describe the function of each ingredient in a washing-up liquid

Questions

1 Three washing powders were compared with each other. They were used to wash three different types of stain: grass stains, gravy stains and tomato ketchup stains. They were used at 40 °C and at 95 °C. Cleaning power was judged on a 0–10 scale, where 0 means no improvement and 10 means complete cleaning. The results are shown in Table 4e.1.

	Sparkle 2		New Bio Dash		Ultimate	
	40 °C	95 °C	40 °C	95 °C	40 °C	95 °C
grass	8	10	7	5	6	9
gravy	6	8	10	4	4	6
ketchup	5	8	10	3	3	6

Table 4e.1

a What combination of washing powder and temperature was best for cleaning off grass stains?

b What is the effect of using a hotter wash if you use Sparkle 2?

c Why might a homeowner prefer not to wash clothes at 95 °C?

d How good is Ultimate at cleaning off food stains at 40 °C?

e How good is New Bio Dash at cleaning off food stains at 40 °C?

f What ingredient might New Bio Dash contain that is missing from Ultimate?

[H] g Explain why New Bio Dash is less effective at cleaning off food stains at 95 °C.

continued on next page

Questions - *continued*

2 Washing powders and washing-up liquids do a very similar job. The ingredients they contain are similar but not identical.

   a Why do they both contain detergents?

   b Name *two* other ingredients contained in both of them.

   c Name *two* ingredients found in washing powders but not in washing-up liquids. What do these ingredients do?

   d Name *two* ingredients found in washing-up liquids but not in washing powders. What do these ingredients do?

3 Sally spilt some copper sulfate solution on the cuff of a clean white shirt. She rinsed the cuff straight away in warm water and fortunately all of the copper sulfate washed out. Describe what happened when she rinsed her shirt cuff, using the following words: *solvent*, *solute*, *solution* and *soluble*.

4 The washing label on a red T-shirt says that it should be washed at 40 °C.

   a Describe *two* ways in which the T-shirt might be damaged if it was washed at a higher temperature.

   b What other advantage is there to be gained by washing clothes at 40 °C instead of boil-washing them?

H 5 Detergent molecules have a hydrophilic head and a hydrophobic tail. Use intermolecular forces to suggest an explanation of why the tail mixes with oil but the head mixes with water.

The Haber process is continuous

Once a Haber process plant starts going, the workforce makes sure it keeps going. More nitrogen and hydrogen are constantly fed in and the ammonia produced is constantly tapped off. This continues 24 hours a day, seven days a week, until the plant stops for maintenance work. The Haber process is therefore called a **continuous process**.

Advantages of continuous processes

There are many advantages to using a continuous process.

- Very large amounts of product are made.
- Some industrial processes take a significant amount of time to get started. A continuous process keeps these 'start-up times' to a minimum. After the process has been started, it will be kept going without stopping for an extended period.
- The Haber process runs at 450 °C so the reactants require heating when the plant first starts up. However, the reaction that makes ammonia is exothermic – it gives out heat. Once the process has started, the heat given out by the reaction is used to heat the next lot of reactants. This reduces energy costs. This is only possible in a continuous process.

Disadvantages of continuous processes

- A continuous process can maintain a high level of production. If demand for the product suddenly changes, this becomes a disadvantage.
- It is difficult to reduce the amount being produced in a continuous process without stopping it altogether.
- It is also difficult to increase output if the plant is already operating near to its full capacity. However, the demand for ammonia is usually steady because it is used to make fertilisers.

SAQ

1 a What is meant by a 'continuous process'.
   b Give *three* reasons for making a product by a continuous process.

Batch processes

The other way of making a chemical product is called the **batch process** method. Drugs and medicines are made by batch processes.

- The right amounts of reactants are mixed and they are allowed to react and make products.
- When the reaction is complete the desired products are purified out. Each individual lot of products made is called a batch.
- After making each batch the equipment used will be thoroughly cleaned and then another batch can be made.

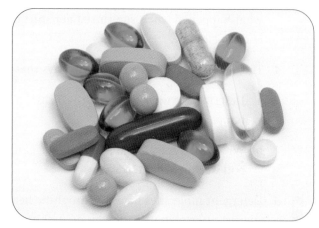

Figure 4f.1 These tablets are used to treat distressing conditions such as depression. They were made by batch processes. Medicines like these, and other well known medicines like aspirin and paracetamol, are also known as **pharmaceuticals**.

Advantages of batch processes

- It is easy to match supply to demand. If demand is high, batches can be made more often. If demand is low, batches can be made less often. It is much more difficult to vary the output of a continuous process if demand is variable.
- A batch process is a good way to make a small amount of product. Medicines are usually made in very much smaller amounts than chemicals such as ammonia.
- Each batch can be tested carefully to make sure it is of good quality. This is very important when manufacturing something people are going to swallow.

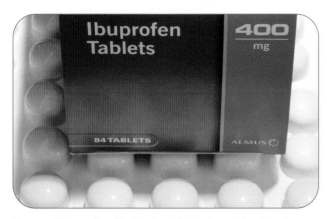

Figure 4f.2 People will take these ibuprofen tablets in order to get relief from pain. It is essential that the tablets are pure and of the highest possible quality.

Disadvantages of batch processes

- Batch processes are often less automated. This means more people are required to work on the process. This is known as being **labour intensive**. Such a process has high wage costs.
- If the reaction requires heating, each new batch will have to be heated. Any heat generated by an exothermic reaction in a previous batch cannot be reused. This increases energy costs.
- Batch processes are relatively small scale. A batch process can often be run in a laboratory. If it becomes necessary to make more of the product, it may be difficult to scale up the process.

SAQ

2 **a** Summarise the advantages and disadvantages of making a product by a batch process.

  b Why is a continuous process suited to making ammonia?

  c Why are most pharmaceuticals made by batch processes?

The cost of pharmaceuticals

Medicines are often very expensive. There are a number of reasons why they are so costly.

- It is necessary to make medicines by batch processes in order to maintain a high quality product. However, these batch processes are difficult to automate and result in high labour costs.
- Batch processes involve high energy costs.

- Every new medicine requires extensive research and testing. This is very expensive and no product can be sold during this period. During the period of developing and testing a new product, costs are high and income is low.
- The raw materials used to make medicines are expensive. They are often costly chemicals that have been made in laboratories – these are known as **synthetic** chemicals. Alternatively, they may be natural chemicals obtained from plants. (Green plants, not chemical plants.) Extracting and purifying such chemicals from plants is costly.

Figure 4f.3 The painkiller we call aspirin was originally extracted from the bark of willow trees. It is now made from synthetic chemicals.

- Marketing and advertising a new medicine is expensive.
- Medicines are swallowed by people. Because of this, the law insists that they are tested thoroughly to ensure that they are safe. Testing new drugs is costly. Costs like this are known as **legislative demands**.

SAQ

3 Summarise the reasons why medicines are expensive.

Extracting chemicals from plants

Useful chemicals can be extracted from plant material by a three-stage process. First, the plant material is crushed, or ground up with an abrasive material like sand. This breaks open the plant cells and releases the chemicals inside the cells.

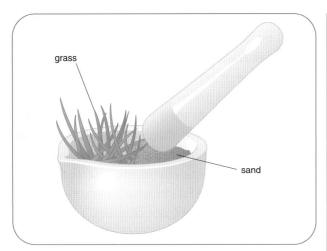

Figure 4f.4 To obtain chlorophyll from grass, the grass is first ground up with sand.

Second, the crushed or ground plant material is mixed with a suitable solvent. The desired chemical dissolves in the solvent.

This solution can be separated from undissolved substances by **filtration**.

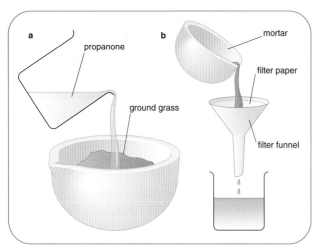

Figure 4f.5 a A solvent called propanone is added to the ground grass. The chlorophyll dissolves in the propanone. **b** The mixture is filtered to separate the solution from the sand and crushed grass.

The solution obtained may contain several dissolved substances, including the one that is wanted. This mixture is separated by **chromatography** so that the desired chemical is obtained in a pure state.

There are several different ways of using chromatography. In column chromatography, the solution is poured through a glass column filled with a packing material. More solvent is then continually poured through the tube. As this solvent drains down through the column, it carries the different dissolved solutes at different speeds. Because of this, the dissolved solutes come out of the bottom of the column at different times.

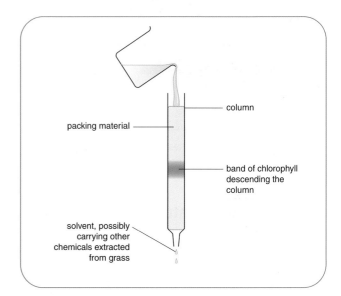

Figure 4f.6 The chlorophyll descends the column, carried by the solvent. When it reaches the bottom of the column, it can be collected in a pure state.

SAQ

4 Explain why each of the following is used in order to obtain a useful substance from plant material:

   a crushing   **b** solvent   **c** chromatography.

H New drugs

When a company manufactures a drug, they must be able to make a profit. When the drug is in the planning stage, many economic decisions must be made. The company estimates how much demand there will be for a new drug. They decide on a price for the drug. They can then estimate how much profit it will generate for them.

This profit must then be compared with the costs incurred by the manufacturer. Some of these costs are incurred in the early stages, before the drug goes on sale.

● Research and development costs are likely to be high with any new drug. The scientists who develop the drug must be highly skilled and may therefore be costly to employ. This part of the process can take a long time – success in research doesn't happen overnight.

Figure 4f.7 New drugs become available every year. The companies that manufacture them have invested huge sums of money by this stage.

- Once the drug has been developed, testing it will take several years. The drug will first of all be tested on tissue culture or on laboratory animals. If it appears safe at this stage, it will then be tested on human volunteers. Time must be allowed to make sure that there are no delayed side effects.
- The drug will cost money to manufacture owing to the costs of the workforce, the energy required and the raw materials.

The research and testing stages involve a large initial investment before the drug can be made. Once the drug is being sold at a profit, the money spent in these stages can be recovered. The time taken to do this is called the **pay-back time**. If the pay-back time is too long, the drug is unlikely to be profitable for the company, unless it proves to be a great success for many years.

SAQ

5 a Why might a drugs company decide on a very high price for a new drug?

   b What advantages are there if a company employs the very best research scientists?

   c Why must the company ensure that new drugs are tested thoroughly?

Quinine

Cinchona calisaya is the Latin name for a tree which is native to Central and South America. There is a story that says that, long ago, there was an earthquake in Peru and many cinchona trees were uprooted and tumbled into a lake. One of the local people was suffering from severe malaria at the time. The malaria gave him a raging thirst. During his illness, he drank a lot of water from the lake. The water quenched his thirst, and it also cured his malaria!

We cannot be sure of the truth of this story, but the people in Peru certainly discovered that something in the bark of the cinchona tree cures malaria. People began to chew the tree bark, or to brew it up in boiling water. They then drank the 'cinchona tea' they had made. These home-made medicines gave them relief from malaria.

We now know that cinchona bark contains a very powerful drug for combating malaria – quinine. Quinine was first isolated in 1820 by the French chemists J. B. Caventous and P. J. Pelletier. The pharmaceutical industry began to manufacture synthetic quinine in 1944.

As quinine became widely used, the parasite that causes malaria became resistant to it. This means the parasites were able to survive even if the patient took quinine. The pharmaceutical companies then developed and manufactured new drugs based on quinine. At first, the new drugs were very effective against malaria, but the parasites again developed resistance. The pharmaceutical companies have to develop new drugs regularly in order to combat this problem.

Figure 4f.8 A young *Cinchona calisaya* tree.

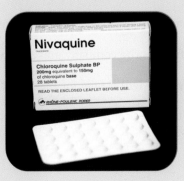

Figure 4f.9 Chloroquine was one of the first of the new quinine-based drugs to be developed, tested, and marketed successfully. Nivaquine is the brand name of these chloroquine tablets.

Summary

You should be able to:

- state that ammonia is made in a large scale, continuous process

- state that medicines are often made in small scale, batch processes

- state and describe the factors affecting the cost of making and developing a new medicine

- H evaluate the advantages and disadvantages of batch and continuous processes

- explain how economic considerations affect the development of new drugs

- know that the raw materials for making chemicals such as medicines may be either synthetic or extracted from plants

- describe how a chemical can be extracted from plant material

Questions

1 Summarise the advantages and disadvantages of batch and continuous processes.

2 a The Haber process is a 'continuous process'. What is its product?

   b Use the Haber process as an example to explain what 'continuous process' means.

   c Why would a sudden increase or decrease in world demand for fertiliser have consequences for a Haber process plant?

   d Why would it be difficult for the Haber process plant to respond to this change in demand?

3 Describe the steps you would use to extract and purify quinine from some pieces of bark from the *Cinchona calisaya* tree.

H 4 A new drug takes 3 years to develop and 8 years to test. Research and development costs are £20 000 000 per year, testing costs are £8 000 000 per year. When the drug is in use, the manufacturing costs total £30 000 000 per year. The money the company receives from selling the drug is £55 000 000 per year.

   a What is the cost of the 3 years research?

   b What is the cost of the 8 years testing?

   c What is the total cost of the 11 years of development?

   d What profit does the company make per year when the drug is in use?

   e How many years will it take for the profit to pay back the cost of developing the drug?

   f Explain why small companies rarely develop new drugs.

Carbon

Carbon is the sixth element in the Periodic Table. Each carbon atom therefore has six protons in its nucleus. Unlike most other elements, pure carbon can exist in three different forms. Different forms of the same element are called **allotropes**. The allotropes of carbon are called **diamond**, **graphite** and **buckminsterfullerene**.

Diamond has been known of since prehistoric times and of course it is used for jewellery.

Figure 4g.1 Diamonds!

Graphite will also have been familiar to our Stone Age ancestors – the soot from a fire is mostly graphite. So is the line produced by a pencil; the writing tip of a pencil is known as 'lead' but it's not made of metal.

Figure 4g.2 A pencil leaves a black line of graphite on paper.

Buckminsterfullerene is the newcomer on the scene. This form of carbon was first discovered in 1985 by a team from the University of Sussex. Diamond and graphite are made of large molecules consisting of many billions of atoms. Buckminsterfullerene is made of much smaller molecules that only measure billionths of a metre across. A billionth of a metre is called a **nanometre**.

The discovery of buckminsterfullerene has led to a whole new branch of chemistry – **nanochemistry**. In the past, chemists were used to working with materials on a large scale. The development of nanochemistry has enabled chemists to work on an atomic scale.

SAQ

1 **a** What are the three allotropes of carbon called?

b Which of these three allotropes was discovered most recently?

c Which one is used for writing?

d Which one is used to make engagement rings?

Diamond

Diamond is made of carbon atoms that are bonded together by strong covalent bonds in a strong three-dimensional structure.

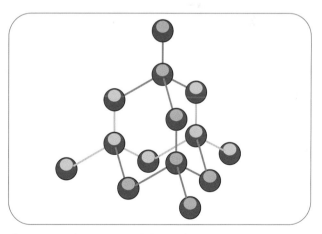

Figure 4g.3 The structure of diamond.

The properties and uses of diamond

- Diamond reflects light, so it sparkles. We say that it is **lustrous**. It is colourless and clear (transparent). Diamonds are used in jewellery because of these properties.
- Diamond does not dissolve in water.
- Diamond does not conduct electricity.

- Diamond has a very high melting point and is very hard. Because of these properties, diamond is used to make many tools. A drill can have diamonds on its tip and cutting tools like saws can be diamond-tipped too. This makes the tools expensive but very hard-wearing.

Figure 4g.4 Diamond-tipped grinding tools.

SAQ

2 a State *five* properties of diamond.

  b Explain why diamonds are used in jewellery.

  c State the advantage of a diamond–tipped drill.

Structure and properties

The properties of a substance can often be explained by its structure. The carbon atoms in diamond are bonded together covalently. A single diamond crystal may consist of over a billion billion atoms bonded together in this way. Diamond has a **giant covalent structure**.

In the diamond structure, all of the outer electrons on each carbon atom are used for covalent bonding. This means they are not free to move. They are not **delocalised**. This is the reason why diamond does not conduct electric current.

The covalent bonds in diamond are very strong. This makes diamond very hard and gives it a very high melting point.

SAQ

3 a Why are diamonds very hard?

  b Why does a diamond have a very high electrical resistance?

Graphite

Graphite is made of carbon atoms that are bonded together by strong covalent bonds into flat sheets.

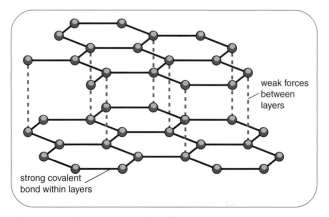

Figure 4g.5 The structure of graphite.

The properties and uses of graphite

- Graphite is very dark black and lustrous. You cannot see through it; it is opaque. These properties are important for its use in pencil leads.
- Graphite is insoluble in water.
- Graphite has a high melting point.
- Graphite can conduct electricity. This means it can be used to make electrodes for electrolysis.
- Graphite is slippery. This helps a pencil lead to slide smoothly over paper. It has also led to graphite being used to make lubricants.

SAQ

4 a State *five* properties of graphite.

  b Explain why graphite is used in pencil lead.

  c Explain why graphite is used in some engine oils.

Structure and properties

Like diamond, graphite has a **giant covalent structure**. However, the structures of the two allotropes are different, and this difference in structure explains their differing properties.

In the graphite structure, only three of the four outer electrons on each carbon atom are used for covalent bonding. The fourth outer electron on each carbon atom is free to move. These electrons are **delocalised**. This is the reason why graphite will conduct electric current.

H In the graphite structure, carbon atoms are held together in the layers by strong covalent bonds. This gives graphite a high melting point. However, the layers of carbon atoms are only held together by weak forces. Because of this, the layers of carbon atoms can slide over each other easily. This makes graphite slippery.

SAQ

5 **a** Why is graphite slippery?

   b Why can graphite conduct electricity?

Buckminsterfullerene

Buckminsterfullerene is made of molecules consisting of 60 carbon atoms. Its formula is therefore C_{60}. The carbon atoms are bonded together into a spherical shape. These tiny spheres are known as 'buckyballs'. Because they are so tiny, they are called **nanoparticles**.

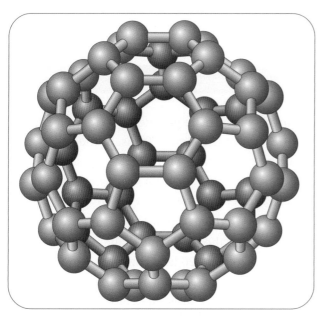

Figure 4g.6 The structure of buckminsterfullerene. Each red circle represents a carbon atom. There are 60 of them.

Buckminsterfullerene is a black solid. It dissolves in petrol to give a red or purple solution. This is a good example of how nanoparticles like buckminsterfullerene have different properties from the materials they are made from. Neither of the other forms of carbon can dissolve in a solvent like petrol.

SAQ

6 **a** What is the formula of buckminsterfullerene?

   b What colour is a solution of buckminsterfullerene in petrol?

It is possible to make similar molecules, called **fullerenes**, containing more than 60 atoms per molecule. These new nanoparticles are shaped like tiny tubes, so they are called **nanotubes**. Nanotubes are tubular structures made of carbon atoms, only a few nanometres in diameter.

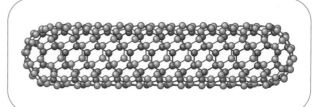

Figure 4g.7 A short nanotube. The red circles represent carbon atoms.

The properties and uses of nanotubes

- Nanotubes can conduct electricity and are very strong for their size. These properties have allowed them to be used as semiconductors in electric circuits.
- Nanotubes are strong, so they are used to reinforce graphite in tennis rackets. This enables the rackets to be both very strong and very light.
- Nanotubes have catalytic properties. Nanotubes are also used in the manufacture of other industrial catalysts. If catalyst particles are attached to nanotubes, this results in a very large surface area of catalyst. This increases the activity and effectiveness of the catalyst.

SAQ

7 **a** What are nanotubes?

   b Give *three* uses of nanotubes.

H Fullerene cages

The positive effect of a drug on the patient who takes it is known as its **therapeutic activity**. Many drugs owe their therapeutic activity to the exact shapes of their molecules.

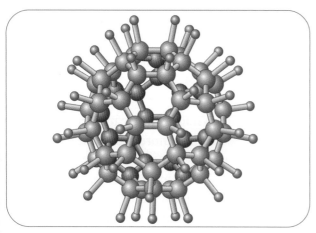

Figure 4g.8 The attachment points on the outside of a buckminsterfullerene molecule are shown here as blue dots.

It is possible to attach other groups to the outside of fullerene molecules. There are many different attachment points on the outside of the 'buckyball', so many possible products can be made. Many differently shaped product molecules can be produced. Some of the product molecules have the correct shapes to work as drugs.

The product molecules can be built up one atom at a time in order to achieve the correct shape.

This is called **molecular manufacturing**. Another molecular manufacturing technique involves starting off with a larger molecule and removing atoms from it. The removal is stopped when a molecule of the desired shape has been produced.

Placing metal atoms or ions inside each fullerene molecule can increase the therapeutic activity of some fullerene-based drugs. It is as if the metal is caged inside the buckyball! Some of the metals used are damaging to body tissues, but they are not released from their fullerene cages so no damage is done. The fullerene cage enables the manufacturer to gain the therapeutic advantage of the metal without any of the disadvantages.

SAQ

8   Describe *two* molecular manufacturing techniques that can produce a product molecule of a particular shape.

9   The shapes of these product molecules are called 'nanoscale features'. What do you think 'nanoscale' means?

Buckminsterfullerene

Richard Buckminster Fuller was one of the leading architects of the 20th century. One of his more radical designs was called the geodesic dome. A geodesic dome is based on a pattern of triangles that form 12 pentagons and 20 hexagons. This shape gives the maximum strength from the minimum amount of material. Geodesic domes have proved their strength by their ability to survive earthquakes and hurricanes unharmed.

In the 1970s, a group of scientists at Sussex University, led by David Walton and Harry Kroto, were interested in molecules called cyanoacetynes. These molecules featured long chains of carbon atoms and appeared to be made in large amounts by astronomical objects called carbon stars.

Kroto decided to use a very high intensity laser beam to reproduce the temperature in the carbon stars. He hoped this would make

cyanoacetynes. The experiments were performed in 1985 at Rice University, Texas. In the products of their experiments, Kroto's team discovered molecules with the formula C_{60}. A new form of the element carbon had been discovered!

The shape of the C_{60} molecule is called a 'truncated icosahedron'. This shape is composed of 12 pentagon and 20 hexagon rings – exactly the same as the shape on which Richard Buckminster Fuller based his geodesic dome. The newly discovered form of carbon was named 'buckminsterfullerene' in his honour.

Figure 4g.9 This greenhouse in Wales is based on Buckminster Fuller's design for a geodesic dome.

Summary

You should be able to:

- recall the three forms of carbon and recognise their structures
- **H** recall that these three forms of carbon are called *allotropes*
- describe the physical properties and uses of diamond and explain how these properties are linked to the uses of diamond
- **H** explain the physical properties of diamond in terms of its structure
- describe the physical properties and uses of graphite and explain how these properties are linked to the uses of graphite
- **H** explain the physical properties of graphite in terms of its structure
- describe the physical properties of buckminsterfullerene
- state the formula of buckminsterfullerene
- recall that fullerenes can be used to make nanotubes, and recall some properties and uses of nanotubes
- **H** describe the use of fullerenes in the drugs industry
- describe the use of nanotubes in improving the effectiveness of catalysts
- compare the scale of normal chemistry with the scale of nanochemistry
- recall that nanoparticles have different properties from the bulk material
- **H** describe two methods of molecular manufacturing

Questions

1 Sketch the structures of graphite, diamond and buckminsterfullerene.

2 Explain why:

a diamond is used to make drill tips but graphite isn't;

b graphite is used to make electrodes but diamond isn't;

c diamonds are used to make jewellery;

d pencil leads are made of graphite;

e some lubricants contain graphite.

3 Why are nanotubes used:

a in electric circuits; b in the chemical industry; c in the sporting industry?

4 **H** Use the structures of the materials involved to explain why:

a diamond and graphite both have very high melting points;

b graphite conducts electricity but diamond doesn't;

c diamond is very hard whereas graphite is soft;

d poisonous metals can be incorporated into drugs manufactured from buckminsterfullerene.

How pure is our water?

Water supplies

Life is impossible without water. Ever since humans first existed we have chosen to live near water. We drink it, cook food in it and wash with it. People in the developed world get treated water, which is safe to drink, piped into their homes. Many people around the world do not have access to safe drinking water. They use untreated water from lakes and rivers, which may be contaminated by humans or animals. Water supplies like this can carry diseases like cholera and typhoid.

Figure 4h.1 a This young girl is collecting water in Moyale, Northern Kenya. The water is polluted by livestock that drink here.

Figure 4h.1 b Collecting clean water. This water project was funded by donations to an international aid agency called World Vision.

The UK is a small country with a large population density. All of its inhabitants need water. As well as being used in the home, very large amounts of water are used by our industries.

- Water can be a cheap raw material. It is used to make sulfuric acid.

  Sulfur trioxide + water → sulfuric acid

- Water is used to cool machinery and prevent it overheating.
- Water is used as a solvent. Many medicines are made by a neutralisation reaction between an acid and an alkali. The acid and the alkali that are used are both dissolved in water first – water acts as a solvent in this reaction.

Water supplies are more important than ever. They must be created, maintained and managed.

Lakes and rivers

Early human settlements were established near lakes, rivers and streams. These natural sources of water are still important worldwide.

Figure 4h.2 The Thames at London. This river provided the city with docks for London's trade links and water for its people.

Aquifers

An **aquifer** is an underground water source. Rain water that has soaked through the soil and through the rocks near the surface collects deep underground. The water in an aquifer is usually found soaked into porous rocks, not as an underground lake. The water is obtained by digging wells or bore holes and then pumping it out. The water in aquifers may be naturally good to drink, containing useful dissolved minerals and very few bacteria.

Reservoirs

Lakes, rivers and aquifers are natural water sources. The UK population in the 21st century is so high that we now rely on an artificial water source too – **reservoirs**. If a dam is built across a river valley, an artificial lake will form. This artificial lake is called a reservoir.

Figure 4h.3 This dam has caused the reservoir on the left.

Water conservation

In many countries around the world the reservoirs and natural sources of water can only just meet the demand for water. If there is a drought – a prolonged period of low rain – water can become in short supply. There is a need for water to be conserved. This means saving water by only using what is necessary.

Some water conservation methods are voluntary. A householder can choose to place a solid object inside their toilet cistern. This results in less water being used with each flush – water is conserved. Other water conservation methods are compulsory. During years of low rainfall, the water companies may ban the use of hosepipes. This stops people using hoses to water their lawns or wash their cars.

SAQ

1 a Describe *four* sources of water that are used for water supplies.

b Which of these four sources is artificial?

c Describe *two* ways of conserving water.

Dissolved substances

The water that comes out of our taps should be fit to drink, but it is not pure. It contains dissolved substances. Some of these are harmless but some are pollutants from several sources.

● Ammonium nitrate fertiliser dissolves in water. The dissolved ammonium nitrate then gets into lakes, rivers, aquifers and reservoirs, and therefore into water supplies.

● Some old houses still have lead water pipes. Lead compounds form in these pipes and can then contaminate the water.

● Farmers sometimes spray their crops with **pesticides** in order to control pests. These chemicals can be carried by rainwater into rivers and eventually into water supplies.

Figure 4h.4 This lead water pipe may have been responsible for polluting domestic water supplies.

SAQ

2 Describe how lead compounds and pesticides can pollute water supplies.

Water treatment

Water companies try to remove these pollutants from domestic water supplies along with other contaminants.

● Insoluble material makes the water look cloudy and unattractive. It is removed by **filtration**. The filtering is usually done using containers of sand or fine gravel, called filter beds.

● Dissolved substances, such as salts, minerals and ammonium nitrate, are more difficult to remove. Some can be removed by **sedimentation**. This is

Figure 4h.5 These filter beds are used to filter domestic water supplies.

done by adding a solution to the water. The solution reacts with the dissolved substances to form an insoluble sediment. This sediment can then be removed by a second filtration.

• Micro-organisms are removed from the water by **chlorination**. The chlorine kills the microbes. This is a good example of how water treatment makes the water less pure – something is added to it. The chlorination makes the water safer to drink.

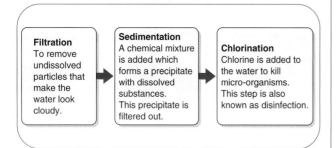

| **Filtration**
To remove undissolved particles that make the water look cloudy. | **Sedimentation**
A chemical mixture is added which forms a precipitate with dissolved substances. This precipitate is filtered out. | **Chlorination**
Chlorine is added to the water to kill micro-organisms. This step is also known as disinfection. |

Figure 4h.6 Flow diagram of water treatment.

The problem with ammonium nitrate

Some dissolved substances, including ammonium nitrate, cannot be removed by sedimentation. This is because they do not react to form an insoluble sediment. Domestic water supplies are often contaminated with ammonium nitrate. High levels of ammonium nitrate are toxic, especially to young babies. For that reason ammonium nitrate levels in drinking water are monitored very closely.

SAQ

3 a Describe how ammonium nitrate can pollute water supplies.

b Why is ammonium nitrate difficult to remove from water supplies?

H Drinking water from the sea

Two-thirds of the world is covered in water, but unfortunately the sea is too salty to drink. Sea water can be turned into fresh water by **distillation**. If sea water is distilled, it leaves the salt and other contaminants behind. The water has to be boiled in order to distil it. This requires a lot of energy and therefore this process is too expensive for most countries.

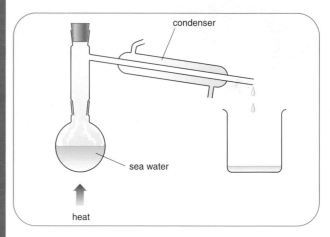

Figure 4h.7 a The sea water is boiled and the steam rises into the condenser, where it is cooled and forms pure water.

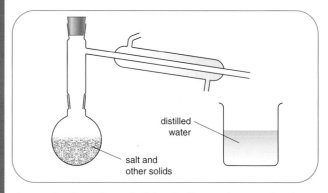

Figure 4h.7 b At the end of the distillation, the water has been separated from all of the dissolved solids.

Domestic water supplies have been obtained by distilling sea water in some middle-eastern countries. These countries are mostly desert and have very little rain. They have a great need for an alternative supply of fresh water. These countries have large oil supplies, so energy is comparatively cheap.

SAQ

4 Explain why a middle-eastern country may use distillation of sea water as a source of drinking water but the UK is unlikely to.

Testing water

There are some simple tests that can be done to identify dissolved substances in water.

Using barium chloride solution

Figure 4h.8 a The white precipitate shows that this water sample contained dissolved sulfate ions.

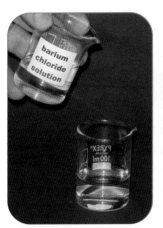

Figure 4h.8 b A different water sample shows no white precipitate. This water sample did not contain dissolved sulfate ions.

When barium chloride solution is added to water containing dissolved sulfate ions, an immediate white **precipitate** is seen. When it is added to water containing no dissolved sulfate ions, there will be no visible change.

This is an example of a **precipitation reaction** because the product is called a precipitate. The precipitation reaction involves the barium ions from the barium chloride solution and the sulfate ions from the water. The precipitate formed is called barium sulfate.

A word equation for this reaction is:

barium ions + sulfate ions → barium sulfate

Using silver nitrate solution

When silver nitrate solution is added to water containing dissolved chloride ions, an immediate white precipitate is seen. When it is added to water containing dissolved bromide ions, an immediate cream precipitate is seen. When it is added to water containing dissolved iodide ions, an immediate yellow precipitate is seen. If none of these ions is present, there will be no visible change. Chloride ions, bromide ions and iodide ions are known as **halide ions**.

The precipitation reaction in Figure 4h.9 involves the silver ions from the silver nitrate solution and the chloride ions from the water. The precipitate formed is called silver chloride.

A word equation for this reaction is:

silver ions + chloride ions → silver chloride

The precipitation reaction in Figure 4h.10 involves the silver ions from the silver nitrate solution and the bromide ions from the water. The precipitate formed is called silver bromide.

Figure 4h.9 The white precipitate shows that this water sample contained dissolved chloride ions.

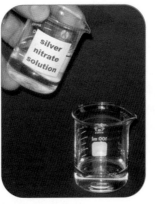

Figure 4h.10 The cream precipitate shows that this water sample contained dissolved bromide ions.

5 Write a word equation for the reaction in Figure 4h.10.

The precipitation reaction in Figure 4h.11 involves the silver ions from the silver nitrate solution and the iodide ions from the water. The precipitate formed is called silver iodide.

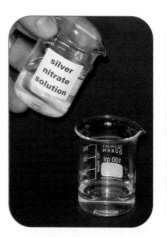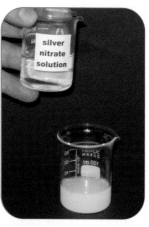

Figure 4h.11 The yellow precipitate shows that this water sample contained dissolved iodide ions.

6 Write a word equation for the reaction in Figure 4h.11.

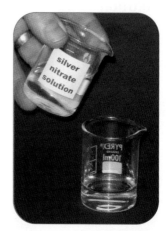

Figure 4h.12 This water sample did not contain dissolved chloride, bromide or iodide ions.

Rutland Water

Rutland used to be the smallest county in England. Then it got even smaller – in land area anyway. The decision was taken to build a dam across the River Gwash in the 1970s and flood half of the county. The result was called Rutland Water – a reservoir large enough to guarantee a water supply to a large area of the English Midlands.

Many people had to leave their homes. Whole villages disappeared under the water. The water company paid compensation to those who lost their homes and land. There was great sadness at the loss of some typical English countryside.

Thirty years later it is clear that a new and rich environment has replaced it. Rutland Water does its job as a supplier of water to homes and industry, but it is also home to up to 20 000 waterbirds. Over 250 different species of bird have been recorded at Rutland Water. It has been designated an SSSI – a site of special scientific interest.

Recently, Rutland Water has been the centre of a project to reintroduce the osprey to this part of England, where it had been extinct. Several of these magnificent birds now nest at the reservoir. Eight chicks were raised between 2001 and 2005.

As well as attracting birdwatchers, Rutland Water attracts anglers, sailors, walkers and cyclists. It is now a major recreational area for the English East Midlands. The county of Rutland is not the same environment as it was in 1970, but it is still a very rich environment and its water supplies do a very important job.

Figure 4h.13 An osprey flies above Rutland Water.

Ionic equations

The ionic equation for the reaction between barium ions and sulfate ions is:

$$Ba^{2+} + SO_4^{2-} \rightarrow BaSO_4$$

Because the charge on the barium ion is 2+ and the charge on the sulfate ion is 2−, there is one barium ion and one sulfate ion in each formula unit of barium sulfate.

The ionic equation for the reaction between silver ions and chloride ions is:

$$Ag^+ + Cl^- \rightarrow AgCl$$

Because the charge on the silver ion is 1+ and the charge on the chloride ion is 1−, there is one silver ion and one chloride ion in each formula unit of silver chloride.

These equations only include the ions that are involved in the precipitation reactions. They do not include the nitrate ions that are part of the silver nitrate, for example, because these ions do not take part in the reaction. The nitrate ions are known as 'spectator ions' in this reaction.

SAQ

7 The formula of a bromide ion is Br^-. The formula of an iodide ion is I^-.

a Write an ionic equation for the reaction between silver ions and bromide ions.

b Write an ionic equation for the reaction between silver ions and iodide ions.

Summary

You should be able to:

◆ describe four different types of water resource

◆ explain the importance of clean water to developing nations

◆ explain the importance of water in industry

◆ explain why it is important to conserve water

◆ state three of the pollutants that may be found in domestic water supplies and recall their origins

◆ state the types of substance present in water before it is treated

◆ describe and explain how these substances are removed by water treatment plants

◆ explain why some soluble substances cannot be removed and recall that these may be poisonous

◆ explain the disadvantages of using distilled seawater as a source of fresh water

◆ explain how barium chloride and silver nitrate can be used to test for the presence of certain ions in water

◆ recall that these reactions are examples of precipitation reactions

◆ write word equations for these precipitation reactions

◆ write ionic equations for these precipitation reactions

Questions

1 Explain *three* ways in which water can be important to an industrial process.

2 What would you deduce about a water supply if:

   a it formed a white precipitate when barium chloride solution was added to it?

   b it formed a yellow precipitate when silver nitrate solution was added to it?

   c it formed a white precipitate when barium chloride solution was added to it and it formed a white precipitate when silver nitrate solution was added to it?

   d it did not form a precipitate when barium chloride solution was added to it, or when silver nitrate solution was added to it?

3 Explain the importance of the following processes to water treatment.

   a Filtration

   b Chlorination

   c Sedimentation

4 **a** Describe what happens in the distillation flask and in the condenser when sea water is distilled.

   b Explain why distilling sea water removes the salt.

   c Why do you think an industrial plant that distils sea water is called a 'desalination plant'?

5 Concentrated sulfuric acid is made from sulfur trioxide and water in two steps.

- In step 1, sulfur trioxide (SO_3) reacts with concentrated sulfuric acid (H_2SO_4) to form 'oleum' ($H_2S_2O_7$).

- In step 2, oleum reacts with water to form concentrated sulfuric acid.

   a Write a balanced symbol equation for step 1.

   b Write a balanced symbol equation for step 2.

   c Find out why concentrated sulfuric acid is made in two steps rather than one.

Describing motion

In this Item, we look at ideas of motion and speed. In later Items, we will look at how physicists came to understand the forces involved in motion and how to control them to make our everyday travel possible. We will also look at how, if we understand the energy changes involved in motion, we can make travel safer and more efficient.

Measuring speed

If you drive on a major highway or through a large city, the chances are that someone is watching you. Cameras on the verge and on overhead gantries keep an eye on traffic as it moves along. Some cameras are there to monitor the flow, so that traffic managers can take action

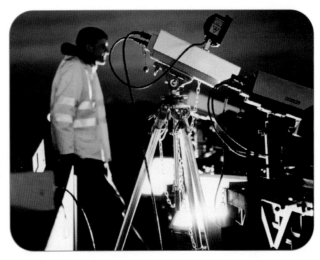

Figure 3a.1 Traffic engineers use sophisticated cameras and computers to monitor traffic. Understanding how drivers behave is important for safety, and also to improve the flow of traffic.

On the move

Today, people are on the move (Figure 3a.2). You may have to travel several kilometres to get to school each day. Members of your family may live in a different part of the country, so that you have to travel hundreds of kilometres to visit them. Holidays may involve flying to another country or even to another continent. You might choose to become a student at a university a long way from home.

For some people, movement is a more desperate business. They may be refugees fleeing from a war zone. They may be economic migrants travelling great distances to find work and a decent standard of living.

Life in today's industrialised countries is very different from how it was two centuries ago. If you had lived then, your life would probably have been tied to farming and the land. You would have lived in a village and married someone from the next village. A journey to the nearest town would have been a rare event. News travelled no faster than a galloping horse. Then came the Industrial Revolution, with work concentrated in the new cities. The railways

made it easier to travel longer distances; cars and motorways followed, and then today's fast and reliable air transport.

Figure 3a.2 We live in a world of contrasts. For some people, movement is limited to a few kilometres each day, on foot or bicycle. For many people from the industrialised world, travelling hundreds of kilometres in a day is commonplace, thanks to cars, trains and planes.

when blockages develop or when accidents occur. Others are equipped with sensors to spot speeding motorists or those who break the law at traffic lights (Figure 3a.1). In the busiest places, traffic police may observe the roads from helicopters circling above.

Drivers should know how fast they are moving – they have a speedometer to tell them their speed at any instant in time. Traffic police can use a radar speed 'gun' to give them an instant readout of another vehicle's speed (such 'guns' use the Doppler effect to measure a car's speed). Alternatively, they may time a car between two fixed points on the road. Knowing the distance between the two points, they can calculate the car's speed.

Distance, time, speed

As we have seen, there is more than one way to determine the speed of a moving car or aircraft. Several methods rely on making two measurements:

- the **distance travelled** between two points;
- the **time taken** to travel between these two points.

Then we can work out the **average speed** between the two points:

$$\text{average speed} = \frac{\text{distance travelled}}{\text{time taken}}$$

or

$$\text{speed} = \frac{\text{distance}}{\text{time}}$$

Notice that this formula tells us the vehicle's *average* speed. We cannot say whether it was travelling at a steady speed, nor whether its speed was changing. For example, you could use a stopwatch to time a friend cycling over a fixed distance – say, 100 m (see Figure 3a.3). Dividing distance by time would tell you their average

Figure 3a.3 Timing a cyclist over a fixed distance. Using a stopwatch involves making judgements as to when they pass the starting and finishing lines, and this can introduce an error into the measurements. An automatic timing system may be better.

speed, but they might have been speeding up or slowing down along the way.

Table 3a.1 shows the different units that may be used in calculations of speed. SI units are the 'standard' units used in Physics; SI stands for *Système Internationale*, or International System. In practice, many other units are used. In US space programmes, heights above the Earth are often given in feet, whereas the spacecraft's speed is given in knots (nautical miles per hour). These awkward units didn't prevent them from reaching the Moon!

SAQ

1 If you measured the distance travelled by a snail in inches and the time it took in minutes, what would be the units of its speed?

2 Which of the following could *not* be a unit of speed? km/h, s/m, mph, m/s, ms.

Quantity	SI unit	Other units	
distance	metre (m)	kilometre (km)	miles
time	second (s)	hour (h)	hour (h)
speed	metres per second (m/s or m s^{-1})	kilometres per hour (km/h)	miles per hour (mph)

Table 3a.1 Quantities, symbols and units in measurements of speed.

3 Table 3a.2 shows information about three cars travelling on a motorway. Which is moving fastest? Which is moving slowest?

Vehicle	Distance travelled in km	Time taken in minutes
Car A	80	50
Car B	72	50
Car C	85	50

Table 3a.2

Worked example 1

A cyclist completed a 1500 m stage of a race in 37.5 s. What was her average speed?

Step 1: start by writing down what you know, and what you want to know.

Distance = 1500 m

Time = 37.5 s

Speed = ?

Step 2: now write down the formula.

Speed = distance / time

Step 3: substitute the values of the quantities on the right-hand side.

Speed = 1500 m / 37.5 s

Step 4: calculate the answer.

Speed = 40 m/s

So the cyclist's average speed was 40 m/s.

SAQ

4 An aircraft travels 1000 m in 4.0 s. What is its speed?

5 A car travels 150 km in 2 hours. What is its speed? (Show the correct units.)

Rearranging the equation

The formula 'speed = distance / time' allows us to calculate speed from measurements of distance and time. We can rearrange the formula to allow us to calculate distance or time. For example, a railway signalman might know how fast a train is moving and need to be able to predict where it will have reached after a certain length of time:

distance = speed × time

Similarly, the crew of an aircraft might want to know how long it will take for their aircraft to travel between two points on its flight path:

$$\text{time} = \frac{\text{distance}}{\text{speed}}$$

A note on learning equations

What is the best way to learn the relationship between speed, distance and time? The most important thing is to understand what the equation is telling you. In the form 'speed = distance / time', it says that speed is a distance divided by a time. You may find it easier to think in terms of units: speed is measured in metres per second, so we need to divide a distance by a time. Remembering a relationship like this will remind you of the underlying physics.

We have seen that there are three forms of the equation; it is probably best to learn one form and be able to rearrange it.

Finally, you may find it helpful to learn the 'formula triangle' that represents this equation (Figure 3a.4). The problem with this is that, in learning the triangle, you are ignoring the important physical relationship between speed, distance and time – so only resort to the formula triangle if you have serious difficulties in rearranging equations.

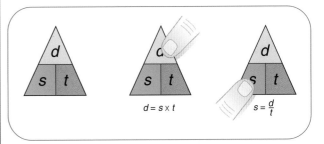

Figure 3a.4 The 'formula triangle' method for rearranging a simple equation. To find the formula for one of the quantities, cover it up; then you can see whether to multiply or divide the other two quantities. Key: *d* = distance, *s* = speed, *t* = time.

Worked example 2

A spacecraft is orbiting the Earth at a steady speed of 8 km/s (Figure 3a.5). How long will it take to complete one orbit, a distance of 40 000 km?

Figure 3a.5 A spacecraft orbiting Earth.

Step 1: start by writing down what you know, and what you want to know.

Speed = 8 km/s
Distance = 40 000 km
Time = ?

Step 2: now choose the appropriate equation, with the unknown quantity as the subject (on the left-hand side).

$$Time = \frac{distance}{speed}$$

Step 3: substitute values; it can help to include units.

$$Time = \frac{40\,000\,km}{8\,km/s}$$

Step 4: perform the calculation.

Time = 5000 s

This is about 83 minutes. So the spacecraft takes 83 minutes to orbit the Earth once.
(This example illustrates the importance of keeping an eye on units. Because speed is in km/s and distance in km, we do not need to convert to m/s and m. We would get the same answer if we did the conversion:

$$Time = \frac{40\,000\,000\,km}{8\,00\,km/s} = 5000\,s)$$

SAQ

6 An interplanetary spacecraft is moving at 20 000 m/s. How far will it travel in one day? (Give your answer in km.)

7 How long will it take a coach travelling at 90 km/h to travel 300 km along a motorway?

Measuring speed

You have probably made measurements of speed in science lessons in the past. You need to choose suitable instruments for these measurements.
● Distance: use a tape measure or a trundle wheel.
● Time: use a stopclock or stopwatch.
    You may have seen surveyors measuring roads or railway tracks using a trundle wheel (Figure 3a.6). Accurate measurements are vital if, for example, the correct amount of tarmac is to be supplied for road repairs.

Figure 3a.6 A surveyor working with a trundle wheel.

Measuring speed – in the laboratory

Here's a method similar to a traffic camera which can be used in a school laboratory. A video camera is linked to a computer. The camera makes a succession of images at equal intervals of time (Figure 3a.9). These are stored by the computer and can be measured to find the position of the object as it moves.

In the street

Flash! A traffic camera records a speeding motorist. The photo is needed to identify the vehicle (from its number plate). It also shows the road markings. These include marks at equal intervals, for checking distances on the photograph. There may also be letters and numbers identifying the individual camera.

A speed camera (Figure 3a.7) uses one of three different methods to determine a vehicle's speed.

- It may take two photographs with a short interval of time between them. The distance travelled in this time can be measured from the photograph, using the markings on the road.
- There may be two sensor strips buried in the road. These are activated as the vehicle passes over them. An electronic circuit records the time taken between the two strips and, knowing their separation, it calculates the vehicle's speed.
- The camera may have a radar gun; this reflects radio waves off the back of the vehicle.

Traffic camera photographs can be used as evidence in court. They may be challenged by defence lawyers, so the police need to be sure that they are functioning correctly. Inside the yellow box are complex circuits to ensure the reliability of the camera (Figure 3a.8).

Figure 3a.7 A speeding van passes a traffic camera. It is about to reach the section of the road which carries the markings needed to check that the camera is functioning correctly.

Figure 3a.8 What's in the box? As well as the camera (top right), this traffic camera has an electronic flash gun, a radar speed gun and circuits to record data and to check that the camera is functioning reliably. A police officer is preparing to change the film.

Another method uses two **light gates** (Figure 3a.10). In a light gate, a beam of (invisible) infra-red radiation is received by a detector. When the moving trolley breaks the beam of the first gate, the electronic timer is started. When the second beam is broken, the timer stops. This tells you the time the trolley has taken to travel between the two gates; if you measure the distance between the two, you can then calculate the trolley's average speed between them.

An alternative method uses a single light gate and an **interrupt card**, fixed to the trolley. The front edge of the card breaks the beam and starts

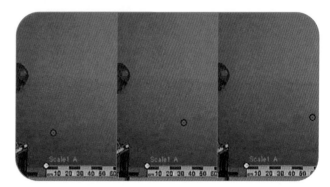

Figure 3a.9 In the laboratory, a video camera can record a series of images of a moving object. The images are made at regular intervals of time, and you can vary the time interval to suit the speed of the object. It helps if the object is moving past a scale so that its position can be read directly from the images.

the timer. When the trailing edge passes the gate, the beam is unbroken and the timer stops. Given the length of the interrupt card, you can calculate the trolley's average speed. (If the timer is connected to a computer, it may do the calculation for you.)

Infra-red gates like this are used to check the positions and speeds of cars on roller-coaster rides at fairgrounds. It is important that the ride's computer system knows where each car is, so that it can take action to avoid a collision if a fault occurs. You may also have noticed an infra-red gate that controls the conveyor belt on a supermarket checkout. When an item reaches the end of the belt, the beam is broken and the belt stops automatically.

SAQ

8 A trolley takes 0.25 s to travel between two light gates, separated by 0.40 m. What is its speed?

Distance–time graphs

You can describe how something moves in words:

'The coach pulled away from the bus stop. It travelled at a steady speed along the main road, heading out of town. After five minutes, it reached the motorway, where it was able to speed up. After ten minutes, it was forced to stop because of congestion.'

We can show the same information in the form of a distance–time graph, as shown in Figure 3a.11. The graph is in three sections, corresponding to the three sections of the coach's journey.

Section A The graph slopes up gently, showing that the coach was travelling at a low speed.

Section B The graph becomes steeper. The distance of the coach from its starting point is increasing more rapidly; it is moving faster.

Section C The graph is flat (horizontal). The distance of the coach from its starting point is not changing; it is stationary.

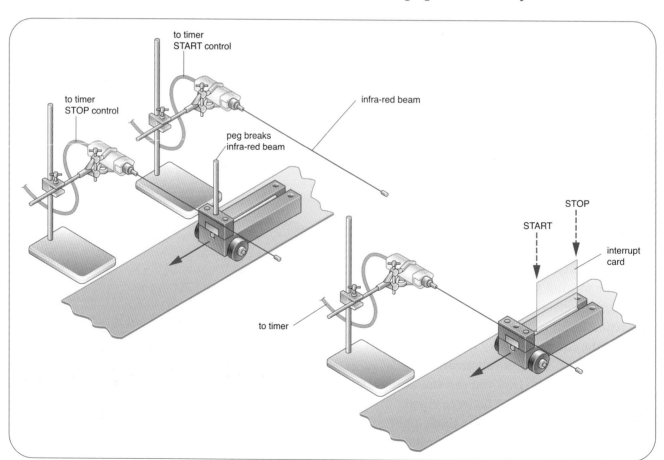

Figure 3a.10 Two ways to use light gates to measure the speed of a moving trolley.

Figure 3a.11 A graph to represent the motion of a coach, as described in the text. The slope of the graph tells us about the coach's speed. The steepest section (B) corresponds to the greatest speed. The horizontal section (C) shows that the coach was stationary.

The slope of the distance–time graph tells us how fast the coach was moving. The steeper the graph, the faster it was moving (the greater its speed). When the graph becomes horizontal, its slope is zero. This tells us that the coach's speed was zero; it was not moving.

SAQ

9 Sketch a distance–time graph to show this:

'The car travelled along the road at a steady speed. It stopped suddenly for a few seconds. Then it continued its journey, at a slower speed than before.'

Distance–time data

The Settle to Carlisle railway follows a very picturesque route across the hills of northern England. It is very popular with tourists, as well as providing a service to local communities. Figure 3a.12 shows a train crossing a viaduct on the line, together with part of the timetable. This suggests another way to represent the motion of a moving object: in the form of a table.

Table 3a.3 shows how far a train has travelled at different points along its journey, together with the time taken. The same information is plotted as a graph in Figure 3a.13. The graph does not always have the same slope; this shows that the train's speed was changing as it went along. It was moving fastest where the graph is steepest.

Station										
Settle	d	1345	-	-	1544	-	-	1853	1853	2023	-
Horton in Ribb.	d	1354	-	-	1552	-	-	1902	1902	2031	-
Ribblehead	d	1402	-	-	1600	-	-	1910	1910	2039a	-
Dent	d	1411	-	-	1610	-	-	1919	1919	-	-
Garsdale	d	1417	-	-	1615	-	-	1925	1925	-	-
Kirby Stephen	d	1429	-	-	1628	-	-	1937	1937	-	-
Appleby	d	1443	-	-	1640	-	-	1950	1950	-	-
Langwathby	d	1457	-	-	1654	-	-	2004	2004	-	-
Lazonby & Kirk.	d	1503	-	-	1700	-	-	2010	2010	-	-
Armathwaite	d	1510	-	-	1708	-	-	2017	2017	-	-
Carlisle	a	1526	1734	1731	1726	1957	1957	2032	2032	-	-
Motherwell	a	1712	1913	1913	1913	2115	2101	2158	2145	-	-
Glasgow Central	a	1729	1935	1935	1935	2132	2119	2216	2205	-	-

Figure 3a.12 a A train crossing the Ribblehead viaduct in the north of England. **b** Part of the timetable for this line. Information from the timetable allows us to draw a distance–time graph like the one shown in Figure 3a.13.

Station	Distance travelled in km	Time taken in minutes
Settle	0	0
Horton	10	8
Ribblehead	18	16
Dent	28.5	26
Garsdale	34	31
Kirkby Stephen	47	43
Appleby	60	56

Table 3a.3 This table represents the motion of a train; the data was used to plot the graph shown in Figure 3a.13.

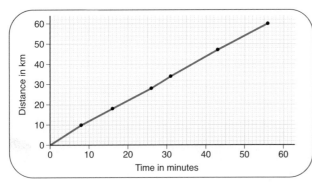

Figure 3a.13 A distance–time graph for a train, based on the data shown in Table 3a.3. From the slope of the graph we can deduce the train's speed.

SAQ

10 From the graph in Figure 3a.13, find when the train was travelling fastest.

H Calculating speed

Figure 3a.14 represents the motion of a runner in a 400 m race. You can see that, for most of the race, the graph is a straight line. This shows that the runner was moving at a steady speed. The graph also shows how to calculate the runner's speed. Here, we are looking at the straight section of the graph, where the runner's speed was constant. We need to find the value of the slope (or gradient) of the graph, which will tell us the speed:

speed = slope of distance–time graph

Worked example 3

Step 1: identify a straight section of the graph.

Step 2: draw horizontal and vertical lines to complete a right-angled triangle.

Step 3: calculate the lengths of the sides of the triangle.

Step 4: divide the vertical height by the horizontal width of the triangle ('up divided by along').

Here is the calculation for the triangle shown in Figure 3a.14:

vertical height = 200 m
horizontal width = 25 s

$$slope = \frac{200\,m}{25\,s}$$

H Worked example 3 - *continued*

So the runner's speed was 8 m/s for most of the race. It is important to include units in this calculation. Then the answer will automatically have the correct units – in this case, m/s. If we had used the data for the train (Figure 3a.13), where distances are in km and times in minutes, the speed would have been in km/minute.

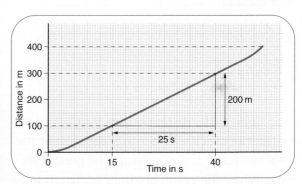

Figure 3a.14 This graph represents the motion of a runner in a race. The short curved section at the beginning represents the runner setting off; for most of the time, the graph is a straight line, indicating steady speed. At the end, the graph becomes a little steeper, indicating that the runner put on a final spurt in a dash for the tape.

SAQ

11 Table 3a.4 shows information about a train journey. Use the data in the table to plot a distance–time graph for the train. Find the train's average speed between Beeston and Deeville.

Station	Distance travelled in km	Time taken in minutes
Ayton	0	0
Beeston	20	30
Seatown	28	45
Deeville	36	60
Eton	44	70

Table 3a.4

Graphs of different shapes

We have seen the following points about distance–time graphs:
● the gradient of the graph tells us the object's speed;
● the steeper the gradient, the greater the speed;

H • a horizontal graph (gradient = 0) indicates that the object is not moving.

Figure 3a.15 shows some further points:

• where the graph is curved, the object's speed is changing – here it is **accelerating** (see Item 3b);

• where the graph slopes downwards (negative gradient), the object is moving in the opposite direction.

For a negative slope, we can still work out the speed from the slope, but we have to remember to include a minus sign.

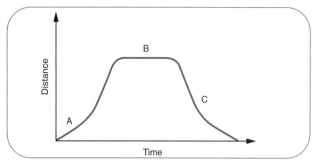

Figure 3a.15 Distance–time graphs are not always straight lines. This graph is for a car moving along a road. The curved section **A** shows it speeding up; section **B** shows where it has stopped; section **C** has a negative slope, showing that the car is returning towards its starting position.

Summary

You should be able to:

♦ state the quantities which must be measured to determine an object's speed

♦ relate this to the operation of a speed camera

♦ use and interpret the formula: $\text{speed} = \dfrac{\text{distance}}{\text{time}}$

H ♦ use the equation to calculate distance and time

♦ draw and interpret distance–time graphs

H ♦ deduce the speed of an object from a distance–time graph

♦ recognise that a curved distance–time graph shows changing speed (acceleration)

Questions

1 A runner travels 400 m in 50 s. What is her average speed?

2 For a Geography project, you have to measure the speed of flow of a river. You have a supply of plastic ducks which will float in the river. Describe how you would find the speed of the river.

3 Table 3a.5 shows the distance travelled by a car at intervals during a short journey. Draw a distance–time graph to represent this data. What does the shape of the graph tell you about the car's speed?

Distance in m	0	200	400	600	800
Time in s	0	10	20	30	40

Table 3a.5

continued on next page

Questions - *continued*

4 The graph in Figure 3a.16 represents the motion of a bus. It is in two sections, A and B. What can you say about the motion of the bus during these two sections?

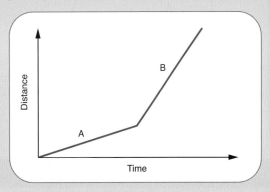

Figure 3a.16

Figure 3a.17 Distance–time graph for a roller-coaster car.

5 The graph in Figure 3a.17 shows the distance travelled by a car on a rollercoaster ride at different times along its trip. It travels along the track and then returns to its starting position. Study the graph and decide which point best fits the following descriptions. In each case, give a reason to explain why you have chosen that point.

a The car is stationary.

b The car is travelling its fastest.

c The car is speeding up.

d The car is slowing down.

e The car starts on its return journey.

6 How far will a bus travel in 30 s at a speed of 15 m/s?

7 Scientists have measured the distance between the Earth and the Moon by reflecting a beam of laser light off the Moon. They measure the time taken for light to travel to the Moon and back. What other piece of information is needed to calculate the Earth–Moon distance? How would the distance be calculated?

8 Table 3a.6 shows information about the motion of a number of objects. Copy and complete the table.

Object	Distance travelled	Time taken	Speed
bus	20 km	0.8 h	
taxi	6 km		30 m/s
aircraft		5.5 h	900 km/h
snail	3 mm	10 s	

Table 3a.6

9 A car is moving along the road. Its distance–time graph gradually becomes steeper. Is the car speeding up, slowing down or travelling at a steady speed?

10 Sketch a distance–time graph which represents the following motion: a trolley moves at a steady speed for a short time; then it moves the same distance, speeding up.

Understanding acceleration

Some cars, particularly high-performance ones, are advertised according to how rapidly they can accelerate. An advert may claim that a car goes 'from 0 to 60 miles per hour (mph) in 6 s'. This means that, if the car accelerates at a steady rate, it reaches 10 mph after 1 s, 20 mph after 2 s, and so on. We could say that it speeds up by 10 mph every second; in other words, its **acceleration** is 10 mph per second.

So we say that an object *accelerates* if its speed increases; its *acceleration* tells us the rate at which its speed is changing; in other words, the change in speed per unit time.

If an object slows down, its speed is changing; this is sometimes described as deceleration.

SAQ

1 Give one word that means the same as 'slowing down'.

Express trains, slow buses

An express train is capable of reaching high speeds – perhaps in excess of 300 km/h. However, when it sets off on its journey, it may take several minutes to reach this top speed. Then it takes a long time to slow down when it approaches its destination. The famous French TGV trains (Figure 3b.1) run on lines which are reserved solely for their operation, so that their high-speed journeys are not disrupted by slower, local trains. It takes time to accelerate (speed up) and decelerate (slow down).

A bus journey is full of accelerations and decelerations (Figure 3b.2). The bus accelerates away from the stop; ideally, the driver hopes to travel at a steady speed until the next stop. A steady speed means that you can sit comfortably in your seat. Then there is a rapid deceleration as the bus slows to a halt. A lot of accelerating and decelerating means that you are likely to be thrown about as the bus changes speed. The gentle acceleration of an express train will barely disturb the drink in your cup; the bus's rapid accelerations and decelerations would make it impossible to avoid spilling the drink.

Figure 3b.1 France's high speed trains, the TGVs (trains à grande vitesse), run on dedicated tracks. Their speed has made it possible to travel 600 km from Marseilles in the south to Paris in the north, attend a meeting and return home again within a single day.

Figure 3b.2 It can be uncomfortable on a packed bus as it accelerates and decelerates along its journey.

Calculating acceleration

An express train may take 300 s to reach a speed of 300 km/h. Its acceleration is 1 km/h per second.

These are not very convenient units, although they may help to make it clear what is happening when we talk about acceleration. To calculate an object's acceleration, we need to know two things:
- its change in speed (how much it speeds up)
- the time taken (how long it takes to speed up).

Then its acceleration is given by:

$$\text{acceleration} = \frac{\text{change in speed}}{\text{time taken}}$$

In the case of the car above, its speed increased by 60 mph in 6 s, so its acceleration was 60 mph / 6 s = 10 mph/s.

Worked example 1

An aircraft accelerates from 100 m/s to 300 m/s in 100 s. What is its acceleration?

Step 1: start by writing down what you know, and what you want to know.

Initial speed = 100 m/s

Final speed = 300 m/s

Time = 100 s

Acceleration = ?

Step 2: now calculate the change in speed.

Change in speed = 300 m/s – 100 m/s

$\qquad\qquad$ = 200 m/s

Step 3: substitute into the formula.

Acceleration = change in speed / time taken

$\qquad\qquad$ = 200 m/s / 100 s

$\qquad\qquad$ = 2 m/s^2

Units of acceleration

In Worked example 1, the units of acceleration are given as m/s^2 ('metres per second squared'). These are the standard units of acceleration. The calculation shows that the aircraft's speed

increased by 2 m/s every second, or 2 metres per second per second. It is simplest to write this as 2 m/s^2, but you may prefer to think of it as 2 m/s per second, because this emphasises the meaning of acceleration.

Other units for acceleration are possible – earlier we saw examples of acceleration in mph per second and km/h per second, but these are unconventional. It is usually best to work in m/s^2.

SAQ

2 Which of the following could *not* be a unit of acceleration? km/s^2, mph/s, km/s, m/s^2.

3 A car sets off from traffic lights. It reaches a speed of 27 m/s in 18 s. What is its acceleration?

4 A train, initially moving at 12 m/s, speeds up to 36 m/s in 120 s. What is its acceleration?

5 A test pilot accelerated from rest to a speed of 280 m/s in just 5 s. What was his acceleration?

6 What was his acceleration when he stopped in 1 s?

Measuring acceleration in the laboratory

If you place a trolley at the top of a sloping ramp, it will run downhill, picking up speed as it travels down the ramp. It is accelerating. To measure its acceleration, you need to make two measurements of its speed (initial and final), and you also need to know the time interval between the two measurements. Here are two ways of doing this, using light gates connected to a timer or computer.

With two light gates (Figure 3b.3), the trolley has a rectangular interrupt card attached. The card breaks the infra-red beam of the first light gate and then it breaks the beam of the second light gate, but for an even shorter time because it is moving more quickly. Knowing the length of the interrupt card, the computer connected to the light gates can then calculate the trolley's speed as it passes the first gate, and then its speed as it passes the second gate. (These are the initial and final speeds.) It can also calculate the time taken by the trolley to travel from one gate to the next. This is enough information for the computer to work out the trolley's acceleration.

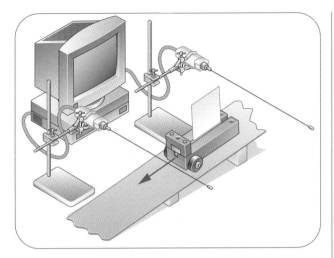

Figure 3b.3 To measure the acceleration of a trolley on a sloping ramp, you need two measurements of its speed. One method uses two light gates. The computer calculates the trolley's speed as it passes through each gate, and then calculates its acceleration.

With a single light gate, the trolley must be fitted with a different interrupt card (Figure 3b.4). This is cleverly designed, with two sections of equal lengths sticking up to break the infra-red beam. The first section breaks the beam for a short time; the second section breaks the beam for a shorter time, because the trolley is speeding up. The computer is programmed with the length of each of the sections; it then calculates the trolley's speed as each section passes the gate. Using the timer interval between these two measurements, it then calculates the acceleration.

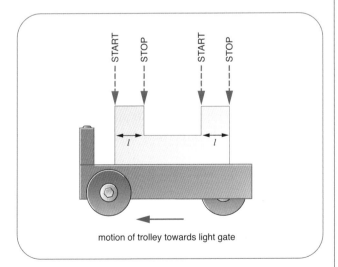

Figure 3b.4 With a single light gate, the trolley must be fitted with an interrupt card that breaks the infra-red beam twice. The computer then records four measurements of time; it is also programmed with the length of the sections of the interrupt card. From this data, it can calculate two speeds and the time interval between them, and hence the trolley's acceleration.

Accelerating downwards

The trolley on the sloping ramp accelerates down the slope. The steeper the ramp, the greater its acceleration. A falling object also accelerates. If you hold a ball at arm's length and let it go, the force of gravity pulls it downwards, causing it to speed up as it falls. You can see the pattern of acceleration in Figure 3b.5. This is a multi-flash photograph, and shows that a ball speeds up as it falls.

Figure 3b.5 The increasing speed of a falling ball is captured in this multi-flash image.

Further calculations

Using the formula for acceleration:

$$\text{acceleration} = \frac{\text{change in speed}}{\text{time taken}}$$

we can calculate change in speed or time taken.

Worked example 2

A stone falls with acceleration $10\,\text{m/s}^2$. How long will it take to reach a speed of $35\,\text{m/s}$?

Step 1: start by writing down what you know, and what you want to know.

Change in speed $= 35\,\text{m/s}$

Acceleration $= 10\,\text{m/s}^2$

Time $= ?$

Step 2: rearrange the formula.

$$\text{Acceleration} = \frac{\text{change in speed}}{\text{time taken}}$$

$$\text{Time taken} = \frac{\text{change in speed}}{\text{acceleration}}$$

continued on next page

H

Worked example 2 - *continued*

Step 3: substitute values into this formula.

$$\text{Time taken} = \frac{35\,\text{m/s}}{10\,\text{m/s}^2} = 3.5\,\text{s}$$

So the stone will take just 3.5 s to reach a speed of 35 m/s.

Changing direction

An object accelerates when it speeds up or slows down. There is also acceleration involved when an object changes direction. Suppose a car turns right; its forward speed has decreased to zero, and its speed to the right has increased.

SAQ

7 A stone falls with acceleration $10\,\text{m/s}^2$. How fast will it be travelling 2.5 s after it was released?

8 A locomotive has a maximum acceleration of $2\,\text{m/s}^2$. How long will it take to accelerate at this rate from 5 m/s to 27 m/s?

Speed-time graphs

Just as we can represent the motion of a moving object by a distance–time graph, we can also represent it by a speed–time graph. (It is easy to get these two types of graph mixed up. Always check out any graph by looking at the axes to see what their labels say.) A speed–time graph shows how the object's speed changes as it moves.

Figure 3b.6 shows a speed–time graph for a bus as it follows its route through a busy town. The graph frequently drops to zero because the bus must keep stopping to let people on and off. Then the line slopes up, as the bus accelerates away from the stop. Towards the end of its journey, it manages to move at a steady speed (horizontal graph), because it does not have to stop. Finally, the graph slopes downwards to zero again as the bus pulls into the terminus and stops.

The slope of the speed–time graph tells us about the bus's acceleration.

● The steeper the slope, the greater the acceleration.

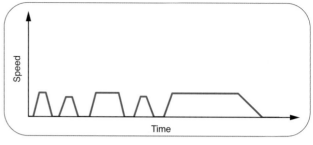

Figure 3b.6 A speed–time graph for a bus on a busy route. At first, it has to halt frequently at bus stops. Towards the end of its journey, it maintains a steady speed.

● A negative slope means a deceleration (slowing down).
● A horizontal graph (slope = 0) means a constant speed.

Graphs of different shapes

Speed–time graphs can show a lot about an object's movement. Was it moving at a steady speed, or speeding up, or slowing down? Was it moving at all? The graph shown in Figure 3b.7 represents a train journey. If you study the graph, you will see that it is in four sections, each of which illustrates a different point.

Section A Sloping upwards – speed increasing – the train was accelerating.

Section B Horizontal – speed constant – the train was travelling at a steady speed.

Section C Sloping downwards – speed decreasing – the train was decelerating.

Section D Horizontal – speed has decreased to zero – the train was stationary.

The fact that the graph lines are curved in sections A and C tells us that the train's acceleration was changing. If its speed had changed at a steady rate, these lines would have been straight.

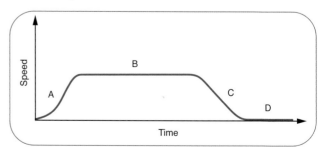

Figure 3b.7 An example of a speed–time graph (for a train during part of its journey). This illustrates how such a graph can show acceleration, constant speed, deceleration and zero speed.

SAQ

9 Sketch a speed–time graph for this journey: a car travels at a steady speed; when the driver sees the red traffic lights ahead, she slows down and comes to a halt.

10 Look at the speed–time graph in Figure 3b.8. Name the sections which represent:

 a steady speed

 b speeding up (accelerating)

 c being stationary

 d slowing down (decelerating).

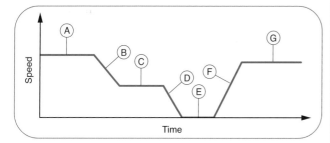

Figure 3b.8

Calculations using speed–time graphs

Acceleration

We can find the acceleration of an object by calculating the gradient of its speed–time graph:

 acceleration = slope of speed–time graph

Worked example 3

A train travels slowly as it climbs up a long hill, then it speeds up as it travels down the other side. Table 3b.1 shows how its speed changes. Draw a speed–time graph to show this data and use the graph to calculate the train's acceleration during the second half of its journey.

Speed (m/s)	6.0	6.0	6.0	8.0	10.0	12.0	14.0
Time (s)	0	10	20	30	40	50	60

Table 3b.1 Speed–time data for a train.

Before starting to draw a graph, it is worth looking at the data in Table 3b.1. The values of speed are given at equal intervals of time (every 10 s). The speed is constant at first (6.0 m/s); then

Worked example 3 - *continued*

it increases in equal steps (8.0, 10.0, etc.). In fact, we can see that the speed increases by 2.0 m/s every 10 s, and this is enough to tell us that the train's acceleration is 0.2 m/s². However, we will follow through the detailed calculation to illustrate how to deduce acceleration from a graph.

Step 1: Figure 3b.9 shows the speed–time graph drawn using the data in Table 3b.1. You can see that it falls into two parts:
- the initial horizontal section shows that the train's speed was constant (zero acceleration);
- the sloping section shows that the train was then accelerating.

Step 2: the triangle in Figure 3b.9 shows how to calculate the slope of the graph. This gives us the acceleration:

acceleration = (14.0 m/s – 6.0 m/s) / (60 s – 20 s)

$$= 8.0 \,\text{m/s} \,/\, 40 \,\text{s}$$

$$= 0.2 \,\text{m/s}^2$$

So, as we expected, the train's acceleration down the hill is 0.2 m/s².

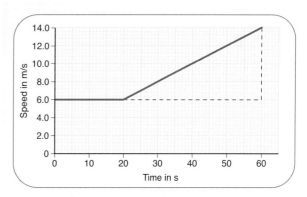

Figure 3b.9 Speed–time graph for a train, based on the data in Table 3b.1. The triangle is used to calculate the slope of the second section of the graph; this tells us the train's acceleration.

Finding the distance moved

A speed–time graph represents an object's movement. It tells us about how its speed changes, and we have already seen how to calculate the object's acceleration (from the slope of the graph). We can also use the graph to deduce how far the object moves. To do this, we have to make use of the formula

distance = speed × time

Here are three worked examples to illustrate how to find the distance moved, using a speed–time graph.

Worked example 4

You cycle for 20 s at a constant speed of 10 m/s (see Figure 3b.10). How far do you move?

Step 1: the distance you travel is

distance moved = 10 m/s × 20 s

= 200 m

Step 2: this is represented by the shaded area under the graph. This rectangle is 20 s long and 10 m/s high, so its area is 10 m/s × 20 s = 200 m.

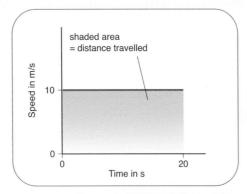

Figure 3b.10 Speed–time graph for constant speed. The distance travelled is represented by the shaded area under the graph.

Worked example 5

(This is a little more complicated.)

You set off down a steep ski slope. Your initial speed is 0 m/s; after 10 s you are travelling at 30 m/s (see Figure 3b.11). Your average speed is 15 m/s. How far do you move?

Step 1: the distance you travel is

distance moved = 15 m/s × 10 s

= 150 m

Worked example 5 - *continued*

Step 2: again, this is represented by the shaded area under the graph. In this case, the shape is a triangle whose height is 30 m/s and whose base is 10 s. The area of a triangle $= \frac{1}{2}$ base × height, so we have

area $= \frac{1}{2}$ × 10 s × 30 m/s

= 150 m

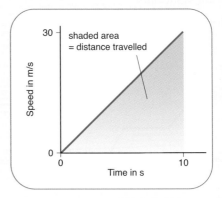

Figure 3b.11 Speed–time graph for constant acceleration from rest. Again, the distance travelled is represented by the shaded area under the graph.

Worked example 6

Calculate the distance travelled in 60 s by the train whose motion was described in Worked example 3 (Figure 3b.9 and Table 3b.1). The graph in Figure 3b.12 is the same as that in Figure 3b.9 but it has been shaded to show the area we need to calculate to find the distance moved by the train. This area is in two parts.

Step 1: a rectangle of height 6 m/s and width 60 s.

Area = 6 m/s × 60 s

= 360 m

(This tells us how far the train would have travelled if it had maintained a constant speed of 6 m/s.)

Step 2: a triangle of base 40 s and height (14 m/s – 6 m/s) = 8 m/s.

continued on next page

Worked example 6 - *continued*

Area $= \frac{1}{2}$ base × height

$\qquad = \frac{1}{2} \times 40\,s \times 8\,m/s$

$\qquad = 160\,m$

(This tells us the extra distance travelled by the train because it was accelerating.)

Step 3: find the total distance.

Total distance travelled $= 360\,m + 160\,m$

$\qquad\qquad\qquad\quad = 520\,m$

So, in 60 s, the train travelled 520 m.

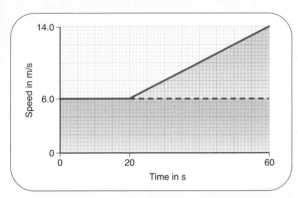

Figure 3b.12 Calculating the distance travelled by a train. Distance travelled is represented by the area under the graph. To make the calculation possible, this area is divided up into a rectangle and a triangle, as shown.

Summing up the use of speed-time graphs

In fact, the distance travelled by a moving object can always be found from the area under the speed–time graph. In more complicated situations, you may have to divide the area under the graph into rectangles and triangles, and add up their separate areas. Alternatively, you can use the technique of counting the squares on the graph paper. Remember:

distance travelled = area under speed–time graph

acceleration = gradient of speed–time graph

SAQ

11 A car travels for 10 s at a steady speed of 20 m/s along a straight road. The traffic lights ahead change to red and the car slows down with a constant deceleration so that it halts after a further 8 s.

a Draw a speed–time graph to represent the car's motion during the 18 s described.

b Use the graph to deduce the car's deceleration as it slows down.

c Use the graph to deduce how far the car travels during the 18 s described.

Summary

You should be able to:

◆ calculate acceleration using the formula

$\qquad$ acceleration $= \dfrac{\text{change in speed}}{\text{time}}$

◆ state that acceleration is measured in m/s^2 (metres per second squared)

◆ state that the gradient of a speed–time graph tells us about an object's acceleration

◆ state that the steeper the gradient, the greater the acceleration

◆ calculate acceleration from the gradient of a speed–time graph

◆ calculate distance moved from the area under a speed–time graph

◆ state that there is acceleration when an object changes the direction in which it is moving

Questions

1 A runner accelerates from rest to 8 m/s in 2 s. What is her acceleration?

2 The speed–time graph for part of a train journey is a horizontal straight line. What does this tell you about the train's speed, and about its acceleration?

3 Sketch speed–time graphs to represent the following:

a an object starting from rest and moving with constant acceleration

b an object moving at a steady speed, then slowing down and stopping.

H 4 Which of the following does *not* involve acceleration?

Speeding up   Slowing down   Steady speed   Changing direction

5 A runner accelerates from rest with an acceleration of 4 m/s² for 2.3 s. What will her speed be at the end of this time?

6 A car can accelerate at 5.6 m/s². How long will it take to reach a speed of 24 m/s, starting from rest?

7 Draw a speed–time graph to show the following motion: a car accelerates uniformly from rest for 5 s; then it travels at a steady speed of 12 m/s for 5 s.

On your sketch graph, shade the area that represents the distance travelled by the car in 10 s.

Calculate the distance travelled in this time.

8 Table 3b.2 shows how the speed of a car changed during a section of a journey. Draw a distance–time graph to represent this data and use it to calculate:

a the car's acceleration during the first 30 s of the journey

b the distance travelled by the car during the journey.

Speed in m/s	0	9	18	27	27	27
Time in s	0	10	20	30	40	50

Table 3b.2

9 The graph in Figure 3b.13 shows how a car's speed changed as it travelled along.

a In which section(s) was its acceleration zero?

b In which section(s) was its acceleration constant?

c What can you say about its acceleration in the other section(s)?

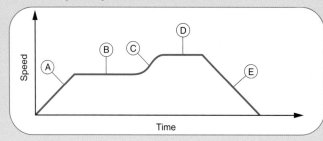

Figure 3b.13 A speed–time graph for a car.

Roller-coaster forces

Some people get a lot of pleasure out of sudden acceleration and deceleration. Many fairground rides involve sudden changes in speed. On a roller coaster (Figure 3c.1), you may speed up as the car runs downhill. Then, suddenly, you veer off to the left – you are accelerated sideways. A sudden braking gives you a large, negative acceleration (a deceleration). You will probably have to be fastened into your seat to avoid being thrown out of the car by these sudden changes in speed.

What are the forces at work in a roller coaster? If you are falling downwards, it's gravity that affects you. This gives you an acceleration of about $10 \, \text{m/s}^2$. We say that the G-force acting on you is 1 (i.e. one unit of gravity). When the brakes slam on, the G-force may be greater, perhaps as high as 4. The brakes make use of the force of friction.

Changing direction also requires a force. So when you loop the loop or veer to the side, there must be a force acting. This is simply the force of the track, whose curved shape pushes you round. Again, the G-force may reach as high as 4.

Roller-coaster designers have learned how to surprise you with sudden twists and turns. You can be scared or exhilarated. However you feel, you can release the tension by screaming.

Figure 3c.1 A roller-coaster ride involves many rapid changes in speed. These accelerations and decelerations give the ride its thrill. The ride's designers have calculated the accelerations carefully to ensure that the car will not come off its track and that the riders will stay in the car.

Figure 3c.2 The space shuttle accelerating away from its launch pad. The force needed is provided by several rockets. The strap-on boosters will be jettisoned once they have used all their fuel to reduce the mass which is being carried up into space.

We have lift-off

It takes an enormous force to lift the giant space shuttle off its launch pad, and to propel it into space (Figure 3c.2). The booster rockets which supply the initial thrust provide a force of several million newtons. As the spacecraft accelerates upwards, the crew experience the sensation of being pressed firmly back into their seats. That is how they know that their craft is accelerating.

Forces change motion. One moment, the shuttle is sitting on the ground, stationary. The next moment, it is accelerating upwards, pushed by the force provided by the rockets.

In this Item, we will look at how forces – pushes and pulls – affect objects as they move. You will be familiar with the idea that forces are measured in newtons (N).

To give an idea of the sizes of various forces, here are some examples.

- You lift an apple. The force needed to lift an apple is roughly 1 newton (1 N).
- You jump up in the air. Your leg muscles provide the force needed to do this – about 1000 N.
- You reach the motorway in your high-performance car and put your foot down. The car accelerates forwards; the engine provides a force of about 5000 N.
- You are crossing the Atlantic in a Boeing 747 jumbo jet. The four engines together provide a thrust of about 1 000 000 N – that's about the same as the thrust provided by each of the space shuttle's booster rockets.

Forces produce acceleration

The car driver in Figure 3c.3a is waiting for the traffic lights to change. When they go green, he moves forwards. The force provided by the engine causes the car to accelerate. In a few seconds, the car is moving quickly along the road.

The arrow in the diagram shows the force pushing the car forwards. If the driver wants to get away from the lights more quickly, he can press harder on the accelerator. The forward force is then bigger, and the car's acceleration will be greater.

The driver reaches another junction where he must stop. He applies the brakes. This provides

another force to slow the car down (Figure 3c.3b). The car is moving forwards, but the force needed to make it decelerate is directed backwards. If the driver wants to stop in a hurry, a bigger force is needed. He must press hard on the brake pedal, and the car's deceleration will be greater.

We have seen several things about forces.

- They can be represented by arrows. A force has a direction, shown by the direction of the arrow.
- A force can make an object accelerate. The acceleration is in the direction of the force.
- The bigger the force acting on an object, the bigger the acceleration it gives to the object.

So the greater the force, the greater the acceleration it will produce. Twice the force produces twice the acceleration; three times the force produces three times the acceleration, etc.

The car driver uses the accelerator pedal to control the car's acceleration; this alters the force provided by the engine. There is another factor which affects the car's acceleration. Suppose the driver fills the boot with a lot of heavy shopping and then collects several children from school. He will notice the difference when he moves away from the traffic lights: the car will not accelerate so readily because its mass has been increased. Similarly, when he applies the brakes, it will not decelerate as readily as before. The mass of the car affects how easily it can be accelerated or decelerated. Drivers learn to take account of this.

To summarise:

- the bigger the mass of an object, the smaller the acceleration it is given by a particular force.

So big (more massive) objects are harder to accelerate than small (less massive) ones. If we double the mass of the object, its acceleration for a given force will be halved. Another way to think of this is to say that we need double the force to give it the same acceleration.

Figure 3c.3 A force can be represented by an arrow.
a A forward arrow represents a forward force; the forward force provided by the engine causes the car to accelerate forwards.
b A backward arrow represents a backward force; the backward force provided by the brakes causes the car to decelerate.

SAQ

1 a John can kick the ball harder than Jack. Who can give the ball a greater acceleration?

b To make the contest fairer, how should the mass of John's ball be altered so that his force gives it the same acceleration as Jack's?

Force, mass, acceleration

These relationships between force, mass and acceleration can be combined in a single, very useful, equation:

force = mass × acceleration

The quantities involved in this equation, and their units, are summarised in Table 3c.1. Worked examples 1 and 2 show how to use the equation.

Quantity	SI unit
force	newton, N
mass	kilogram, kg
acceleration	metres per second squared, m/s^2

Table 3c.1 The three quantities related by the equation force = mass × acceleration.

Worked example 1

When you strike a tennis ball which another player has hit towards you, you provide a large force to reverse its direction of travel and send it back towards your opponent. You give the ball a large acceleration. What force is needed to give a ball of mass 0.1 kg an acceleration of 500 m/s^2?

Step 1: we have

mass = 0.1 kg

acceleration = 500 m/s^2

force = ?

Step 2: substituting in the equation to find the force gives

force = mass × acceleration

$= 0.1 \, \text{kg} \times 500 \, \text{m/s}^2$

$= 50 \, \text{N}$

SAQ

2   What force is needed to give a car of mass 600 kg an acceleration of 2.5 m/s^2?

3   A stone of mass 0.2 kg falls with an acceleration of 9.8 m/s^2. How big is the force that causes this acceleration?

Worked example 2

A Boeing 747 jumbo jet has four engines, each capable of providing 250 000 N of thrust. The mass of the aircraft is 400 000 kg. What is the greatest acceleration which the aircraft can achieve?

Step 1: the greatest force provided by all four engines working together is

4 × 250 000 N = 1 000 000 N

Step 2: now we have

force = 1 000 000 N

mass = 400 000 kg

acceleration = ?

Step 3: the greatest acceleration the engines can produce is then given by

acceleration = force / mass

$= 1\,000\,000 \, \text{N} / 400\,000 \, \text{kg}$

$= 2.5 \, \text{m/s}^2$

SAQ

4   What acceleration is produced by a force of 2000 N acting on a person of mass 80 kg?

5   One way to find the mass of an object is to measure its acceleration when a force acts on it. If a force of 80 N causes a box to accelerate at 0.1 m/s^2, what is the mass of the box?

Pairs of forces

Imagine being stationary in a swimming pool. If you can swim, imagine how you can get moving (Figure 3c.4). You use your hands to push the water backwards. This provides the force to move you *forwards*.

Here we have two forces: your push on the water and the water's push on you. These forces are equal in size and opposite in direction. One is exerted by you and acts on the water. The other is exerted by the water and acts on you. Without the force of the water on you, you would not start moving forwards.

Figure 3c.4 Forces are created in pairs. The swimmer pushes backwards on the water, and the water pushes forwards on the swimmer. These forces are equal and opposite.

Isaac Newton realised that forces are always created in pairs. A single force cannot be created out of nothing. Here are some more examples of these equal-and-opposite pairs of forces; some are more surprising than others:

If you hold two bar magnets so that the north pole of one faces the south pole of the other, they will attract one another (Figure 3c.5). You will feel that they attract each other equally strongly. Even if one magnet is more strongly magnetised than the other, you will feel that they are pulled equally. Reverse one of the magnets, and you will feel that they now repel each other with equal forces.

Figure 3c.5 Magnets attract and repel one another with equal and opposite forces. Even if one magnet was replaced by a piece of unmagnetised iron, the forces of attraction would be equal and opposite.

Friction is necessary for walking. To walk, you push backwards on the ground. Your foot is tending to slide backwards on the ground, so a frictional force pushes forwards to oppose you. There are thus two frictional forces acting: one which acts backwards on the ground, and the other which acts forwards on your foot.

We are used to the idea that the Earth's **gravity** pulls on us. If we fall down, it is because of gravity. The force of gravity on us is our weight – say, 500 N. But at the same time, we pull on the Earth. We have a gravitational pull, too. We pull upwards on the Earth with a force of about 500 N. However, because the Earth's mass is so great, our pull has very little effect on it.

SAQ

6 Draw a diagram to show the two frictional forces which act when you walk (as described above). Which of the two forces causes you to accelerate forwards?

Newton's third law

The idea that forces are always created in pairs is known as **Newton's third law of motion**.

When bodies interact, they exert equal and opposite forces on each other.

You may sometimes hear a pair of forces like this referred to as *action* and *reaction*. The idea is that one force, the action, always results in a second force, the reaction. So the third law is sometimes stated as 'for every action, there is an equal and opposite reaction.' However, you need to be careful that you understand what the terms *action* and *reaction* mean.

Investigating 'force = mass × acceleration'

In the laboratory, you can investigate the equation *force = mass × acceleration* using a linear air track (see Figure 3c.6). (Alternatively, you could use a trolley.) The track has a row of tiny holes along either side. Air is blown out through these holes and this provides a cushion of air on which a glider can float smoothly from one end of the track to the other. The feet of the track can be adjusted so that the track is horizontal; then, when the glider is given a gentle push to start it moving, it moves at a steady speed from one end of the track to the other.

To provide a force to accelerate the glider, a thin string is attached to one end and passed over a pulley. Masses are hung on the end of the string. Each 100 g mass provides a pulling force of (approximately) 1 N. When the masses are released, they pull the glider along the track, causing it to accelerate. The acceleration can be measured as described in Item P3b, using an interrupt card and a single light gate.

The force pulling the glider can be varied by adding more masses to the hanger on the end of the string. In practice, this is done using ten 100 g masses as follows. Initially, one mass is hung on the end of the string. The other nine are attached to the glider and its acceleration is measured. Then one mass is transferred to the end of the string, leaving eight on the glider. The acceleration is measured again, another mass transferred to the end of the string, and so on. To understand the reason for doing this, it is important to realise that it is not just the glider that is being accelerated; the masses are also accelerating. By transferring masses from the glider to the hanger, the *total mass* remains constant. Only the force increases; the greater the force, the greater the glider's acceleration.

To investigate how acceleration depends on mass, the accelerating force must be kept constant. A fixed number of masses (say, four) is hung on the end of the string. Gradually, masses are added to the glider and its acceleration measured for each value of mass. Again, it is important to include the mass of the masses hanging on the end of the string, since they are also being accelerated by the force.

Figure 3c.7 shows the shapes of the graphs which can be drawn to represent the results of these experiments.

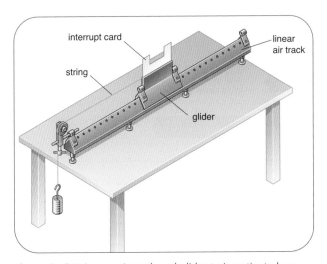

Figure 3c.6 Using an air track and glider to investigate how acceleration depends on force and mass. There is no friction to slow the glider because it is supported on a cushion of air. Acceleration is measured using an interrupt card and light gate. It is a good idea to make repeat measurements of each value of acceleration, to see how closely the results can be repeated. Take an average of the values found.

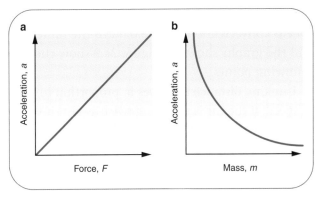

Figure 3c.7 Graphs to represent the results from the experiment shown in Figure 3c.6. Graph **a** is a straight line passing through the origin. This shows that acceleration is proportional to force. Graph **b** shows that, as the mass of the glider increases, its acceleration decreases (for a given force).

Thinking and braking

The *Highway Code* is a booklet which tells drivers and others how to use Britain's roads safely. On its back cover, it carries a diagram (Figure 3c.8) showing the shortest distances in which a car can come to a halt if the driver has to stop in a hurry. A car travelling at 20 mph, for instance, requires at least 12 m in which to stop. From the diagram, you can see that the **stopping distance** is made up of two parts.

Thinking distance   This is the distance which the car travels in the time it takes between the instant that the driver realises he or she must stop, and the instant when they apply the brakes. It takes a fraction of a second for the driver to react and for their brain to send a message to their foot. The car travels at a steady speed in this time.

Braking distance   This is the distance travelled by the car as it slows down to a halt, once the brakes have been applied. Its speed decreases steadily (constant deceleration) during this time.

stopping distance =
    thinking distance + braking distance

The distances in the chart assume that the conditions of both the road and the car are good, and that the driver's reactions are quick. The braking distance, in particular, may be greater if the road surface is slippery or if the car's brakes or tyres are faulty. The thinking distance may be greater if the driver has been drinking alcohol, for example, or if they are sleepy.

Table 3c.2 shows the same information, with speeds converted to m/s. The data in the table, and the graphs shown in Figure 3c.9 show the following points.

● Thinking distance increases in proportion to speed. If the car is travelling twice as fast, it will travel twice as far during the **thinking time**. From the graph (Figure 3c.9a), we can see that the thinking time is about two-thirds of a second.

● Braking distance increases more and more rapidly as the car's speed increases. Look at the values for 20 mph and 40 mph. Braking distances are 6 m and 24 m. Doubling the speed gives four times the braking distance. Similarly, trebling the speed from 20 mph to 60 mph gives nine times the braking distance. This gives the curved graph shown in Figure 3c.9b.

The *Highway Code* also advises drivers to maintain a good distance from the car ahead. Some drivers habitually 'tailgate' the car in front – that is, they drive very close to its back bumper. The advice suggests that drivers should maintain a gap of at least 2 s, considerably more than the minimum thinking time of $\frac{2}{3}$ s; at a speed of 70 mph (31 m/s) on a motorway, this means a gap of 62 m. In the UK, distance markers are positioned every 100 m on the motorway verge, and these can help drivers to keep to a safe separation.

SAQ

7 Study the data shown in Table 3c.2.

   a Which is greater for a car travelling at 40 mph, thinking distance or braking distance?

   b At what speed are these two distances equal?

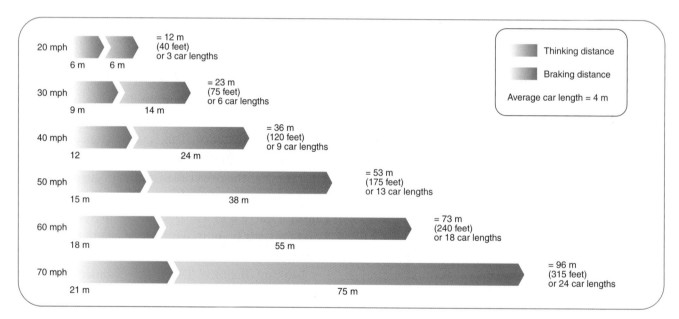

Figure 3c.8 A chart of stopping distances, like the one in the *Highway Code*. These distances assume an alert driver in good road conditions with a well-maintained car.

Speed in mph	Speed in m/s	Thinking distance in m	Braking distance in m	Stopping distance in m
20	8.9	6	6	12
30	13.3	9	14	23
40	17.8	12	24	36
50	22.2	15	38	53
60	26.7	18	55	73
0	31.1	21	75	96

Table 3c.2 Data on stopping distances, taken from the *Highway Code*. Stopping distance increases more and more rapidly at high speeds.

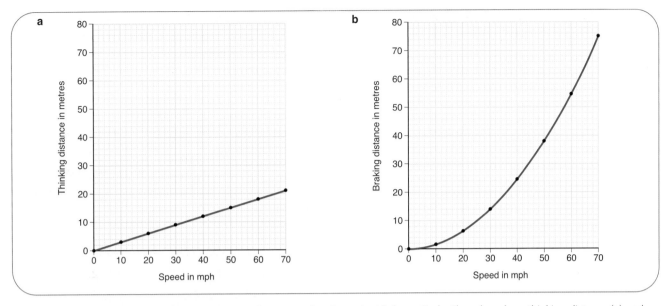

Figure 3c.9 Graphs to represent data on stopping distances, taken from the *Highway Code*. They show how thinking distance (**a**) and braking (**b**) distance depend on a car's speed.

Keeping out of trouble

You can drive more safely if you think about how different factors may affect your stopping distance.

Thinking distance will be greater if a driver's thinking is slower (Figure 3c.10). This might be because:

- the driver is tired;
- the driver is affected by alcohol or other drugs;
- the driver is distracted.

Distractions can come from something outside the car or from passengers, music or phone calls in the car.

Braking distance will be greater if driving conditions are not ideal. This might be because:

- the road surface is slippery, owing to ice or rainfall;
- the car is defective, with bald (smooth) tyres or faulty brakes.

All of these factors will have a greater effect at higher speeds because the car will travel further before it stops. So increased speed increases both thinking distance and braking distance.

SAQ

8   When a car stops suddenly, the stopping distance is the sum of the thinking distance and the braking distance. Which of these two factors might be affected by a slippery road surface? Explain why both factors might be affected if the driver had consumed alcohol or drugs.

Figure 3c.10 A driver being breath tested. A high proportion of drivers involved in accidents are found to have traces of drugs in their blood.

Your life in a driver's hands

The police spend a lot of time investigating road traffic accidents (RTAs); in Figure 3c11, what are they looking for?

They hope to be able to reconstruct the events leading up to the accident. This is partly because they may collect enough evidence to prosecute a dangerous driver who has caused the accident. It is also because there may be a problem with the road which has led to the accident, and which can be put right in the future.

So the police investigate the scene of the crash. They look for marks on the road, which will tell them whether any of the vehicles involved skidded. They can deduce whether anyone was speeding and whether they braked later than might have been expected. They look at the condition of the road surface – is it slippery, perhaps because of water or an oil spill?

At the same time, they investigate the people involved. Were any of the drivers affected by drink or drugs? Were they using their mobile phones? (This can be checked from the phone companies' records.) Perhaps one of the drivers had been at the wheel for too long and was suffering from tiredness, which can cause lapses in concentration.

Children may be the victims of accidents (Figure 3c.12), and this seems particularly tragic. There is increasing pressure to develop more 'home zones', areas of towns where the speed limit is low and where streets are designed to exclude motorists who are passing through, so that the only motorists will be residents. Perhaps this will help us to return to the situation where children could safely play out in the street.

Figure 3c.11 Police investigations at the scene of a road traffic accident in Derbyshire.

Figure 3c.12 Lowered speed limits help to reduce the number of child victims on the roads.

Summary

You should be able to:

◆ state that, when a force acts on an object, a greater force produces a greater acceleration

◆ state that a greater mass requires a greater force to produce a given acceleration

◆ use the equation

force = mass × acceleration

◆ state that *thinking distance* is the distance travelled between the driver seeing the need to stop and applying the brakes; and that it is affected by the condition of the driver

◆ state that *braking distance* is the distance travelled by the car whilst the brakes are applied until it stops, and that it is affected by the condition of the road and the car

◆ interpret the formula

stopping distance = thinking distance + braking distance

◆ state that both thinking distance and braking distance are greater at higher speeds

Questions

1   What are the units of mass, force and acceleration?

2   Why is it sensible on diagrams to represent a force by an arrow? Why should mass not be represented by an arrow?

3   Which will produce a bigger acceleration: a force of 10 N acting on a mass of 5 kg or a force of 5 N acting on a mass of 10 kg?

4   What force is needed to give a mass of 20 kg an acceleration of 5 m/s^2?

5   A drunken driver takes a long time to stop on a wet road. Explain why, by referring to the formula

stopping distance = thinking distance + braking distance

6   Look at the data from the *Highway Code* shown in Table 3c.2. Use it to calculate the thinking time (the time between the driver realising that he or she must stop and when the brakes are applied).

7   A train of mass 800 000 kg is slowing down. What acceleration is produced if the braking force is 1 400 000 N?

8   A car speeds up from 12 m/s to 20 m/s in 6.4 s. If its mass is 1200 kg, what force must its engine provide?

9   An astronaut on the Moon drops a stone. It accelerates at 1.6 m/s^2. If the force on the stone is 80 N, what is its mass?

Uphill struggle

Figure 3d.1 Racing uphill is very demanding on your legs. They have to push your body upwards, against the downward pull of the Earth's gravity. The higher you go, the thinner the atmosphere and the harder you will find the climb.

Running up a mountain is hard work, although some people do it for pleasure (Figure 3d.1). As you climb upwards, gravity is trying to pull you back down. Your legs have to push hard on the ground if you are to move quickly upwards.

If you are running uphill, your legs have the job of pushing the weight of your body upwards. This takes energy. You have to draw on the energy reserves stored in your body. Your blood supply brings the necessary oxygen to your muscles. If you are climbing at high altitudes, you may run short of oxygen. The air gets thinner the higher you climb, and you may suffer from altitude sickness. People who live at high altitudes have more haemoglobin in their blood so that more oxygen gets carried to their muscles. When the Olympic Games were held in Mexico City, at a height of 2200 m above sea level, athletes had to spend several weeks acclimatising themselves to the conditions before they could compete.

It is much easier coming downhill. Now gravity is on your side. If you let yourself go, you can simply roll down to the bottom. Gravity will make you go faster and faster; friction will tend to slow you down. In practice, your legs will be working hard again, pushing down on the ground to stop you from going out of control.

Push, pull, lift

In this Item, we will look at the energy changes involved when you move things – by pushing, pulling, lifting or whatever. These energy changes involve forces.

Doing work

Figure 3d.2 shows one way of lifting a heavy object. Pulling on the rope raises the heavy box. As you pull, the force on the box moves upwards.

To lift an object, you need a store of **energy** (as chemical energy, in your muscles). You give the object more **gravitational potential energy (GPE)** (because it is getting higher). The force is your means of transferring energy from you to the object. The name given to this type of energy transfer by a force is **doing work**.

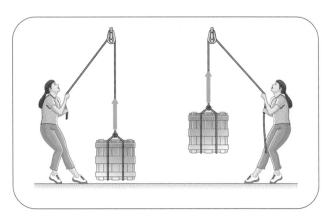

Figure 3d.2 Lifting an object requires an upward force, pulling against gravity. As the box rises upwards, the force also moves upwards. Energy is being transferred by the force to the box.

Figure 3d.3 Three examples of forces doing work. In each case, the force moves as it transfers energy.

Some further examples of forces doing work are shown in Figure 3d.3.

a Pushing a shopping trolley to start it moving. The pushing force does work; it transfers energy to the trolley; the trolley's kinetic energy increases.

b An apple falling from a tree. Gravity pulls the apple downwards. Gravity does work, and the apple's kinetic energy increases.

c Braking to stop a cycle. The brakes produce a backward force of friction which slows the cycle. The friction does work and reduces the cycle's kinetic energy. Energy is transferred to the brakes, which get hot.

Calculating work done

Think about lifting a heavy object, as shown in Figure 3d.2. A heavy object needs a big force to lift it. The heavier it is and the higher it is lifted, the more its GPE increases. This suggests that the amount of energy transferred by a force depends on two things:

● the *size of the force* – the greater the force, the more work it does;

● the *distance moved* by the force – the further it moves, the more work it does.

We can write:

work done = force × distance moved by the force

When you lift an object, it gains energy and you lose energy. Some of your energy has been transferred to the object. You can see that *energy is needed to do work.*

The amount of work done is measured in joules (J). This is the same unit as for energy, because the work done by a force is equal to the energy it transfers. Note:

1000 J = 1 kJ (1 kilojoule)

1 000 000 J = 1 MJ (1 megajoule)

Worked example 1

A crane lifts a crate upwards through a height of 20 m. The lifting force provided by the crane is 5 kN. How much work is done by the force?

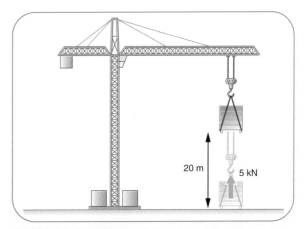

Figure 3d.4 A crane provides the upward force needed to lift a crate. The force transfers energy from the crane to the crate; the crate's GPE increases.

continued on next page

Worked example 1 - *continued*

Step 1: write down what you know and what you want to know.

Force = 5 kN = 5000 N

Distance moved = 20 m

Work done = ?

Step 2: write down the equation for work done, substitute values and solve.

Work done = force × distance moved

$$= 5000 \, \text{N} \times 20 \, \text{m}$$

$$= 100\,000 \, \text{J}$$

So the force does 100 000 J of work, or 100 kJ.

SAQ

1 How much work is done by a 50 N force when it moves through 2 m?

2 A girl lifts a heavy box. She provides a force of 200 N and the box is lifted 1.5 m. How much work has the force done?

Forces doing no work

If you sit still on a chair (Figure 3d.5a), there are two forces acting on you. These are your weight, acting downwards, and the upward contact force of the chair, which stops you from falling through the bottom of the chair.

Neither of these forces is doing any work on you. The reason is that neither of the forces is moving, so they do not move through any distance. Hence, from *work = force × distance*, the amount of work done by each force is zero. When you sit still on a chair, your energy doesn't increase or decrease as a result of the forces acting on you.

Figure 3d.5b shows another example of a force which is doing no work. A spacecraft is travelling around the Earth in a circular orbit. The Earth's gravity pulls on the space craft to keep it in its orbit. The force is directed towards the centre of the Earth. However, because the spacecraft's orbit is circular, it does not get any closer to the centre of the Earth. There is no movement *in the direction*

of the force and so gravity does no work. The spacecraft continues at a steady speed (its KE is constant) and at a constant height above the Earth's surface (its GPE is constant). Of course, although the force is doing no work, this doesn't mean it isn't having an effect; without the force, the spacecraft would escape from the Earth and disappear into the depths of space.

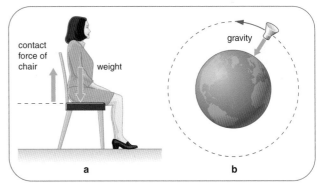

Figure 3d.5 a When you sit still on a chair, there are two forces acting on you. Neither transfers energy to you. **b** The spacecraft stays at a constant distance from the Earth. Gravity keeps it in its orbit without transferring any energy to it.

You will by now understand that 'work' is a word which has a specialised meaning in Physics, different from its meaning in everyday life. If you are sitting thinking about your homework, no forces are moving and you are doing no work. It is only when you start to write that you are doing work. To make the ink flow from your pen, you must push against the force of friction, and then you really are 'doing work'.

Worked example 2

A boy can only provide a pushing force of 200 N. To move a box weighing 400 N onto a table, he uses a plank as a ramp (Figure 3d.6). How much work does he do in raising the box?

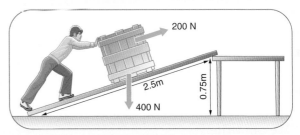

Figure 3d.6 A ramp can allow you to lift a heavy load, but you do more work than if you could raise it unaided.

continued on next page

Worked example 2 - *continued*

This example illustrates the importance of using the correct distance when calculating work done.

Step 1: from the diagram, you can see that the box is raised 0.75 m vertically, but the boy has to push it 2.5 m up the slope.

Step 2: calculate the work done in pushing the box up the slope.
We must take distance = 2.5 m.

Step 3: work done by pushing force up slope = force × distance.

Work done = 200 N × 2.5 m

$\qquad$ = 500 J

So the boy does 500 J of work.

SAQ

3 If the boy in Figure 3d.6 was stronger, he could lift the box vertically. He would need to provide a force of 400 N. How much work would he do?

Worked example 3

A weightlifter does 16 800 J of work in lifting a load of 800 N. How high does he lift the load?

Step 1: write down what you know and what you want to know.

Force = 800 N

Work done = 16 800 J

Distance moved = ?

Step 2: write down the equation for work done, and rearrange it.

Work done = force × distance moved

$$\text{Distance moved} = \frac{\text{work done}}{\text{force}}$$

Worked example 3 - *continued*

Step 3: substitute values and calculate the answer:

$$\text{Distance moved} = \frac{16\,800\,\text{J}}{800\,\text{N}}$$

$$= 2.1\,\text{m}$$

So the load was lifted 2.1 m upwards.

SAQ

4 A car moves 1500 m along the road. Its engine does 600 000 J of work against friction. Calculate the force of friction opposing the car.

Power

Exercising in the gym (Figure 3d.7) can put great demands on your muscles. Your trainer might ask you to find out how many times you can lift a set of weights in one minute. Equally, speeding up a treadmill means that you have to work harder to keep up. These exercises are a test of how powerful you are. The faster you work, the greater your power.

In Physics, the word **power** is often used with a special meaning. It means the rate at which you do work (that is, how fast you work). The more

Figure 3d.7 It's hard work down at the gym. It's easier to lift small loads, and to lift them slowly. The greater the load you lift and the faster you lift it, the greater the power required. It's the same with running on a treadmill; the faster you have to run, the greater the rate at which you do work.

work you do and the shorter the time in which you do it, the greater your power. We can write this as an equation:

$$\text{power} = \frac{\text{work done}}{\text{time taken}}$$

(You may have studied this relationship previously, in Module P2 in Book 1.)

Units of power

Power is measured in watts (W). One watt is the power when one joule of work is done in one second. So

1 watt = 1 joule per second (1 W = 1 J/s)
1000 W = 1 kilowatt (kW)
1 000 000 W = 1 megawatt (MW)

Worked example 4

A car accelerates from rest to a speed of 25 m/s in 10 s. In this time, the force of the engine does 250 000 J of work. What is its power?

Step 1: write down what you know and what you want to know.

Work done = 250 000 J

Time taken = 10 s

Step 2: calculate the power.

$$\text{Power} = \frac{\text{work done}}{\text{time taken}}$$

$$= \frac{250\,000\,\text{J}}{10\,\text{s}}$$

$$= 25\,000\,\text{W}$$

So the energy is being transferred to the car (from its engine) at a rate of 25 kW, or 25 kJ per second. In fact, car engines are not very efficient. In this example, the car's engine may transfer energy at the rate of 100 kW or so, although most of this is wasted as heat energy.

SAQ

5 A car engine does 50 000 J of work each second. What is its power?

6 An electric motor does 480 J of work in 1 minute. What is its power?

Power in general

We can apply the idea of power to any transfer of energy. For example, electric light bulbs transfer energy supplied to them by electricity. They produce light and heat. Most light bulbs are labelled with their power rating (40 W, 60 W, 100 W and so on) to tell the user about the rate at which it transfers energy. (You learned about electrical power in Item P2b in Book 1.)

We can even think about a human being as an energy transfer device. Our energy comes from our food – about 10 MJ per day. Unless we are storing lots of energy, we must be transferring 10 MJ to our surroundings each day. There are 86 400 seconds in a day, so our power is 10 MJ / 86 400 s = 115 W approximately. So 10 people produce heat at a similar rate to a 1 kW heater. You can understand why a large theatre or conference hall holding 1000 people or more soon gets warm and needs an effective air-conditioning system to avoid the audience becoming overheated.

Worked example 5

An electric motor does work at the rate of 20 W. If it operates for 10 minutes, how much work does it do?

Step 1: to solve this, we will use the equation

$$\text{power} = \frac{\text{work done}}{\text{time}}$$

Rearranging gives:

work done = power × time

Step 2: power is the rate at which work is done.

Power = 20 W

Time = 10 minutes

= 600 s

Step 3: substituting gives

work done = 20 W × 600 s

= 12 000 J

So the motor does 12 000 J of work; it transfers 12 000 J of energy.

Summary

You should be able to:

- ◆ state that, when a force moves, it transfers energy; it does work

- ◆ state that work is measured in joules (J)

- ◆ use the formula: work done = force × distance moved

- ◆ state that power tells us how quickly work is done

- ◆ state that power is measured in watts (W)

- ◆ use the formula: power = $\dfrac{\text{work done}}{\text{time taken}}$

Questions

1 In what units is the work done by a force measured? What are the units of power?

2 A fast-moving car has 0.5 MJ of kinetic energy. The driver brakes and the car comes to a halt. How much work is done by the force provided by the brakes?

3 How much work is done by a force of 1 N moving through 1 m? How much work is done by a force of 5 N moving through 2 m?

4 Which does more work, a force of 500 N moving through 10 m or a force of 100 N moving through 40 m?

5 It is estimated that the human brain has a power requirement of 40 W. How many joules is that per second?

6 A light bulb transfers 1000 J of energy in 10 s. What is its power?

7 An electric motor transfers 100 J in 8 s. If it then transfers the same amount of energy in 6 s, has its power increased or decreased?

H **8** A force does 20 000 J of work in lifting a box 5 m upwards. How big is the force?

9 A car engine provides a force of 5000 N. In accelerating the car, it does 200 kJ of work. How far does the car move?

10 A steel ball of weight 50 N hangs at a height of 5 m above the ground on the end of a chain 2 m in length. How much work is done on the ball by gravity and how much is done by the upward pull of the chain?

11 How many watts are there in a kilowatt and in a megawatt?

12 A crane lifts a load of 500 N through a vertical distance of 8 m in 10 s. What is the power of the crane?

13 A light bulb has a power rating of 60 W. How long will it take to transfer 5400 J of energy?

Kinetic energy

It takes energy to make things move. You transfer energy to a ball when you throw it or hit it. A car uses energy from its fuel to get it moving. Elastic potential energy stored in a stretched piece of rubber is needed to fire a pellet from a catapult. So a moving object is a store of energy; this energy is known as **kinetic energy** (KE) (Figure 3e.1).

IRCAM

Figure 3e.1 The word 'kinetic' means 'related to movement'. Many sculptors have had fun making kinetic sculptures – you have probably made a mobile. These kinetic sculptures are at the Centre Georges Pompidou, an arts centre in Paris. Water squirts out of jets on the sculptures, pushing them around so that the water makes great sprays in the air.

The KE of an object depends on two factors:
- the object's mass – the greater the mass, the greater its KE;
- the object's speed – the greater the speed, the greater its KE.

SAQ

1 What can you say about the kinetic energy of a bus:

   a when it is speeding up;

   b when it is travelling at a steady speed;

   c when it is slowing down?

2 A bus, moving at 50 km/h, carries 60 people to their work. Then it returns empty to the depot, also at 50 km/h. How does its KE change between the two halves of its journey?

H Calculating KE

Here is the formula for KE:

$$\text{kinetic energy} = \tfrac{1}{2} \times \text{mass} \times \text{speed}^2$$

Worked example 1 shows how to use the formula to calculate the KE of a moving object.

Worked example 1

A van of mass 2000 kg is travelling at 10 m/s. Calculate its kinetic energy. If its speed increases to 20 m/s, by how much does its kinetic energy increase?

Step 1: Calculate the van's KE at 10 m/s.

$$KE = \tfrac{1}{2} \times \text{mass} \times \text{speed}^2$$
$$= \tfrac{1}{2} \times 2000 \text{ kg} \times (10 \text{ m/s})^2$$
$$= 100\,000 \text{ J}$$
$$= 100 \text{ kJ}$$

Step 2: calculate the van's KE at 20 m/s.

$$KE = \tfrac{1}{2} \times \text{mass} \times \text{speed}^2$$
$$= \tfrac{1}{2} \times 2000 \text{ kg} \times (20 \text{ m/s})^2$$
$$= 400\,000 \text{ J}$$
$$= 400 \text{ kJ}$$

Step 3: calculate the change in the van's KE.

$$\text{Change in KE} = 400 \text{ kJ} - 100 \text{ kJ}$$
$$= 300 \text{ kJ}$$

So the van's KE increases by 300 kJ when it speeds up from 10 m/s to 20 m/s.

Comments on Worked example 1

It is worth looking at the worked example in detail, because it illustrates several important points.

When calculating KE using $\tfrac{1}{2} \times \text{mass} \times \text{speed}^2$, take care! Only the speed is squared. Using a calculator:
- start by squaring the speed
- then multiply by the mass
- finally divide by 2.

When the van's speed doubles from 10 m/s to 20 m/s, its KE increases from 100 kJ to 400 kJ. In other words, its speed doubles but its KE increases by a factor of 4. This is because KE depends on speed squared. If the speed trebled, the KE would increase by a factor of 9 (see Figure 3e.2).

When the van starts moving from rest and speeds up to 10 m/s, its KE increases from 0 kJ to 100 kJ; when its speed increases by the same amount again, from 10 m/s to 20 m/s, its KE increases by 300 kJ, three times as much. It takes a lot more energy to increase your speed when you are already moving quickly.

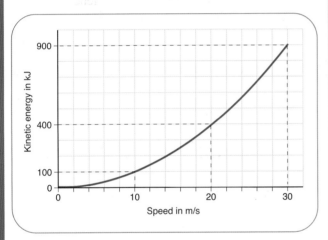

Figure 3e.2 The faster the van travels, the greater its kinetic energy – see Worked example 1. Double the speed means four times the kinetic energy, because KE depends on speed squared. The graph shows that KE increases more and more rapidly as the van's speed increases.

SAQ

3 Calculate the KE of a girl of mass 50 kg running at 8 m/s.

4 A car of mass 2000 kg increases its speed from 10 m/s to 30 m/s. By how much has its KE increased?

Decreasing KE

When a car stops, it loses its KE. Where does this energy go? The car's brakes provide the force needed to stop the car. There is a frictional force in the brakes, and this does the work needed to stop the car:

work done by friction in brakes =

decrease in KE of car

The frictional force generates heat in the brakes, which can become red hot if the driver brakes violently. That's where the car's KE goes – it appears as heat energy in the brakes.

We can relate this to the idea of braking distance (Item P3c). A fast-moving car has more KE; the force of the brakes must do a lot of work to bring it to a halt, which means that the force must act over a long distance. (Remember: work done = force × distance moved.) Because a car moving at 20 m/s has four times as much KE as one moving at 10 m/s, the force must act for four times the distance (so that it does four times as much work). You can see this pattern in the data in Table 3c.2 and in the graph of Figure 3c.9 (on page 239).

SAQ

5 Explain why the braking distance of a car moving at 30 m/s is nine times that of a car moving at 10 m/s.

Fuel figures

Most cars run on fossil fuels. In the UK, that mostly means petrol and diesel. A few cars are adapted to run on liquefied petroleum gas (LPG). When fossil fuels are burned (which is what happens in a car's engine), carbon dioxide is produced, along with other polluting gases.

Demand for fossil fuels is rising worldwide, so prices are rising. That means it is desirable to make the most of the fuel we use. By economising on fuel, we also contribute less pollution to the atmosphere.

Table 3e.1 shows fuel consumption figures for typical petrol and diesel versions of several makes of car. You can see that the diesel versions get more miles per gallon from their fuel – we say that they have better fuel economy.

Make and model	Diesel	Petrol
Ford Focus	51.4 mpg	42.2 mpg
Peugeot 307	54.1 mpg	43.4 mpg
Toyota Corolla	58.9 mpg	42.2 mpg
Vauxhall Corsa	61.4 mpg	53.3 mpg
Volkswagen Golf	53.3 mpg	43.5 mpg

Table 3e.1 Fuel economy data *(figures from VCA)*.

From Table 3e.1, you can see that a diesel engine gives more miles per gallon of fuel. That should make it cheaper to run a diesel car – but they usually cost more to buy in the first place, so a motorist who is looking for the cheaper alternative will have some sums to do.

SAQ

6 Table 3e.2 shows the fuel consumption figures for three different cars. They are given in litres per 100 km; in other words, they show how many litres of fuel are used in driving 100 km. Which car has the best fuel economy? Which has the worst?

Car	Fuel consumption in litres per 100 km
Batmobile	5.8
Herbie	4.9
Chittychitty-Bangbang	5.5

Table 3e.2

H Where does the energy go?

The fuel which a car engine burns is its energy source. Where does the energy stored in the fuel go to?

Think about a car starting from rest. The driver presses on the accelerator pedal and the car moves off.

● Some of the energy in the fuel becomes the kinetic energy of the moving car. The faster the car moves, the more fuel must be burned.

● As the car moves along, it has to overcome air resistance, and there is also friction within the car engine itself. Some energy is required to do work against friction.

Now the driver presses on the brake pedal.

● The car's KE decreases and its brakes get hot.

You can see that using both pedals (accelerator and brake) leads to energy from the fuel being used up. A driver can reduce their car's fuel consumption by thinking about the way they drive.

● They can avoid speeding up and slowing down; a steady speed wastes less fuel.

● They can rely on drag to slow the car down, rather than using the brakes.

H Although high speeds lead to a waste of fuel, so do low speeds. Car engines are designed to be efficient at intermediate speeds. At low speeds, the engine makes poor use of the fuel. That's why sitting in traffic jams can waste a lot of fuel, money and time.

Figure 3e.3 A vehicle undergoing fuel-efficiency testing.

Cars of the future

Fossil fuels cause pollution. Wouldn't we be better off with electric cars? Most car manufacturers have produced electric models (see Figure 3e.4). These have large batteries under the bonnet instead of a petrol engine. The owner charges up the batteries by plugging into the mains supply; when fully charged, the battery can power the car for a reasonable distance – perhaps 200 km. That is easily enough for most people's daily requirements, and the car can be recharged overnight.

There's a problem; where does the mains electricity come from? It comes from a power station and, at least in the UK, most of our electricity comes from burning fossil fuels. So, although the car can drive through the city streets without adding to the pollution, the power station is producing polluting gases somewhere else. So there is an advantage – less pollution where the car is used, but the pollution has simply been transferred to the power station.

Figure 3e.4 An electric car being charged up. This is part of an experimental scheme in Malaysia.

In 2005, Honda announced a new record for car fuel economy. Its Insight model (Figure 3e.5) achieved a fuel economy rating of 103 mpg. The car concerned is described as a hybrid vehicle – part petrol powered, part electric. How does it work?

On the open road, the car uses its petrol engine, just like a conventional car. However, there is a difference when the driver brakes. The car has 'electric brakes'; instead of generating heat, which is wasted energy, the brakes turn dynamos which charge up the car's batteries. So that saves energy.

In town, where pollution is a greater problem, the car switches to using its electric motor, powered by the car's batteries. So the car's fuel economy is improved because it makes better use of its fuel, and it is also less polluting.

Figure 3e.5 The Honda Insight, the first car to break through the 100 mpg barrier.

SAQ

7   Explain why the system shown in Figure 3e.6 can be described as *renewable*.

Hydrogen cars

Engineers at Surrey University (see www.hydrogensolar.com) are devising a system which uses sunlight to produce hydrogen, which can be used as a carbon-free fuel. Figure 3e.6 shows how it works.

● You install solar cells on the roof of your house.

● Sunlight falling on the cells produces electricity.

● The electricity is used to electrolyse (split) water into hydrogen and oxygen.

● The hydrogen is collected and stored, under pressure, in a tank.

● You fill your car's fuel tank with hydrogen and drive off.

Because hydrogen burns with oxygen to produce water, there is no pollution. There is another benefit from this system. Any spare hydrogen can be burned for heating; it can even be used to generate electricity for lighting etc. And if you have any left over after that, you could sell it to your neighbours.

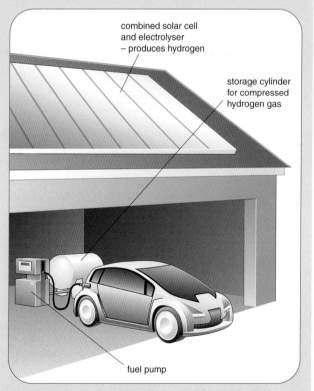

Figure 3e.6 A scheme for powering hydrogen-fuelled cars from sunlight.

Summary

You should be able to:

♦ describe kinetic energy as the energy of a moving object

♦ state that the greater an object's mass and the greater its speed, the greater its kinetic energy

♦ state that most cars use fossil fuels (petrol, diesel) as their energy source and that this results in pollution

♦ state that electric cars are cleaner in use but that pollution is produced when the electricity used to charge their batteries is generated

H ♦ use the equation
kinetic energy = $\frac{1}{2}$ mass × speed^2

♦ state that a car uses fuel to increase its KE and to overcome friction

Questions

1 Which object in Table 3e.3 has the greatest KE? And which has the least KE?

Table 3e.3

Object	Mass in kg	Speed in m/s
rugby ball	2.5	10
football	2.5	15
javelin	4.0	15
tennis ball	0.2	10

2 John has an electric car. He charges its batteries at night from the mains electricity supply. John claims that his car is environmentally friendly.

  a Explain why John's car can help to reduce pollution on the roads.

  b Explain why John is still responsible for some pollution when he uses his car.

H 3 How much KE is stored by a 1 kg ball moving at 1 m/s?

4 Which has more KE, a 2 g bee flying at 1 m/s or a 1 g wasp flying at 2 m/s?

5 A car of mass 750 kg accelerates away from traffic lights. At the end of the first 100 m, it has reached a speed of 12 m/s. During this time, its engine provides an average forward force of 780 N, and the average force of friction on the car is 240 N.

  a Calculate the work done on the car by the force of its engine.

  b Calculate the work done on the car by the force of friction.

  c Calculate the increase in the car's kinetic energy as it accelerates.

  d Can you account for all of the energy supplied to the car's engine?

Smashing fun

Figure 3f.1 This test at the TRL shows a head-on collision between a car, with dummy driver, and a lorry. You can see that both vehicles are wired up with sensors and recording equipment. At the back, engineers are filming the collision.

The Transport Research Laboratory (TRL) is based at Wokingham in Berkshire. Here, scientists and engineers work to design and test improvements in vehicle design. The photograph (Figure 3f.1) shows a vehicle undergoing crash testing. In this case, the lorry has been fitted with an 'under-run guard', the black strip of metal just above the level of the road. These are now a standard feature of trucks; before they were fitted, it was much easier for a car to go under the truck, with fatal consequences for the driver.

However, the TRL isn't solely concerned with vehicle safety. Many accidents can be avoided by improved road design. The TRL is also interested in improving the attitudes and behaviour of drivers, and this work is carried out by a team of psychologists.

Crash, bang, wallop

A car may have a mass of 1 tonne (1000 kg). A large truck may be 44 tonnes. When those vehicles are moving at 70 mph (almost 30 m/s), they have a lot of kinetic energy.

The impact of a collision may last only a second or two. During this time, the vehicles come to rest, so that they no longer have KE. Where does this energy go?

Under normal circumstances, a driver uses the brakes to slow down. The vehicle's KE is reduced by the frictional force within the brakes, which causes a lot of heating. So the brakes are a very effective way of transforming kinetic energy to heat energy.

In a collision, the driver may not have time to apply the brakes. In old-fashioned cars, a lot of the vehicle's KE might then be transferred to its occupants, with disastrous consequences. Modern cars are designed to absorb the energy without transferring it to the occupants.

At the front and back of a car are the **crumple zones**. These parts are designed to crush easily on

impact. They are made of metal which is easily bent, and this takes up energy. The occupants of the car are safely contained within the central section of the car, described as a **safety cage** (see Figure 3f.2). The safety cage is a strong, stiff metal framework around the passenger compartment. So the crumple zone can be squashed on impact while the occupants remain protected.

Two other features are important for absorbing energy: **seat belts** and **air bags** (Figure 3f.3). Think of

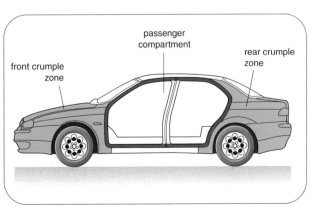

Figure 3f.2 A car has crumple zones at the front and at the back, and a safety cage in between.

a passenger in a moving car. They, too, have kinetic energy which must be absorbed during a collision.

- The seat belt allows the passenger to pitch forward during the collision, but then pulls them back into their seat. Without a seat belt, they would be flung forward; they might even fly out through the windscreen. Because the seat belt has a certain amount of 'give' in it, it slows down the rate at which the passenger comes to a halt, so that they are not jerked suddenly to a halt.
- The air bag inflates almost instantaneously on impact. It explodes outwards, in front of the passenger, and is another way of cushioning the impact.

After a crash, a seat belt may have been permanently stretched and so it must be replaced.

Figure 3f.3 Seat belts and air bags help to cushion the impact of a collision.

SAQ

1 Old-fashioned seat belts kept their wearers firmly restrained in their seats. How would this affect the force of impact on a motorist, compared with a modern belt which 'gives' slightly during a collision?

H Greater distance, smaller force

We can use the idea of a force doing work to understand these car safety features. Remember

that the idea is to reduce the car's kinetic energy while transferring as little as possible of it to the people in the car.

Think first about the crumple zones at the front and back of the car. In a collision, work is done by the force of impact in crumpling the metal of the crumple zone. That uses up a lot of the car's KE, so much less is transferred to the passengers.

Inevitably, some energy is transferred to the occupants of the car. The equation that tells us about this energy transfer is:

work done = force × distance moved

Rearranging gives:

force = work done / distance moved

This tells us that, for the force of impact to be small, we want it to act over a longer distance. So, for example, a modern seat belt allows the wearer to move forward for perhaps 50 cm during the collision, thus reducing the force on them. A tight seat belt holding the person rigidly in place would produce a sudden, jerking force.

There are other safety features which work like this. Motorway crash barriers are designed to crumple when a car hits them. A weak barrier might let a car pass through; a rigid one might cause the car to bounce off into the traffic.

Also, it is a good idea to make the collision happen more slowly, by increasing the length of time it takes to happen. A sudden, violent collision involves a bigger force than a slower collision. This is how seat belts and air bags work; they bring the passenger to rest in a longer time than a sudden jolt.

SAQ

2 Explain how the crumple zone at the front of a car increases the collision distance (the distance the car moves during a collision). Why does this reduce the force of impact?

Safety features, active and passive

You are cycling along. Suddenly, a ball flies out into the road in front of you, followed by a child. You jam on your brakes – and you skid to a halt.

Skidding is tricky. Your bike is out of your control; it may slew round sideways, as the back end overtakes the front. You may tip over and fall to the ground. That is bad enough on a bike, and it could be much worse in a car. In order to avoid a collision, a driver needs to be in control of their car, so it is vital that they avoid getting into a skid.

The only parts of a car in contact with the road are the tyres. These provide the traction (pulling force) to move the car along, and that depends on **friction**. If there is not enough friction, the tyres do not grip and the wheels start to spin (see Figure 3f.4).

Figure 3f.4 Skidding can be fun, in the right circumstances. The wheels have lost their grip on the road surface, throwing mud everywhere.

Active safety

Ideally, when a driver needs to stop in a hurry to avoid a collision, they need to apply the brakes firmly, to stop as quickly as possible, but not so firmly that the car skids. That is difficult to do and, in the past, drivers needed to practise avoiding skidding. Today, many cars are fitted with **ABS brakes**. ABS stands for anti-lock braking system. How does it work?

There are sensors on the car's axles (Figure 3f.5). When the driver brakes hard, these sensors check that the wheels are not skidding round. If they are, the pressure of the brakes is reduced slightly until the sensors detect that the wheels are pulling properly on the road.

A car's wheels may also skid round if the driver tries to accelerate too much. **Traction control** is

designed to avoid this. It is similar to ABS, but it works when the driver is accelerating. Sensors detect if the car's wheels are spinning because they have lost their grip on the road. The brakes are automatically applied until traction is restored.

ABS and traction control are **active safety features** because they involve a computer detecting what is going wrong and acting to correct the problem. A third active control is **stability control**, which comes into effect when a car starts to 'yaw', so that it twists round whilst braking or accelerating. The controlling computer applies the brakes to just two wheels so that the car returns to pointing in the correct direction.

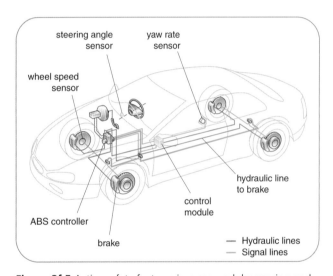

Figure 3f.5 Active safety features in a car work by sensing and controlling the car's brakes.

Passive safety

There are many ordinary features of cars which help to improve safety, simply because they are designed to make it easier for the driver to operate them. Drivers can be easily distracted, so the simpler a car is to drive, the better.

Electric windows. Wind-down windows require the driver to take one hand off the wheel for several seconds, and it would be dangerous to reach across to wind down the passenger window.

Adjustable seating allows the driver to find a comfortable position from which to operate the car's controls and to see all around.

Cruise control. On a long, safe stretch of road, the car reaches a safe 'cruising' speed. The driver flicks a switch and the car knows that it must maintain this speed. The driver can relax.

Paddle controls are the 'stalk' controls which stick out behind the steering wheel. These may control the indicators or the stereo system, or they may allow the driver to operate the gears without letting go of the steering wheel.

All of the above are described as **passive safety features** because the driver doesn't have to take any special action to operate them. Their very design ensures that the driver will be able to focus more attention on the road ahead.

SAQ

3 Which safety feature is described here? 'The car's brakes are automatically applied if the wheels start to skid while the driver is attempting to accelerate.'

H Safety features save lives

ABS brakes can help to stop a car in a shorter distance than might otherwise be the case. How is this?

- A driver wants to stop in a hurry. If the car goes into a skid, there is no frictional force between the tyres and the road. So the car moves forward without slowing down. That increases the braking distance.
- To avoid the car skidding, the driver may press cautiously on the brakes. The braking force is thus smaller than it might be and, again, the braking distance is longer than necessary.
- With ABS, the driver brakes hard and the automatic system applies just the right force to the brakes without going into a skid. This minimises the braking distance.

Summary

You should be able to:

◆ state that, when a vehicle slows down, its kinetic energy must be transformed; it may become heat in the brakes or it may result in the deformation of the crumple zone

◆ state that seat belts and air bags cushion an impact, reducing the force on driver and passengers

◆ describe a range of safety features, active and passive, which help to reduce injuries on the road

H ◆ explain that, to reduce the forces when a vehicle stops suddenly, the stopping time and stopping distance should be increased

◆ explain how, by controlling the frictional force, ABS brakes bring a vehicle safely to a halt in the minimum distance.

Questions

1 Divide the following car safety features into two lists, active and passive:

ABS braking   electric windows   safety cage   paddle shift controls   traction control

2 Why must seat belts be replaced after a collision?

H 3 On some steep hills, there is an 'escape lane' for vehicles which have gone out of control going down the hill. Imagine a large truck, running downhill; its brakes fail. It enters the escape lane at the side of the carriageway; at the end of this lane is a large trough filled with sand or gravel. This causes the truck to grind to a halt.
   Use the ideas you have studied in this chapter to give a scientific explanation of this.

4 Calculate the KE of a driver of mass 80 kg moving at 25 m/s. In a collision, the driver is brought to a halt in a distance of 50 cm. What is the average force acting on the driver?

The force of gravity

The Earth's gravity is a familiar force that we experience every day. Jump up in the air and you fall back down. Let go of a book and it falls to the floor. When we are standing up, our muscles keep working to ensure that our bodies remain upright; if our muscles stopped working, we would collapse in a heap on the ground. The force of gravity attracts everything towards the centre of the Earth.

The force caused by gravity on any object is called its **weight**. Like any other force, gravity can make things accelerate. You can see this in the multiflash photograph of Figure 3g.1. A ball has been bounced, and its position is shown at regular intervals of time. As the ball accelerates downwards, the images become increasingly far apart.

If you simply drop an object onto the floor, you won't really be conscious of the fact that it speeds up as it falls. But jump out of a high-flying aircraft and you will soon notice that you go faster and faster, at least to start off with. Don't forget to pull the cord on your parachute!

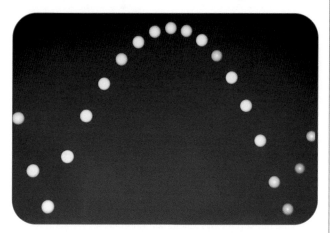

Figure 3g.1 A multiflash photograph of a ball bouncing. The images of the ball are made at equal intervals of time. Gravity makes the ball accelerate, so the images become further apart as it falls.

Measurements show that the acceleration of a falling object is about $10 \, \text{m/s}^2$. (A more accurate value is $9.8 \, \text{m/s}^2$. The precise value of the acceleration caused by gravity depends on just where you are on the Earth's surface.)

Acceleration caused by gravity = $10 \, \text{m/s}^2$

SAQ

1 How can you tell from the photo in Figure 3g.1 that the ball's speed is changing? How can you tell whether the ball is accelerating or decelerating?

Friction, drag and air resistance

Friction is the force which tends to slow down objects when they are moving. It comes about because surfaces cannot be perfectly smooth. On a microscopic scale, all surfaces are rough. This means that they tend to interlock with one another, as shown in Figure 3g.2. A force is needed to push the irregularities on one surface past the irregularities on the other. As they are pushed past one another, the surfaces may be damaged – this is why surfaces tend to wear away as they slide over one another.

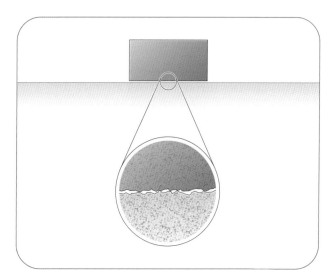

Figure 3g.2 Friction comes about because of the roughness of surfaces, often on a microscopic scale. Pushing one surface over the other causes the surfaces to lock together. A force is needed to push one surface past the other.

You would find it very difficult to move without friction. We all rely on friction in order to walk about – think about the difficulty of walking on an icy surface. In the same way, it is difficult for a car to start moving on an icy or oily surface. Figure 3g.3 shows the part played by friction when a car moves along the road.

The bottom of the car's wheel, where it touches the road surface, is pushing backwards. Friction

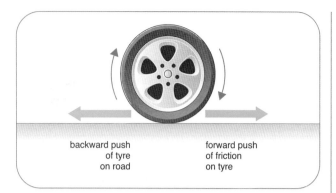

Figure 3g.3 We need friction to get us moving. The backward push of the car tyre on the road produces the frictional force which pushes the car forwards.

backward push of tyre on road

forward push of friction on tyre

pushes in the opposite direction, so it is the force of friction on the car which makes it move.

Drag is a similar force to friction, when an object is moving through a fluid (a liquid or gas). If a whale or a boat, for example, is moving through water, it has to push the water aside. At the same time, the water rubs against its sides. You will have experienced the same thing if you have ever run into the sea; the deeper the water, the greater its resistance to your motion. In deeper water, it is easier to swim forwards than to walk.

Air resistance is the same as drag, when an object is moving through air. An aircraft, for example, moves rapidly through the air and the air resistance tends to slow it down. Air resistance has much less effect than the drag of water because air is much less dense than water, so air is more easily pushed aside.

Air resistance is much more significant for small creatures than for people. Cats are lighter than people, so they do not reach as high a speed as a free-fall parachutist. A cat can survive a fall from a high building, as can other small creatures such as mice. For insects, air resistance is very significant. They can fall a long way without being damaged.

We make use of air resistance in the game of badminton. The shuttlecock is designed to have considerable air resistance; without its feathers, it would fall much more rapidly and the game would be very frustrating indeed.

Falling through the air

Gravity makes you fall, but air resistance can save you. That is the idea used by parachutists (Figure 3g.4). Free-fall parachutists jump from an

Figure 3g.4 Free-fall parachutists, before they open their parachutes. They can reach a terminal speed in excess of 100 mph (45 m/s).

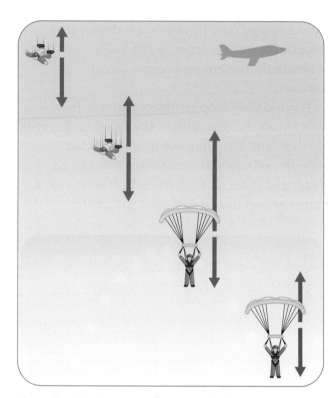

Figure 3g.5 The forces on a falling parachutist. Notice that his weight is constant. When air resistance equals weight, the forces are balanced and the parachutist reaches a steady speed. The parachutist is always falling, although his acceleration is upwards when he opens his parachute.

aircraft. At first, they accelerate downwards with an acceleration of 10 m/s^2, just like any other falling object. However, air resistance gradually comes into play. Air resistance acts upwards on the parachutists who are falling downwards, and this tends to slow them down. The faster they go, the greater the force of air resistance; eventually, air resistance equals weight and the forces are balanced (Figure 3g.5). The parachutists have reached their top speed, known as their **terminal speed**. Terminal speed for a person in free fall is over 50 m/s, considerably faster than the speed limit for traffic on a motorway.

Balanced and unbalanced forces

It is worth taking a close look at Figure 3g.5. You can see that the parachutist's weight is constant all the way down. It doesn't change. However, the faster the parachutist moves, the greater the force of air resistance.

There are several other points to note as well.

- At first, there is little air resistance, because the parachutist is moving slowly. The parachutist's weight is greater than air resistance and so the forces are *unbalanced*. He accelerates.

- Eventually, there is enough air resistance to balance his weight. The forces on the parachutist are *balanced*, so he no longer accelerates. He falls at a steady speed.

- When the parachute opens, there is suddenly a much bigger upward force. The parachutist continues falling, but now the forces on him are *unbalanced* and he decelerates.

- Air resistance decreases as he moves more slowly. Eventually, air resistance = weight again and the forces are *balanced*. The parachutist falls at a steady, safe speed.

SAQ

2   What can you say about the forces on a parachutist falling with terminal speed – are they balanced or unbalanced?

H **Speed control**

Greater speed means greater air resistance. However, there is another factor which affects air resistance: the size and shape of the moving object. Think about a free-fall parachutist – he can adjust his speed by changing the position of his body. Head-first is a streamlined shape, reducing air resistance and giving a greater terminal speed. Spread-eagled, with legs and arms spread out horizontally, is a slower way to travel. This is because a greater area is presented to the air, and so there is more drag.

H Opening the parachute greatly increases the area and hence the air resistance. Now there is a much bigger force upwards; the forces on the parachutist are unbalanced and he slows down. The idea is to reach a new, slower terminal speed of about 10 m/s before reaching the ground.

At this point, weight = drag, and so the forces on the parachutist are balanced.

The graph in Figure 3g.6 shows how the parachutist's speed changes during a fall.

- When the graph is horizontal, speed is constant and forces are balanced.

- When the graph is sloping, speed is changing; the parachutist is accelerating or decelerating and forces are unbalanced.

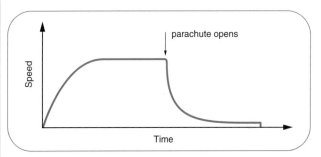

Figure 3g.6 A speed–time graph for a falling parachutist.

SAQ

3   Look at the graph of Figure 3g.6. Find a point where the graph is sloping upwards.

  a Is the parachutist accelerating or decelerating? Which of the two forces acting on the parachutist is greater?

  b Explain the shape of the graph after the parachute has opened.

Reducing friction

Because friction can result in overheating and wear, a great deal of effort has been put into reducing its effects. If you have a bicycle, you may have noticed the benefit of oiling the axles so that the wheels can turn more freely. This is an example of **lubrication**. The oil provides a smooth layer of liquid between the two surfaces so that they can slide more readily over one another.

Drag can be reduced by making sure that the surfaces moving through a fluid are smooth. The hull of a boat may need to be cleaned to remove barnacles and rust, which may be slowing it down. An alternative approach is to reduce the amount of the boat which is in contact with the water. This is used in the modern catamaran-style ferry, whose speed causes it to rise up and ride with most of its hull above the water.

Streamlining is another way to reduce drag. Sports cars and racing cars have a wedge shape to cut through the air. Similarly, many large trucks have a wind deflector fitted to the roof (Figure 3g.7) so that they don't present a flat surface to the air. (Some caravan owners fit a similar deflector.)

Thinking about drag can help drivers to save money. For example, a roof box fitted on top of a car can add greatly to air resistance, and so the car uses much more fuel. Even an unladen roof rack should be removed if the driver wants to economise on petrol.

Figure 3g.7 Streamlining and wind deflection can reduce air resistance and save fuel.

Top speed

Every car has a top speed, the fastest it can go. The speedometer may go up to 150 mph, but the top speed of a typical family car is usually much less than this, perhaps 100 mph. What stops the car from going any faster? The answer, of course, is air resistance. The driver can press on the accelerator to make the car go faster. However, the faster the car moves, the greater the air resistance. Eventually, the car reaches a speed where the backward force of air resistance equals the forward force provided by the engine (Figure 3g.8). At this speed, the forces on the car are balanced and it can go no faster.

A balance of forces

A racing car is designed to have a much higher top speed than a family car. It has a streamlined shape (to reduce drag) and a powerful engine. This means that it has a high top speed.

A large **driving force** means that the car can reach a high speed before the driving force is balanced by air resistance:

driving force = air resistance

SAQ

4 Imagine a racing car setting off at the start of a race. The driver presses on the accelerator to give a large, steady driving force. When driving at top speed, the driver has to brake suddenly, increasing the frictional forces on the car.

Sketch a speed–time graph for this journey. Indicate where the forces on the car are balanced and where they are unbalanced.

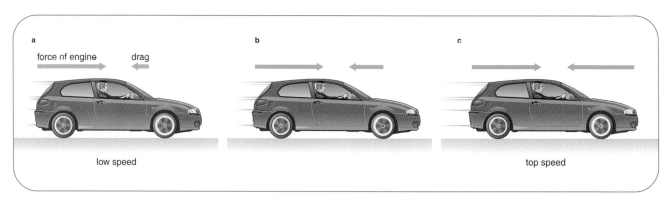

Figure 3g.8 a At low speeds, a car can accelerate because the force of its engine is greater than the drag opposing its motion. **b** As the car's speed increases, the drag on it also increases. **c** At top speed, the drag has increased further until it equals the engine force.

Summary

You should be able to:

◆ state that the force of gravity (weight) causes objects to accelerate as they fall

◆ state that unbalanced forces cause an object to speed up or slow down

◆ state that balanced forces result in steady speed

◆ describe how frictional forces (drag, air resistance) act against movement, and how lubrication and streamlining can reduce these forces

◆ explain how air resistance causes an object falling through air to reach a terminal speed

H ◆ state that air resistance (drag) increases with speed and area

◆ use the idea that, at terminal speed, weight = drag

◆ use the idea that, for a car at top speed, driving force = drag

Questions

1 An apple falls from a tree.

   a What force causes it to fall?

   b How does its speed change as it falls?

2 A car is travelling along at a steady speed. The driving force from its engine pushes it forward.

   a What force acts in the opposite direction to the driving force?

   b Are the forces on the car balanced or unbalanced? Explain how you know.

3 Sketch two cars, one with a more streamlined shape than the other. Explain why a streamlined shape means that a car makes better use of its fuel.

H 4 List as many ways as you can in which a car's design can be altered to give it a greater top speed.

5 A sky diver of weight 1000 N is falling through the air. Sketch a diagram to show the forces acting on her when she reaches terminal speed. What is the force of air resistance acting on her? What resultant force acts on her?

6 Explain why the opening of a parachute causes a parachutist to slow down.

7 Sketch a distance–time graph for a parachutist. Indicate the following stages:

   ● parachute opened   ● terminal speed

   ● falling freely   ● falling slowly and steadily to ground.

8 Imagine that you are playing a game of badminton. Your opponent has surreptitiously added a small metal weight to the shuttlecock. Describe how you would expect the shuttlecock to behave, and explain why using the ideas of air resistance and terminal speed.

Gravitational potential energy

At the start of a roller-coaster ride, your car is hauled up to the top of the slope (Figure 3h.1). After that, it runs freely downhill and uphill until its brakes bring it to a halt at the bottom of the ride. The car has no motor; it is able to complete the run because it has been given energy by that initial pull to the highest point of the ride.

Figure 3h.1 The roller-coaster car has been given energy when it is pulled to the top of the ride.

At the top of the ride, we say that the car has **gravitational potential energy (GPE)**, or simply **potential energy (PE)**. Any object which has been lifted up has GPE. The word *potential* suggests that the object has a store of energy which *could* be used to do something.

- A raised hammer has GPE; it could be used to bang a nail into wood.
- A diver on a high board has GPE; it could be used to plummet into the water below.
- Water behind a hydroelectric dam has GPE; it could be used to generate electricity.

More GPE

The amount of GPE which an object has depends on its height above the ground. The higher it is, the greater its GPE. An object moving upwards has increasing GPE; a falling object has decreasing GPE. An object moving along horizontally isn't changing its height, so its GPE is not changing.

If you lift an object upwards, you provide the force needed to increase its GPE. The greater the mass of the object, the greater the force needed to lift it and hence the greater its GPE.

This suggests that an object's gravitational potential energy depends on two factors:

- the height of the object above ground level – the greater its height, the greater its GPE;
- the object's mass – the greater its mass, the greater its GPE.

This is illustrated in Figure 3h.2. When the box is lifted onto the table, its GPE has increased. When it is lifted onto the top of the filing cabinet, its GPE has increased further.

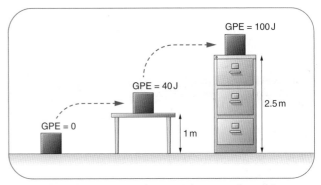

Figure 3h.2 The gravitational potential energy of an object increases as it is lifted higher. The greater its mass, the greater its GPE.

SAQ

1 In the following examples, is the object's GPE increasing, decreasing or remaining constant?

   a An apple falls from a tree.

   b An aircraft flies horizontally at a height of 9000 m.

   c A skyrocket is fired into the sky.

2 Whose rock has the greatest GPE? Whose has the least?

   - John lifts a rock of mass 5 kg on to a table 1.2 m high.
   - Jane lifts a rock of mass 5 kg on to a table 1.5 m high.
   - Jack lifts a rock of mass 4 kg on to a table 1.2 m high.

The strength of gravity

Mountaineering on the Moon should be easy (see Figure 3h.3). The Moon's gravity is much weaker than the Earth's, because the Moon's mass is one-eightieth of the Earth's. This means that the weight of an astronaut on the Moon is a fraction of his or her weight on the Earth. In principle, it is possible to jump several times as high on the Moon. Unfortunately, because an astronaut has to carry an oxygen supply and wear a cumbersome suit, this is not possible.

Experiments on the Moon have shown that a golf ball can be hit much farther than on Earth because it travels a much greater distance horizontally before gravity has pulled it back to the ground.

Figure 3h.3 An astronaut on the Moon.

There are planets where gravity is stronger than here on Earth – Jupiter, for example. An astronaut on the surface of Jupiter would find that gravity is about two and a half times as strong as on Earth, so getting about would be much harder.

The strength of gravity varies from planet to planet and from moon to moon. We can give an idea of the strength of gravity by stating the gravitational field strength (symbol g). Table 3h.1 shows values of g for different places in the solar system.

Because the gravitational field strength on the Moon is weak, it is easier to lift an object there. It takes less work to raise an object and so it has gained less GPE.

Place	Gravitational field strength in N/kg
Earth	9.8
Moon	1.6
Venus	8.9
Mars	3.7
Jupiter	23.1

Table 3h.1 Gravitational field strength on the surface of various planets and the Moon.

SAQ

3 Which planet in Table 3h.1 has the weakest gravitational field? Which has the strongest?

⊞ Calculating weight

The GPE of an object depends on its weight, so we need to know how to calculate the weight of an object if we know its mass.

Weight = mass × gravitational field strength

(You could have guessed this from the units of gravitational field strength, newtons per kilogram.)

For objects close to the Earth's surface, we can use, as an approximation:

$g = 10\,\text{N/kg}$

so that an object of mass 1 kg has a weight of 10 N.

SAQ

4 Calculate the weights of the following objects on the surface of the Earth:

   a a box of mass 0.5 kg

   b a person of mass 80 kg

   c a car of mass 1200 kg.

5 An astronaut on the Moon has a weight of 240 N. What is his mass? (Gravitational field strength on the Moon = 1.6 N/kg)

Calculating GPE

Figure 3h.4 is the same as Figure 3h.2 but with data added. From the values given in the diagram, you can see that GPE is calculated simply by multiplying weight and height.

H Gravitational potential energy = weight × height

We can write this in symbols:

GPE = *mgh*

where *m* = mass, *g* = gravitational field strength and *h* = height.

(Here, we are assuming that an object's GPE is zero when it is at ground level.)

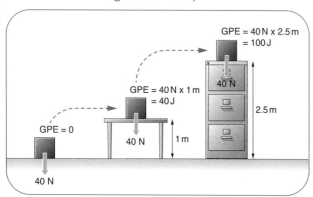

GPE = 40 N x 2.5 m = 100 J

40 N

2.5 m

GPE = 40 N x 1 m = 40 J

40 N

1 m

GPE = 0

40 N

Figure 3h.4 Calculating GPE.

Worked example 1

An athlete of mass 50 kg runs up a hill. The foot of the hill is 400 m above sea level; the summit is 1200 m above sea level. By how much does the athlete's GPE increase? (Acceleration due to gravity g = 10 m/s².)

Step 1: calculate the increase in height.

h = 1200 m – 400 m = 800 m

Step 2: write down the equation for GPE, substitute values and solve.

Change in GPE = weight × change in height = *mgh*

$$= 50\,kg \times 10\,N/kg \times 800\,m$$

$$= 400\,000\,J$$

$$= 400\,kJ$$

So the athlete's GPE increases by 400 kJ.

A note on height

We have to be careful when measuring or calculating the change in an object's height. Firstly, we have to consider the *vertical* height through which it moves. A train may travel 1 km up a long

H and gentle slope, but its vertical height may only increase by 10 m. A satellite may travel around the Earth in a circular orbit. It stays at a constant distance from the centre of the Earth and so its height doesn't change. Its GPE is constant.

Secondly, it is the *change* in height of the object's centre of gravity that we must consider. This is illustrated by the high-jumper shown in Figure 3h.5. As she jumps, she must try to increase her GPE enough to get over the bar. In fact, by curving her body, she may pass over the bar while her centre of gravity passes under it!

Figure 3h.5 This high-jumper adopts a curved posture to get over the bar. She cannot increase her GPE enough to get her whole body above the level of the bar. Her centre of gravity may pass under the bar, so that at no time is her body entirely above the bar.

SAQ

6 A girl of weight 500 N climbs on top of a 2 m high wall. By how much does her GPE increase?

7 A stone of weight 1 N falls downwards. Its GPE decreases by 100 J. How far has it fallen?

KE–GPE transformations

The ski jumper shown in Figure 3h.6 has travelled uphill on a chairlift. This increases her gravitational potential energy. Now she can ski downhill; her kinetic energy increases as her GPE decreases. GPE is being transformed (converted) into KE. As she jumps from the ramp, she starts to rise in the air. Now KE is being transformed into GPE.

There are many other situations where such transformations are going on.

Figure 3h.6 Energy transformations are going on as a ski jumper jumps. Accelerating downhill, GPE is transformed to KE. As she rises from the end of the ramp, KE is transformed back to GPE.

- An apple falls from a tree. It accelerates downwards; GPE is being transformed to KE.
- On a roller-coaster ride, a car runs downhill, then back up the next slope. GPE is being transformed to KE and then back to GPE.
- A pendulum is swinging from side to side (Figure 3h.7). At its highest point, it is momentarily stationary. It has GPE but no KE. As it swings downwards, it speeds up; GPE is transformed to KE. At the lowest point of its swing, it is moving fastest, so its KE is greatest and its GPE is at its least. It slows down again as it swings back up; KE is transformed back to GPE.

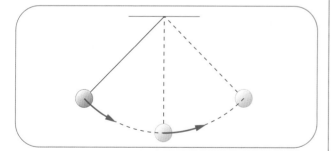

Figure 3h.7 As a pendulum swings back and forth, there is a constant interchange of energy between GPE and KE. The mass on the end of the pendulum has more KE at the lowest point of its swing, where it is moving fastest. Its GPE increases as it rises upwards towards the ends of its swing.

Let's take a closer look at the case of the car on the roller-coaster ride (Figure 3h.8).

- Initially, it must be given GPE. A force does work to pull the car to the highest point.
- As the car runs down the first slope, GPE is being transformed to KE. The car goes faster and faster all the time because, the lower it is, the less its GPE and the greater its KE.
- Now the car starts to run back uphill. Its KE decreases, so it goes slower. Its GPE increases as it goes higher.
- As the car travels along, some of its energy is lost as heat, because of friction. This means it can never get back all of the GPE it started off with and so it cannot reach its original height.

SAQ

8 Figure 3h.9 shows the Oblivion ride at Alton Towers theme park. The car with its passengers plunges down the final drop into a smoke-filled pit. The track then curves so that the car ends up running horizontally; its brakes bring it to a sudden halt.

   a What form of energy does the car have when it is waiting to drop?

   b What energy change occurs as the car drops?

   c What form of energy does the car have as it travels along the horizontal track?

   d At the end, the car is stationary. Where has the car's energy gone?

Figure 3h.9 Oblivion.

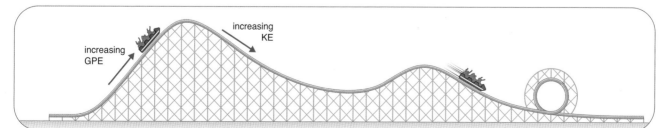

Figure 3h.8 Energy changes along a roller-coaster ride.

Terminal speed

What is going on when a falling object reaches its terminal speed? It is falling, so it is losing GPE. However, it isn't going any faster, so its KE isn't increasing. Where does its GPE go?

Recall that it is friction (air resistance) which causes the object to reach terminal speed. The falling object is doing work against friction, and that generates heat. So, at terminal speed, GPE is being converted directly to heat energy; none of it is becoming KE.

Energy calculations

We can use the idea of energy transformation to perform some useful calculations.

Galileo and gravity

No one knows for sure whether Galileo (Figure 3h.10) performed the experiment described in Worked example 2. In his old age, he mentioned it to a young student when discussing his work in Pisa, and he certainly had a heated debate with other scholars at the time. The question was this: which falls faster, a heavy object or a light one? Galileo claimed that they would fall at the same rate, provided there was no friction to slow them down. The opposing view was that heavy objects fell faster; in particular, if one object weighed ten times as much as the other, it would fall ten times as fast.

Figure 3h.10 Galileo Galilei (known as Galileo) at the age of 42 in 1606.

Today, we agree with Galileo. However, when he was working (in the early 1600s), scientists did not look for the answers to questions in experiments; rather, they studied the writings of long-dead Greek and Arabic philosophers. Aristotle had said that heavy objects fall faster, so they stuck to that view.

It is said that Galileo tried to convince them that they were wrong by dropping large and small cannonballs from the top of the tower in Pisa. The larger one touched the ground momentarily before the smaller one – it was less affected by air resistance. 'Just as we predicted,' said his opponents. Of course, their prediction would have been that the heavier one would reach the ground *much* sooner than the lighter one. Galileo grumbled that they were splitting hairs, and that the experiment justified his view.

Galileo argued like this: suppose a small mass falls slowly and a large one falls more quickly. Now imagine tying a small mass to a larger one. Now we have an even larger mass, which should fall even faster. However, we can imagine the small mass trying to fall more slowly, because it is pulled more weakly by gravity. It will slow down the larger one. Similarly, we would expect the larger one to speed up the smaller one. Their speed, when joined, should be in between their separate speeds, not greater than either.

The solution to this disagreement is to suppose that they fall at the same rate. Then it makes no difference whether they are tied together or not. The large and small masses will fall at the same rate.

This shows the value, in Science, of posing 'What if...?' questions. Sometimes you can reach a sensible answer to a problem without carrying out an experiment; an understanding of the underlying science can help you to work out what the outcome will be.

Today, we would explain Galileo's idea as follows: a larger mass needs a bigger force to make it accelerate, but gravity provides this larger force because it pulls more strongly on a larger mass. Hence, objects fall at the same rate, regardless of their different masses.

A falling object starts off with GPE and ends up with KE:

> decrease in GPE = increase in KE

We can write this as a formula:

$$mgh = \tfrac{1}{2}mv^2$$

This equation can be solved to give v when we know h, or h when we know v (where v = speed).

Figure 3h.11
The Leaning Tower of Pisa. Galileo is said to have carried out experiments on the effect of gravity on falling objects by dropping them from the top of the tower.

Worked example 2

Galileo dropped a steel ball of mass 2 kg from the top of the Leaning Tower of Pisa, 55 m above the ground below (Figure 3h.11). How fast was it moving when it reached the ground?

Step 1: write down the equation which represents the transformation of GPE to KE.

$$mgh = \tfrac{1}{2}mv^2$$

Step 2: substitute values and simplify.

$$2\,\text{kg} \times 10\,\text{m/s}^2 \times 55\,\text{m} = \tfrac{1}{2} \times 2\,\text{kg} \times v^2$$

$$1100\,\text{J} = 1\,\text{kg} \times v^2$$

Step 3: rearrange, and solve for v.

$$v^2 = 1100\,\text{J} / 1\,\text{kg}$$

$$v = (\sqrt{1100})\,\text{m/s} = 33\,\text{m/s}$$

So the ball will be moving at 33 m/s when it reaches the ground. Note that, to get this answer, we have assumed that all of the ball's GPE will be transformed to KE; none is lost as heat due to friction with the air.

Summary

You should be able to:

- state that raising an object gives it gravitational potential energy (GPE)

- state that the GPE of an object depends on:

    - its mass

    - the height through which it is raised

    - the gravitational field strength g

- describe how, as an object falls, GPE is converted to KE

- use the formula: weight = mg

- use the formula: GPE = mgh

- explain that, at terminal speed, work is done against friction and GPE is converted to heat

Questions

1 A ball rolls down a smooth slope. What energy transformation is taking place here?

2 A crane lifts two loads of bricks to the top of a tall building. Load A has a mass of 180 kg; load B has a mass of 200 kg. Which has the greater GPE when they reach the top of the building, and why?

3 Copy the diagram of the roller-coaster in Figure 3h.12. On it, mark the following points:

   a a point where the car's KE is decreasing.

   b a point where GPE is being converted to KE.

   c the point where the car has most GPE.

   d the point where the car has least GPE.

Figure 3h.12

4 What is the weight on the Earth's surface of a brick of mass 3.5 kg?

5 What is the gain in GPE of a box of mass 150 kg when it is lifted 30 m vertically upwards?

6 Calculate the gain in GPE of the box when it is lifted as shown in Figure 3h.13.

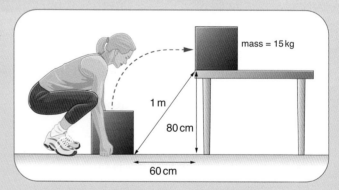

Figure 3h.13

7 A skier comes downhill. We could calculate his speed using $mgh = \frac{1}{2} mv^2$. What assumption are we making if we use this formula?

8 Use the formula $mgh = \frac{1}{2} mv^2$ to explain why an object dropped on the Moon is moving more slowly when it hits the ground than if it had been dropped from the same height on the Earth.

9 Think of a car, moving at a steady speed. It is burning fuel, so that it is using up energy stored in the fuel. However, it is not going any faster, so that its KE is not increasing. Where does the energy of the fuel go to?

Bringing lightning down to Earth

Benjamin Franklin was an American, born in Boston in 1706. He was a scientist, as well as many other things – politician, printer, economist, musician, publisher. His most famous experiment (Figure 4a.1) involved him in a most dangerous activity, flying a kite in a thunderstorm. He was investigating lightning as part of his studies of static electricity.

Figure 4a.1 Benjamin Franklin, flying a kite in an attempt to capture a bolt of lightning. Franklin showed that lightning is similar to the sparks produced in experiments on static electricity. In July 1753, the scientist Georg Richmann was killed in St Petersburg when he tried to repeat Franklin's experiment. Richmann's body was dissected to discover the effect of electricity on his organs.

In the middle of the 18th century, static electricity was a subject of much popular interest (see Figure 4a.2). Scientists gave demonstration lectures and sold equipment to members of the public so that they could try the experiments for themselves. Benjamin Franklin believed that lightning was simply a form of static electricity. He pointed out that a lightning flash was similar in shape and colour to the sparks which could be produced in the laboratory. They seemed to travel with similar speeds and both were known to kill animals. He devised an experiment in which a man stood inside a tall 'sentry box' hut on top of a high tower. Using a long iron rod protruding from the top of the box, the man would draw sparks from storm clouds passing overhead.

In a further demonstration, Franklin attached a sharply pointed metal wire to the top of a kite. He expected to draw down a spark from a lightning bolt; to avoid being electrocuted, he included a metal key at the bottom of the kite string and attached a length of ribbon to the key. Holding the ribbon, he felt he was relatively safe from electrocution. As a bolt of lightning struck the kite, Franklin saw the fibres of the kite string stand on end and a spark jumped from the key to the ground. (Other people who repeated Franklin's experiment, were killed, so don't try this at home.)

Figure 4a.2 Interest in static electricity made it something of a hobby for people in the 18th century. These Americans are rubbing rods to produce an electric charge.

Static electricity

A lightning flash is similar to the flashes you might notice if you pull off an acrylic jumper in a darkened room. We say that the jumper has become charged with **static electricity**.

We experience static electricity in a number of ways in everyday life. There are the tiny sparks you can see and even hear when taking off clothes made of synthetic fibres. You may have felt a small shock when getting out of a car; static electricity builds up on the car and then discharges through you when you touch the metal door. You may get a similar shock if you drag your feet as you walk across a floor which is covered with an insulating material such as vinyl. You have probably rubbed a balloon on your clothes and seen how it will stick to a wall or ceiling.

Before Benjamin Franklin and other scientists started carrying out their systematic experiments on static electricity, little was known about it. It had been known for centuries that amber, when rubbed, could attract small pieces of cloth or paper. Amber is a form of resin from trees, which has become fossilised. It looks like clear, orange plastic. The Greek name for amber is electron and this is where we get the name of the tiny charged particles which account for electricity.

Franklin, and those who worked on the problem at the same time as him, had no idea about electrons; these particles were not discovered until a hundred years later. However, that didn't stop them from developing a good understanding of static electricity. In the discussion which follows, we will talk about **electrons**; after all, they were discovered over a century ago and they make it much easier to understand what is going on in all aspects of electricity.

Charging up

If you rub a plastic ruler with a cloth, both are likely to become electrically charged by the force of **friction**. You can tell that this is so by holding the ruler and then the cloth close to your hair; they attract the hair. (If your hair is not attracted, try some tiny scraps of paper or cork instead.) Some dusters and dusting brushes make use of this – the brushes have fine nylon bristles which become charged up, so that they attract dust. Notice that the materials involved, plastic, cloth and nylon, are all **insulators**.

We can sum up these observations:
● static electricity is generated by rubbing;
● a charged object may attract uncharged objects.

Static explanations

Now we have to think systematically about how to investigate this phenomenon. Firstly, how do two charged objects affect one another? Figure 4a.3 shows one way of investigating this. A plastic rod

is rubbed with a cloth so that both become charged. The rod is hung in a cradle so that it is free to move; when the cloth is brought close to it, the rod moves towards the cloth (Figure 4a.3a).

If a second rod is rubbed in the same way and brought close to the first one, the hanging rod moves away (Figure 4a.3b). Now we have seen both attraction and repulsion, and this suggests that there are two types of static electricity. Both rods have been treated in the same way, so we expect them to have the same type of electricity; the cloth and the rod must have different types.

The two types of static electricity are referred to as **positive charge** and **negative charge**. We can explain the experiments shown in Figure 4a.3 by saying that the process of rubbing gives the rods one type of electric charge (say, negative) while the cloth is given the opposite type (say, positive). Figures 4a.3c and d show the two experiments with the charges marked.

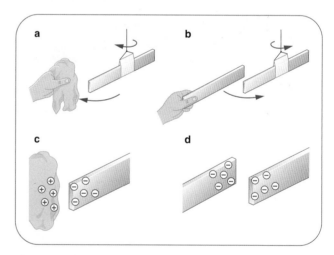

Figure 4a.3 Two experiments, to show the existence of two opposite types of static electricity. **a** The charged rod and cloth attract one another. **b** The two charged rods repel one another. **c** The rod and the cloth have opposite electrical charges. **d** The two rods have electrical charges of the same sign.

From these experiments, we can also say something about the forces which electrical charges exert on each other.
● Like charges repel.
● Unlike charges attract.

('Like charges' means both are positive or both are negative. 'Unlike charges' means opposite charges: one positive and the other negative. People often remember this rule as 'opposites attract'.)

Friction and charging

It is the force of friction which causes charging. When a plastic rod is rubbed on a cloth, friction transfers tiny particles called **electrons** from one material to the other. If the rod is made of polythene, it is usually the case that electrons are rubbed off the cloth and onto the rod.

The two objects have equal amounts of charge, but of opposite signs (positive and negative).

SAQ

1 What force is involved when two objects are rubbed together and become charged?

2 Two positively charged polystyrene spheres are held close to one another. Will they attract or repel one another?

Electrons, more or less

Electrons are a part of every atom. They are negatively charged and they are found on the outside of the atom. Because they are relatively weakly held in the atom, they can be readily pulled away by the force of friction. An atom has no electric charge – we say that it is **neutral**. When an atom has lost an electron, it becomes positively charged.

A polythene rod becomes negatively charged when it is rubbed with a cotton cloth, so we can imagine electrons being rubbed from the cloth onto the rod (see Figure 4a.4).

● A material which has an excess of electrons has a negative charge.

● A material which has a lack of electrons has a positive charge.

It is difficult to explain why one material pulls electrons from another. The atoms which make up polythene must attract electrons more strongly than the atoms of the cotton cloth. Experiments have shown that materials can be ranked in order, according to how strongly they attract electrons when rubbed. The results are shown in Table 4a.1, which shows a competitive ranking: a material near the bottom of the table attracts electrons more strongly than a material higher up. For example, a nylon rod rubbed on silk will become positively charged but, if the

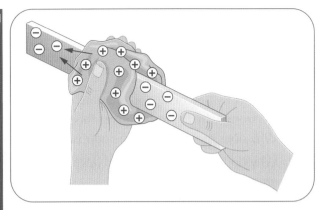

Figure 4a.4 When a polythene rod is rubbed with a cotton cloth, electrons are transferred from the cotton to the polythene. The cotton is left with a positive charge.

same rod is rubbed on human hair, it will become negatively charged.

When Benjamin Franklin was studying static electricity 250 years ago, he imagined that both positive and negative charges could move around. Now we know that, when a material is charged by rubbing, it is only the negatively charged electrons that are mobile.

Material	
Perspex	*most weakly*
glass	
human hair	
nylon	
wool	
silk	
paper	
cotton	
polyester	
polythene	
polyvinylchloride	
Teflon	*most strongly*

Table 4a.1 A list of materials ranked according to how strongly they attract electrons when rubbed. This allows us to say that, for example, polythene will become negatively charged when rubbed with wool; the wool will become positively charged.

SAQ

3 a What charge does an electron have, positive or negative?

   b Would two electrons attract or repel one another?

H **4** A glass rod is rubbed on a cotton cloth.

    a Using the information contained in Table 4a.1, decide whether the rod will become positively or negatively charged.

    b Explain, by describing the movement of electrons, why the rod acquires this charge. Why does the cloth have the opposite charge?

    c Why do the cloth and the rod have equal *amounts* of charge?

Static hazards

Static electricity can be a nuisance and it can also be a serious hazard. Here are some minor problems which come from the effects of static electricity.

- Clothing, particularly if made from synthetic fabrics, tends to become charged up as one item rubs against another. A skirt may not hang straight, a jumper may cling. You may feel tiny shocks when removing a charged-up garment.
- Static electricity can make your hair stand on end if you brush it vigorously after washing it with a lot of conditioner.
- Insulating materials which become charged up will attract dust and become dirty. For example, plastic food containers lose their initial ultra-clean appearance. TV and computer screens become charged because of the electric currents which make them operate and they gather dust from the air.

Getting dangerous

Disasters can happen when static electricity builds up as a result of friction. For example, when oil is pumped on board a shipping tanker, friction between the oil and the pipe can result in a build up of charge. A spark may jump between the negatively charged oil and the positively charged pipe. This spark may ignite the mixture of vapour and air in the hold of the tanker.

The same problem can arise when a petrol tanker is refilling the storage tanks at a petrol station (Figure 4a.5) or where there are high concentrations of oxygen gas (such as at a steelworks). A small spark could be enough to set off a violent explosion.

Figure 4a.5 Before this petrol tanker starts to deliver fuel to the storage tank, a metal wire is connected from the tanker to the tank. This allows any static electricity to flow away safely, avoiding the danger of sparks.

The Van de Graaff generator (Figure 4a.6) is a useful device for demonstrating the effects of static electricity. The woman in the photograph has been charged up to a high voltage; you can tell this from the way her hair is standing on end. She is standing in a plastic tray so that she is insulated and the electric charge cannot get away from her. However, if she were to let go of the dome and touch the man behind her, they would both feel a shock as the static electricity flowed away, into the ground.

The shock experienced here would not be very great, certainly not dangerous, because the amount of charge flowing is very small. However,

Figure 4a.6 The dome of the Van de Graaff generator shares its charge with the woman. She doesn't feel any shock – unless she touches something connected to earth.

it is possible for much larger amounts of charge to build up when electrical machinery (such as motors or generators) is in use and anyone coming close could get a harmful shock. It might even prove fatal. So such equipment must be designed to avoid any build-up of charge.

Avoiding problems

When lightning strikes the ground, electric charge travels safely down into the Earth. The Earth acts as a giant reservoir for electric charge and we can make use of this to get rid of dangerous accumulations of static electricity. For example, if you have been charged up using a Van de Graaff generator, you can get rid of the charge by holding one end of a wooden ruler and touching the other end to the ground or to a bench. Wood is a poor conductor of electricity, but it is good enough for the static charge to flow into the ground. (If you used a rod of metal, which is a much better conductor, the charge would flow much more quickly and you would feel the shock of it.)

So, to avoid hazardous build-ups of charge, we connect things to earth. For example, equipment in use near to computers should be earthed, because any excessive concentration of charge could produce sparks which could destroy the computer chips.

Another way to reduce the possibility of a build-up of charge is to use an antistatic spray or cloth. The spray coats the surface of insulating objects with a thin film of conducting material, so that charge spreads out over the surface rather than accumulating at one place.

People need protecting, too. If there is a risk that you might come into contact with static electric charge, it is advisable to wear shoes with insulating soles and to cover the floor with insulating mats. Then any static charge can flow safely away down an earth connection rather than using you as a convenient path to earth.

We have ignition

Figure 4a.7 shows how opposite charges can build up inside an oil tanker when its hold is filled. The atmosphere inside is a mixture of air and fuel vapour; a single spark could cause an explosion. To avoid this, a good electrical connection must be made between the oil and all the metal parts of the tanker, including the pipes. This is similar to the problem which can arise when fuel is being pumped into an aircraft, or when the storage tanks at a petrol station are being refilled. The tanker supplying the fuel must be connected with a thick cable to the tank it is filling. Then any static charge which builds up as the fuel runs through the pipe will be discharged. Look out for the cable the next time you see a tanker refuelling a petrol station.

Similar problems can arise when dry powders flow along pipes. Friction produces static charges; small amounts of powder in the air can be inflammable and a spark can set off an explosion. This is a particular problem in the food industry – breakfast cereals and custard powder can both be ignited if care isn't taken to provide a method for electric charges to discharge themselves harmlessly.

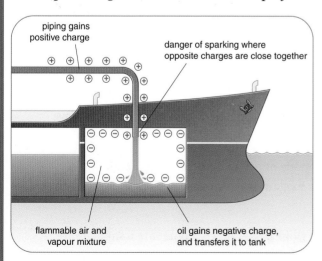

Figure 4a.7 The hold of an oil tanker is likely to contain a dangerously inflammable mixture of oil vapour and air. Sparks must therefore be avoided. Problems arise when the oil flows through the pipe; friction gives rise to static electricity. The positive and negative charges must be given the chance to neutralise one another or a spark may result.

SAQ

5 A factory where custard powder is made blew up in a freak explosion. As the custard powder flowed along a pipe, it gained a negative charge.

   a What charge did the pipe gain?

   The air in the storage chamber was filled with a cloud of fine custard particles.

   b Explain what caused the explosion.

H 6 Figure 4a.7 shows how positive and negative charges can build up in the hold of an oil tanker.

a Explain this in your own words.

b One way to overcome this is to add a substance such as graphite to the oil, so that it becomes conducting. How will this prevent the problem from arising?

Summary

You should be able to:

♦ recognise that, when one object is rubbed against another, the force of friction may transfer electrons from one object to the other; the objects are charged with static electricity

H ♦ state that electrons have a negative charge, so the object which gains electrons becomes negatively charged; the object which loses electrons becomes positively charged

♦ state that charged objects exert forces on each other

♦ state that like charges repel and unlike charges attract

♦ state that a charged object may also attract an uncharged object

♦ explain how static electricity can be hazardous

H ♦ explain how, by connecting an object to earth, any charge which builds up can flow away safely

Questions

1 When a Perspex rod is rubbed on a woollen cloth, the rod acquires a negative electric charge.

   a What type of electric charge does the cloth acquire?

   b What can you say about the amounts of charge on the two charged items?

   c What force gives rise to the charge on the rod and the cloth?

   d If you had two Perspex rods charged up in this way, how could you show that they both have electric charges of the same sign?

2 You walk across a floor which is covered in carpet made of nylon, an insulating material. You touch a metal water pipe and get an electric shock. Explain why you have become charged with static electricity and why you got a shock.

H 3 Explain how a danger of explosion may arise when flour passes down a pipe into a storage silo. How can this danger be avoided?

Making more of static electricity

There are many uses of static electricity. It is at work in photocopiers, attracting the black ink (toner) to the paper. It is vital in the imaging devices used to capture pictures electronically in a digital camera. It can help to direct the printing ink to the right points on a sheet of paper.

Here, we look in detail at three examples of the uses of static electricity.

Charge to the heart

You have probably seen a defibrillator being used in TV medical dramas. A patient collapses – they have had a heart attack. Their heart has stopped beating with the correct rhythm and death will follow if something isn't done quickly to restart it.

The **defibrillator** has two pads connected to the control box by wires. These pads are attached to the patient's bare chest (see Figure 4b.2) and the control box builds up a strong electric charge.

Stand back!

The UK health service has a scheme to provide defibrillators in public places; these are the electrical machines which medical crew use when they attempt to restart a patient's heart after a heart attack (see Figure 4b.1). In this account, Lincoln Abbotts describes how the life of his father, Terry, was saved.

My parents had travelled down from Stoke-on-Trent to the BBC's Maida Vale studios to hear a live broadcast concert, which I had produced.

It was as they began their return journey that events took a dramatic turn. Walking down Platform 13 at Euston Station to catch the 11:55 Virgin Express, all seemed fine. As the whistle blew, however, my father fell to the floor. A fellow passenger quickly got him into the recovery position and the guard's call over the train's Tannoy system led to a young lady doctor quickly arriving on the scene.

In the blur of time that then followed, three members of the Euston station team arrived, carrying the defibrillator unit and took immediate action that saved my father's life.

Blue flashing lights signalled the arrival of the motorcycle paramedic who took over and immediately congratulated the train team on their prompt and decisive action. With my father (Terry) on a stretcher and my mother (Sylvia) on the station's mobility vehicle, we found our way to the waiting ambulance which took us all at speed to University College Hospital. What a short ride – to my mind, it seemed to take forever.

A triple bypass operation was carried out at 8:30 the following morning. My dad is now recovering at home, doing well, and is looking forward to his next trip south, which will be as an outpatient in a few weeks' time. The Euston defibrillator team saved my dad's life and we will be forever grateful.

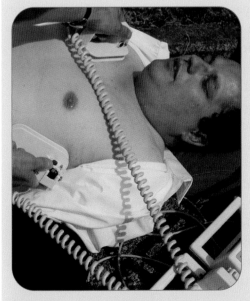

Figure 4b.1 A heart defibrillator in use.

Once the medical crew are clear, the charge is sent down the wires to the pads, where it passes through the patient's chest. The hope is that this will jolt the heart back into its normal pace.

(Some defibrillators have 'paddles', which are pads with insulating handles. The patient's chest is smeared with gel to ensure good conduction and the paddles are pressed against the chest.)

Defibrillation doesn't always work, so second and third shocks may be attempted. Today, for less than £1000, it is possible to buy a defibrillator to keep at home; once put into use, it gives instructions on how it is to be used, assesses the patient's condition and gives the appropriate shock.

A patient whose heart frequently shows signs of losing its regular rhythm may have a miniature defibrillator fitted in their chest. It monitors the heart's beating and supplies a small shock if ever this is needed. (This is different from a pacemaker, which uses electrical signals to keep the two sides of the heart beating in the correct sequence.)

SAQ

1 What features of the use of a defibrillator help to ensure that the medical crew do not get a shock?

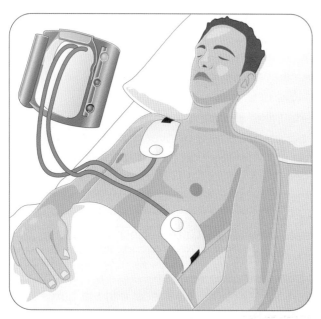

Figure 4b.2 How the pads of a defibrillator are applied to the patient's chest.

Electrostatic smoke precipitators

Figure 4b.3 Power stations are not the only industrial plants whose chimneys produce dirty emissions. These chimneys could be cleaned up using electrostatic precipitators.

Power stations and other industrial plants can produce a great deal of smoke, polluting the environment for hundreds of kilometres around (Figure 4b.3). The smoke consists of very fine particles which may be carried high into the atmosphere, falling to ground a long way off. These particles cause damage to plants and animals living on land and in water. Removing as much dust as possible from the output of power stations is an expensive business, but important for preserving valuable ecosystems.

Static electricity helps to capture the smoke particles. When a power station has an **electrostatic smoke precipitator** (Figure 4b.4), the smoke is passed through a special chamber before it goes up the chimney. The chamber is fitted with a grid of metal wires and its walls are covered with metal plates. The wires are connected to a high voltage supply, so that they become negatively charged. The dust particles become negatively charged and are attracted to the plates lining the walls, where they stick.

The inside of the chamber becomes caked with lumps of dust and, periodically, large hammers strike the plates, causing the dust to fall to the bottom. Here it is collected up and removed for safe disposal. In this way, over 99% of the dust is removed from the exhaust gases coming from the power station's furnaces.

Some homes, restaurants and other public buildings have air conditioning systems which are fitted with dust precipitators. These remove dust from the air entering the buildings and this can be of great benefit to asthma sufferers.

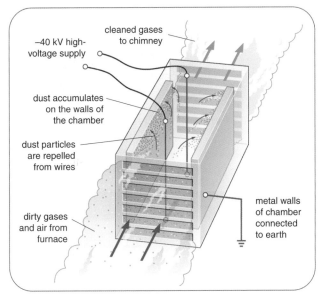

Figure 4b.4 At a power station, an electrostatic precipitator uses static electricity to remove dust from the exhaust gases produced by the furnace.

SAQ

2 If the particles of dust are repelled by the grid of wires in the electrostatic precipitator, what charge must they have?

3 Here is a little more detail of what goes on inside a smoke precipitator.

The wires (negative) and walls (positive) of the chamber have opposite charges. This creates a very strong electric field within the chimney and molecules in the gas passing through the chamber become charged as electrons are ripped from them. The electrons tend to stick to dust particles, so that these are then charged.

a What charge do the dust particles have?

b Why are they attracted to the walls of the chamber?

c What would happen if the wires were given a positive charge and the walls a negative charge?

Paint spraying

In industry, paint is often sprayed electrostatically (Figure 4b.5). The idea is to ensure that paint reaches all sides of a metal object and as little paint as possible is wasted. Here is how it works.

- A spray gun uses high-pressure air to blow a fine spray of paint droplets towards the object to be painted, such as part of a car body.
- The nozzle of the gun is connected to a high-voltage power supply, so that each droplet becomes electrically charged.
- The metal object is connected to the other terminal of the power supply. This ensures that the droplets of paint are attracted to the metal.
- The paint reaches all parts of the metal object, including the back. There are no 'shadows' where the paint does not reach.

This helps to reduce the amount of paint which is wasted and the coating is more uniform. Droplets that fail to reach the metal target fall into a gutter, where they are collected up, ready for re-use.

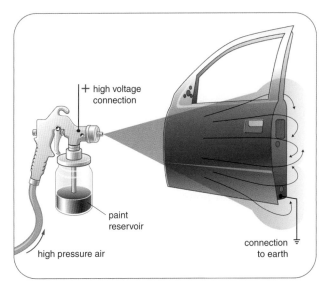

Figure 4b.5 Electrostatic paint spraying. The electrostatic force guides paint to all sides of the object being painted and ensures that very little misses its target.

SAQ

4 What electric charge do the paint droplets have in Figure 4b.5?

5 Explain why the paint droplets move apart from each other. How does this help to ensure an even coating of paint?

Summary

You should be able to:

◆ describe how static electricity can be used in many applications

◆ describe how a defibrillator passes an electric charge through a patient's chest in an attempt to restart the heart's normal rhythm

◆ describe how an electrostatic dust precipitator removes polluting dust from the smoke passing up chimneys

◆ describe how electrostatic paint spraying ensures good coverage of an item and less wastage

Questions

1 What is the purpose of:

   a a defibrillator?

   b an electrostatic dust precipitator?

2 In electrostatic paint spraying:

   a how do the droplets of paint become charged?

   b why is the object to be painted connected to the power supply?

H 3 Figure 4b.6 shows the principle of one type of inkjet printer. (This type of printer is used to add markings to items such as tins on a conveyor belt.)

   Explain how the printer can make marks at different points on the paper.

4 Insect pests often hide on the underside of a plant's leaves. When farmers spray insecticide, it lands on the top surface of the leaf and the insects escape. How could static electricity be used to help overcome this problem? (Think about the example of paint spraying shown in Figure 4b.5. Don't suggest giving the insects an electric shock!)

5 Explain how the way in which a defibrillator is used helps to ensure that the medical crew do not get a shock.

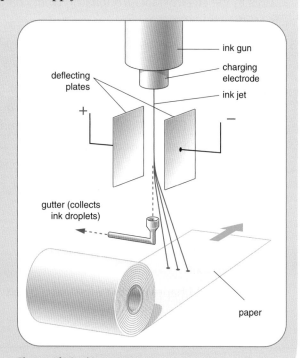

Figure 4b.6 Inkjet printer.

Safe electricals

About circuits

No doubt, you have made electric circuits in the lab and looked at some real-life circuits. The circuits which you have experimented with are **models** for circuits which have real purposes in the world. They are models because they present a simplified view of how circuits work. Practical circuits are usually more complex, but it makes sense to start with simple circuits to build up a picture of how electric current flows.

In this Item, you will learn about how we can control the current which flows in a circuit and how we can use mains electricity safely.

What we do with electricity

The photographs show two rather different types of electric circuit.

Figure 4c.1 shows part of the electric circuit which carries power from a large generating station to the industrial complex where it is used. Electric current is flowing through thick metal cables, held above the ground by tall pylons.

Figure 4c.2 shows the electrical circuits inside a computer. There are several 'chips' (integrated circuits) in the computer. In these, electric current flows through silicon, a material which is not such a good conductor as a metal. Engineers design chips to be as small as possible. This is because, although electric current flows quickly, it is not instantaneous, so current takes less time to flow around a small component than a larger one.

Figures 4c.1 and 4c.2 illustrate the two general uses we have for electric circuits.

- Electricity can be used to transport **energy** from place to place. A circuit contains devices for transforming energy. Think of a simple circuit like a torch. Energy is transferred electrically from the battery to the bulb, where it is transformed into light and heat.

- Electricity can be used to transport **information** from one place to another. Digital information comes into the computer, and its circuits then manipulate the information to produce pictures, sounds and new data. We even have electrical circuits in our bodies for handling information; our brain and nerves work electrically and it is possible to trace the flow of electricity around our bodies.

Figure 4c.1 These cables carry large electric currents. Energy is being transferred from a power station to an industrial plant where it will be used to turn machinery, split chemicals and so on.

Figure 4c.2 The electric circuits of a computer are highly engineered. Each of the chips (the rectangular objects) contains many millions of electric circuits which work to process information at high speed.

Current in electric circuits

If an electric current is to flow, two things are needed:

- a complete circuit for it to flow around
- something to push it around the circuit.

The 'push' might be provided by a battery or power supply. In most familiar circuits, metals such as copper or steel provide the circuit for the current to flow around. Figure 4c.3 shows how a simple circuit can be set up in the laboratory. Once the switch is closed, there is a continuous metal path for the current to flow along. Current flows from the positive terminal of the battery (or cell); it flows through the switch and the filament lamp, back to the negative terminal of the battery.

In this circuit, we know that a current is flowing because the bulb lights up. To know *how much* current is flowing, we would need to include an **ammeter** in the circuit. You should recall that an ammeter is connected *in series* in a circuit, so that the current can flow through the meter.

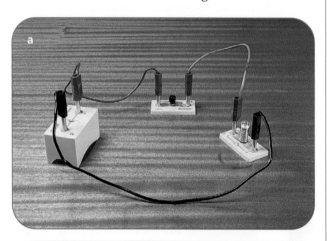

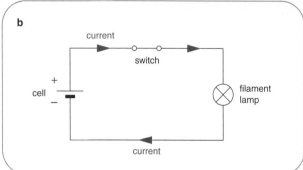

Figure 4c.3 a A simple electrical circuit, set up using laboratory components. **b** The same circuit, represented as a circuit diagram.

Resistors

We can change the current in a circuit using a resistor. Resistors (Figure 4c.4) have two terminals so that the current can flow in one end and out the other. They may be made from metal wire (usually an alloy – a mixture of two or more metals with a high resistance) or from carbon. Carbon (like the graphite 'lead' in a pencil) conducts electricity, but not as well as most metals. Hence high-resistance resistors tend to be made from graphite, particularly as it has a very high melting point.

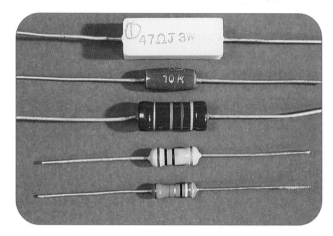

Figure 4c.4 A selection of resistors. Some have colour-coded stripes to indicate their value; others use a number code.

A **variable resistor** can be used to alter the current flowing in a circuit. Figure 4c.5 shows a type of variable resistor called a rheostat. It consists of a coil of resistance wire and a slider. Here is how it works.

- The current flows in at one terminal (bottom left).
- It flows through the wire coil (lots of resistance) until it reaches the metal slider.
- Then it flows through the thick metal bar (almost no resistance) and out at the other terminal (top right).

You can change the resistance of the rheostat by moving the slider. The more wire the current must flow through, the greater the resistance of the rheostat.

Figure 4c.6 shows two different circuit symbols for a variable resistor, together with an example of a circuit which makes use of one.

Figure 4c.5 A laboratory rheostat, used to control the current in a circuit.

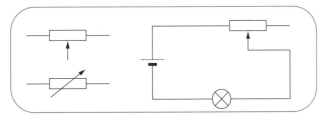

Figure 4c.6 Circuit symbols for a variable resistor and how to connect one in a circuit.

SAQ

1 Look at the circuit shown in Figure 4c.6. The symbol for a variable resistor is like a rheostat; you can imagine sliding the arrow along to change the resistance.

   a If you slide the arrow to the right, will the resistance in the circuit increase or decrease?

   b How will this affect the current in the circuit?

   c How will the brightness of the lamp change?

Electrical resistance

If you use a short length of wire to connect the positive and negative terminals of a battery together, you have created a short circuit and this can do a lot of damage. The wire and the battery may both get hot, because a large current will flow through them. There is very little **electrical resistance** in the circuit, so the current is large. Power supplies are protected by trip switches which cause them to cut out if too large a current flows.

Resistance reduces the current flowing in a circuit. The greater the resistance, the smaller the

current that will flow. Figure 4c.7 shows a circuit in which a cell pushes a current through a resistor. The cell provides the voltage or **potential difference (p.d.)** needed to push the current through the resistor. Potential difference is another word for voltage.

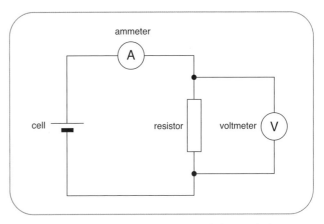

Figure 4c.7 The cell provides the p.d. needed to push the current around the circuit. The amount of current depends on the p.d. and the resistance of the resistor. The ammeter measures the current flowing through the resistor; the voltmeter measures the p.d. across it. This circuit can thus be used to find the resistance of the resistor.

How much current can a cell push through a resistor? This depends on the resistance of the resistor – the greater its resistance, the smaller the current that will flow through it. The resistance of a component is measured in ohms (Ω) and can be calculated using this equation:

$$\text{resistance } (\Omega) = \frac{\text{voltage (V)}}{\text{current (A)}}$$

If you want to make a bigger current flow in a circuit with a resistor, what can you do? Increase the voltage. For a given resistor, the greater the p.d., the greater the current that will flow. This is shown by the graph in Figure 4c.8.

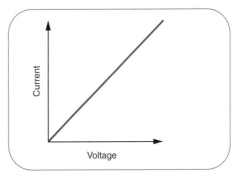

Figure 4c.8 This graph shows that the greater the p.d. across a resistor, the greater the current that flows through it.

Measuring and calculating resistance

Using the circuit shown in Figure 4c.7, we can measure the resistance of a resistor (or any other component). We need to know two things:

- the current flowing through the resistor, measured by the ammeter
- the p.d. across it, measured by the voltmeter.

$$\text{Resistance} = \frac{\text{voltage}}{\text{current}}$$

A **voltmeter** is always connected *in parallel* with the relevant component, because it is measuring the potential difference *between the two ends* of the component.

SAQ

2   What is the resistance of a lamp if a voltage of 10 V makes a current of 2.5 A flow through it?

What is an ohm?

If we think about the formula 'resistance = voltage / current', which tells us what we mean by resistance, we can see that it takes a p.d. of 10 V to make a current of 1 A flow through a 10 Ω resistor, a p.d. of 20 V to make 1 A flow through a 20 Ω resistor, and so on. Hence resistance (in Ω) tells us how many volts are needed to make 1 A flow through that resistance. To put it another way:

an ohm is a volt per amp   1 Ω = 1 V/A

So, it would take 500 V to make 1 A flow through the 500 Ω resistor.

SAQ

3   How many volts are needed to make a current of 1 A flow through a 100 Ω resistor?

Further calculations

There are three arrangements of the equation for resistance which you will find useful:

$$\text{resistance} = \frac{\text{voltage}}{\text{current}}$$

$$\text{current} = \frac{\text{voltage}}{\text{resistance}}$$

voltage = current × resistance

Worked example 1

What current will flow when a p.d. of 240 V is connected across a 60 Ω resistor?

Step 1: write down what you know and what you want to know.

Voltage = 240 V

Resistance = 60 Ω

Current = ?

Step 2: choose the appropriate form of the formula.

$$\text{Current} = \frac{\text{voltage}}{\text{resistance}}$$

Step 3: substitute values and calculate the answer.

$$\text{Current} = \frac{240\,\text{V}}{60\,\Omega}$$

$$= 4\,\text{A}$$

So a current of 4 A will flow.

(You could have done this in your head: 60 V will make 1 A flow; 240 V is 4 times 60 V, and this will make 4 A flow.)

SAQ

4   What p.d. is needed to make a current of 3 A flow through a 120 Ω resistor?

5   What current will flow through a wire of resistance 1000 Ω when the p.d. across its ends is 240 V?

Safe wiring

In the UK, the electricity supply to a house is usually at a voltage of 240 V. Anyone coming into contact with such a high voltage will get a shock, possibly fatal, so domestic wiring systems must be designed for safety.

The cables which pass around a house consist of three separate wires. These are:

- the **live wire** – this is the wire which is at a high voltage;
- the **neutral wire**, which completes the circuit – its voltage is close to 0 V;

- the **earth wire** – this is a safety feature which stops the appliance from becoming live, in the event of a fault in its wiring.

When we refer to 'the 240 V mains', we are talking about the p.d. between the live wire and the neutral wire. When you plug an appliance into a wall socket, you are connecting it between these two wires, so that current flows from the live wire through the appliance and out into the neutral wire.

Electricity enters a house near the fuse box and meter (Figure 4c.9). From there, cables run around the house, connecting each socket to the supply. Usually, these cables are connected in a particular way, called a **ring main circuit**. The cable runs from the fuse box, around part of the house and back to the fuse box. Each of the three wires (live, neutral and earth) thus forms a loop, with several sockets around the loop. When an appliance is connected to a socket, current in the live wire can flow both ways around the loop to reach the appliance. This means that there are two routes for the current, so there is less resistance in the circuit. This allows the wires to be made thinner than if the cable was a single-ended 'spur' from the fuse box. Similarly, the current has two routes along which to flow back to the fuse box, along the neutral wire.

Lighting circuits are generally not ring circuits. A cable leads out from the fuse box to supply several lights, but it does not make a complete loop. This is because lights do not require high currents. This means that the cable can be thinner, too.

SAQ

6 Which *two* wires in a house are at or close to 0 volts?

Plugs

Just as different national electricity supply systems have their own choice of mains voltage and frequency, so they have their own design of plug and socket. Figure 4c.10 shows a typical 13 A plug, as used in the United Kingdom. You can see that the wires which lead from the plug to the appliance are colour-coded, for ease of identification.

- Live: brown.
- Neutral: blue.
- Earth: green and yellow stripes.

A plug with the wires connected to the wrong pins could be extremely dangerous! The plug has several other safety features.

- The pins are made of brass, a metal which is unlikely to corrode. The live and neutral pins have plastic shrouds, so that you cannot touch the live metal when plugging in.
- The case is made of plastic, a good insulating material.
- The wires are screwed firmly into place.
- The flex is held by a cable grip so that it is impossible to pull the wires from the pins.
- An appropriate fuse is fitted, to protect the wiring from high currents.

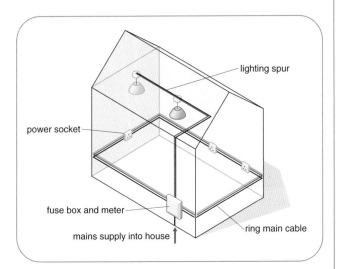

Figure 4c.9 This drawing shows part of the electricity supply cabling in a house. The supply enters the house via the meter and fuse box. A ring main then runs around the ground floor. Each socket is connected to all three wires in the cable. Plugging in an appliance connects to all three wires. Current flows to the appliance along the live wire, around both halves of the ring.

Figure 4c.10 A 13 A plug from the UK, designed to maximise safety for the user. If you live in or visit another country, compare the design of the local plugs with this one. How have they been designed for safety?

Today, manufacturers of electrical appliances generally fit plugs so that it is unnecessary for the user to do so.

SAQ

7 Which wire in an appliance's connecting cable (flex) has two colours?

Fuses and earthing

Fuses are included in circuits to prevent excessive currents from flowing. If the current gets too high, cables can burn out and fires can start. A fuse contains a thin section of wire that is designed to melt and break if the current gets above a certain value (see Figure 4c.11). This makes a break in the circuit, so the current can no longer flow.

Usually, fuses are contained in cartridges which make it easy to replace them, but some fuses use fuse wire (see Figure 4c.11). The thicker the wire, the higher the current that is needed to make it 'blow'. A fuse represents a weak link in the electricity supply chain. Replacing a fuse is preferable to having to rewire a whole house.

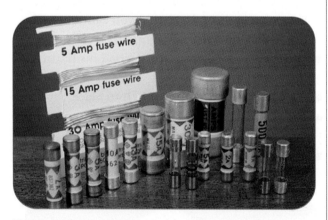

Figure 4c.11 Cartridge fuses and fuse wire. The thicker the wire, the higher the current which causes it to blow.

Many electrical appliances have metal cases: washing machines and fridges, for example. If the case came into contact with the live wire, the user could get a shock simply by touching the case. For this reason, the metal case is **earthed** – it is connected to the earth wire. If the case becomes live, the electric current flows straight to earth. Any metal object can be protected in this way; if it is connected to earth, it cannot become live.

Some appliances, including stereos and hair dryers, have plastic cases (which cannot possibly become live). These are designed so that it is very difficult for the user to come into contact with the mains electricity inside. Such appliances are described as being **double insulated** and they do not require an earth connection. Look for the 'double square' symbol on the plastic case (Figure 4c.12).

Figure 4c.12 The symbol which indicates that an appliance is double insulated.

SAQ

8 What happens inside a fuse when it 'blows'?

Protecting wiring and appliances

The cables which carry electric current around a house are carefully chosen. Figure 4c.13 shows some examples. Each is labelled with the maximum current it is designed to carry. A 5 A cable is relatively thin; this might be used for a lighting circuit, because lights do not require much power so the current flowing is relatively small. The wires in the 30 A cable are much thicker. This might be used for an electric cooker, which requires much bigger currents than a lighting circuit.

Figure 4c.13 Cables of different thicknesses are chosen according to the maximum current they are likely to have flowing through them: 5 A, 15 A and 30 A. Each cable has live, neutral and earth wires, colour-coded.

The wires in each cable are insulated from one another and the whole cable has more protective insulation around the outside. If an excessive current flows in the wires, they will heat up and the insulation may melt. It may emit poisonous

A choice of voltage

The electricity supply to most homes around the world is at a fairly high voltage, usually between 220 V and 250 V. Table 4c.1 shows the supply voltages used in different parts of the world.

Although the mains voltage in the EU is 'officially' 230 V, in practice this is not the case. The UK traditionally uses 240 V, whereas continental Europe has standardised at 220 V. These are the values which you are likely to find using a voltmeter. However, manufacturers of electrical appliances require a standard value so that the same appliance can be used in all parts of the EU. They therefore design their appliances to work in the range 220–240 V. A particular heater, for example, will give out more heat in the UK than in France, because it is connected to a slightly higher voltage.

You may notice that the supply voltage in North America and the Caribbean is only 110 V, half that of most other countries. Why is this?

A mains supply at 230 V is hazardous and must be handled with care. A 110 V supply is less hazardous – you are much less likely to die if you get a shock from the live wire. However, to supply the same amount of electrical power at half the voltage, the current which flows must be twice as great.

To carry a higher current, the wiring in a house must be thicker and this is expensive. If thin wire is used, its resistance may be too great and it may overheat, leading to a fire. Hence there is a balance to be struck between the dangers of high voltages and the dangers of high currents; different countries have chosen different ways of striking this balance.

Country or region	Domestic mains voltage in V	Mains frequency in Hz
European Union (including the UK), Switzerland	230	50
India, China, Hong Kong	220	50
Singapore	220–240	50
South Africa	220–230	50
Australia	240–250	50
New Zealand	230	50
United States, most of the Caribbean	110–120	60
Japan	100	50 or 60

Table 4c.1 The mains electricity supply is different in different countries.

fumes or even catch fire. Thus it is vital to avoid using appliances which draw too much current from the supply. The fuses in the fuse box help to prevent this from happening, by 'blowing' (melting) if the current exceeds the rating of the fuse.

In many houses, the fuses in the fuse box have been replaced by **circuit breakers** (also known as trip switches) (see Figure 4c.14). These are electromagnetic devices which trip and break the circuit when the current gets too high. Once the problem has been sorted out, they can be easily reset – they are a type of resettable fuse.

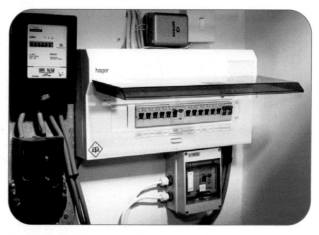

Figure 4c.14 A fuse box without fuses – this one uses circuit breakers, which are easy to reset.

H The fuse in a plug is there to protect the appliance and its flex. An excessive current can cause overheating, leading to a fire; it can also damage the appliance. It is important to choose the correct value of fuse to protect an appliance. Its current rating should be just above the value of the current which flows when the appliance is operating normally.

Protecting people

A **residual current device** (RCD) is different from a circuit breaker; it protects the user rather than an appliance or cable (see Figure 4c.15). In normal circumstances, the currents flowing in the live and neutral wires are the same. However, there may be a fault. Someone may have accidentally touched the live wire, perhaps by running over the flex of the lawnmower. Some current flows through the user, rather than along the neutral wire. Now more current flows in the live wire than the neutral and the RCD detects this and switches off the supply. A difference of 30 mA is enough to trip the RCD. Houses often have RCDs fitted next to the fuse box. School laboratories usually have one too, to protect students and teachers.

H As we have seen above, the earth connection is another safety feature which protects the user. It is particularly important for appliances with metal cases, including heaters and washing machines. If these become live, any dangerous current is conducted away to earth. If the current is large enough, the fuse will blow and the current stops flowing – the hazard is gone.

SAQ

9 It takes a small current – perhaps 100 mA – to kill a person. Explain why it is no good relying on the circuit breakers or fuses in the fuse box to protect you if you accidentally touch a live wire. Why is an RCD needed as well?

Figure 4c.15 An RCD is particularly useful when using electrical appliances outdoors.

Summary

You should be able to:

- state that the greater the resistance in a circuit, the smaller the current that flows (for a given voltage)

- state that increasing the p.d. (voltage) in a circuit causes the current to increase

- state and use the formula: $\text{resistance} = \dfrac{\text{voltage}}{\text{current}}$

H - state and use the formula: $\text{voltage} = \text{current} \times \text{resistance}$

- state and use the formula: $\text{current} = \dfrac{\text{voltage}}{\text{resistance}}$

- explain that the live wire carries the high voltage whereas the neutral wire completes the circuit

- describe that earthing protects the user; an earthed conductor can never become live

- describe how a wire fuse melts when the current becomes too high, protecting the circuit and the appliance

H - explain how fuses and circuit breakers protect wiring and appliances; RCDs and earthing protect people

Questions

1 Copy the table and complete it to show the names and colours of the three wires described.

Name	Colour of wire in flex	Function
		carries the high voltage
		completes the circuit
		stops an appliance becoming live

Table 4c.2

2 a If the resistance in a circuit is increased, how does the current flowing change?

  b Which of the three graphs in Figure 4c.15 shows this correctly?

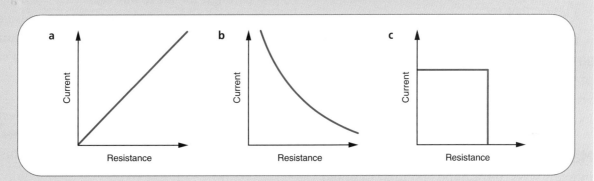

Figure 4c.15

3 Which of these equations shows correctly how to calculate resistance?

$$\text{Resistance} = \text{current} \times \text{voltage} \qquad \text{Resistance} = \frac{\text{current}}{\text{voltage}} \qquad \text{Resistance} = \frac{\text{voltage}}{\text{current}}$$

4 A torch bulb is connected to a 3 V battery. A current of 0.6 A flows through it. Calculate its resistance.

5 a Draw the symbol which indicates that an appliance is double insulated.

  b Which two wires are connected in the plug of a double-insulated appliance?

6 What current will flow through a 40 Ω resistor when it is connected to a 5 V supply?

7 What p.d. is needed to make a current of 8 A flow through a 25 Ω resistor?

8 Explain why it is more dangerous to touch the live wire of a mains supply, rather than the neutral wire or the earth wire.

9 Why are fuses fitted in the fuse box of a domestic electricity supply? What device could be used in place of the fuses?

10 In normal use, a current of 3.5 A flows through a hair dryer. Choose a suitable fuse from the following: 3 A, 5 A, 13 A, 30 A.

Too high to hear

You should recall that **ultrasound** is sound which is too high-pitched for our ears to hear. We can hear sounds whose frequencies lie roughly in the range 20 Hz to 20 kHz; as we get older, our hearing deteriorates and it becomes increasingly hard to hear the highest frequency sounds.

You should also recall that the **frequency** of a sound wave or ultrasound wave is the number of vibrations per second; it is measured in hertz (Hz) or kilohertz (kHz).

1 kHz = 1000 Hz

Ultrasound waves have frequencies higher than 20 kHz, which is known as the **upper threshold of hearing**. Here we will look at how ultrasound waves are made and how they travel; we will also look at how they are used in medicine.

Making waves

In many ways, ultrasound is similar to sound.
- Both are produced by vibrating objects.
- Both travel as waves.
- Both can travel through solids, liquids and gases, but not through a vacuum.

How can we picture the movement of the molecules of the air as sound or ultrasound travels through? Figure 4d.1 shows how the vibrations of a tuning fork are transmitted through the air. As the prong of the fork moves to the right, it pushes on the air molecules on that side, squashing them together. These molecules push on their neighbours, which become compressed (squashed together) and which in turn compress their neighbours. In the

meantime, the prong has moved back to the left, compressing the air molecules on the other side. As the prong vibrates back and forth, **compressions** are sent out into the air all around it. In between the compressions are **rarefactions**, areas in which the air molecules are rarefied (less closely packed together).

Notice that the individual air molecules do not travel outwards from the vibrating fork. They are merely pushed back and forth; it is the compressions and rarefactions which travel through the air to our ears.

This picture of how sound travels also explains why sound cannot travel through a vacuum; there are no molecules or other particles in a vacuum to vibrate back and forth.

Ultrasound works in the same way. It is usually produced by an electronic device; an electronic circuit produces a current which vibrates back and forth with a high frequency, and this is converted into a mechanical vibration.

SAQ

1 **a** What do we call the region of a sound wave where the particles are squashed together?

b What is the opposite of this, and what is it called?

Why waves?

The compressions and rarefactions of a sound wave do not seem much like the waves we see on the sea. So why do we talk about sound and ultrasound 'waves'?

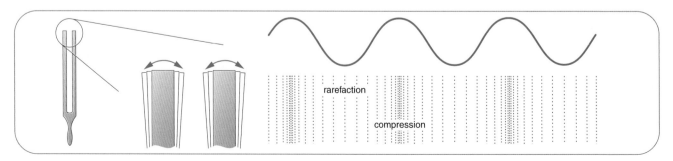

Figure 4d.1 A vibrating tuning fork produces a series of compressions and rarefactions as it pushes the air molecules back and forth. This is how a sound travels through the air (or any other material). We can relate this to the wavy trace on an oscilloscope screen.

Figure 4d.1 includes another way of representing a sound, as a wavy line rather like the trace on an oscilloscope screen. The **crests** on the wave match the compressions; the **troughs** match the rarefactions. Thus the wave represents the changes in air pressure as the sound travels from its source. It is much easier to represent a sound as an up-and-down wave like this, rather than drawing lots of air molecules pushing each other back and forth.

Here we have used two different **models** to represent sound.

● **Model 1** Vibrations travelling through a material; the particles of the material are alternately compressed together and then rarefied as the sound passes through.

● **Model 2** Sound as a wave; a smoothly varying up-and-down line, like the trace on an oscilloscope screen.

The first of these models gives a better picture of what we could see if we could observe the particles of the material through which the sound is passing; the second is easier to draw. It also explains why we talk about **sound waves**. The wavy line is rather like the shape of waves on the sea.

Sound waves and ultrasound waves are examples of **longitudinal waves**. These are waves in which the particles move back and forth along the direction in which the wave is travelling. In Figure 4d.1 the wave is travelling to the right and the air molecules are moving from left to right and back again.

H ## Longitudinal and transverse waves

There is a second type of wave, called a **transverse wave**. In this type, the vibrations are at right angles to the direction in which the wave is travelling – this type is correctly represented by a series of crests and troughs. You can send a wave like this along a stretched rope. Move the end of the rope up and down and the wave travels along the rope.

Figure 4d.2 shows how both types of wave can be created on a stretched spring.

All electromagnetic waves (light, radio waves etc.) travel as transverse waves. You might think that ripples or waves on water are transverse; this

H is not quite true. If you sit in a boat on a wavy sea, you may notice that you go up and down as the waves pass under the boat; at the same time, you also move back and forth. So water waves are half-transverse, half-longitudinal.

SAQ

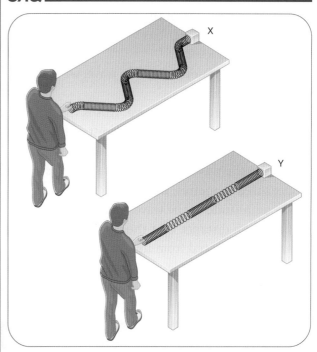

Figure 4d.2 Waves along a stretched spring.

2 Look at Figure 4d.2.

 a Which part shows a transverse wave? Which shows a longitudinal wave?

 b Explain how the person must move the end of the spring to create each type of wave.

Wavelength and amplitude

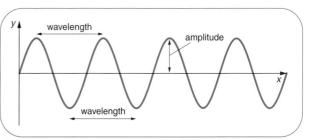

Figure 4d.3 Representing a wave as a smoothly varying wavy line. This shape is known as a sine graph; if you have a graphics calculator, you can use it to display a graph of $y = \sin x$, which will look like this graph.

Figure 4d.3 shows a wave travelling from left to right. The horizontal (*x*) axis shows the distance travelled horizontally by the wave. The vertical (*y*) axis shows how far the surface of the water has been displaced from its normal level. If this was a wave on water, we could think of the *x*-axis as the level of the surface of the water when it is undisturbed. The line of the graph shows how far the surface of the water has been displaced from its undisturbed level.

From this representation of a wave, we can define two quantities.

● The **wavelength** of the wave. This is the horizontal distance from one crest of the wave to the next (or from one trough to the next). Because the wavelength is a distance, it is measured in metres, m.

● The **amplitude** of the wave. This is the maximum distance that the surface of the water is displaced from its undisturbed level; in other words, the height of a crest. Notice that the amplitude is measured from the undisturbed level up to the crest; it is not measured from trough to crest. For ripples on the surface of water, the amplitude is a distance, measured in metres, m.

For a longitudinal wave like a sound wave, there are compressions and rarefactions.

● The **wavelength** is the distance from one compression to the next (or from one rarefaction to the next).

● The **amplitude** is the greatest distance a particle moves from the undisturbed position as the wave passes.

SAQ

3 Draw a diagram to show two waves:
  ● both having the same amplitude
  ● one with a longer wavelength than the other; label them 'long' and 'short'.

4 What is the frequency of the lower threshold of human hearing?

Animals, ultrasound and infrasound

Just because we cannot hear ultrasound does not mean that other creatures cannot. Bats (Figure 4d.4) are well known for using ultrasound to find their way around in the dark. The bat emits a sharp burst of ultrasound, then its ears pick up the reflections from all the surfaces around it. It can deduce two sorts of information about its surroundings:

● how close objects are (from the time taken for the echo to return to it)

● the nature of the surface, hard or soft, smooth or rough (from whether the echo is clear or more diffuse).

Elephants are thought to be able to hear at **infrasound** frequencies, below 20 Hz. They may be able to communicate with one another over long distances by stamping on the ground. Low-frequency sound waves then travel through the ground, to be picked up by other elephants who detect the waves with their legs.

Figure 4d.4 Bats are not blind, but their hearing is much better than their eyesight. They have large ears to pick up ultrasound, reflected from the surfaces of surrounding objects. Although we cannot hear them, the pulses of ultrasound emitted by a bat are very intense. The bat is in danger of deafening itself, but it has a way of avoiding this. It switches its ears off for the short duration of its squeak. It does this by separating the ossicles (tiny bones) in its ears, so that the sound does not reach its cochlea. In a tiny fraction of a second, it switches its hearing back on to listen for the reflected ultrasound.

Using ultrasound in medicine

Doctors use ultrasound to check on the condition of babies before they are born. A probe is moved over the mother's stomach, beaming ultrasound waves into her body (Figure 4d.5). The waves reflect off the baby, the probe detects the reflected ultrasound and a computer analyses the signal to produce an image on the screen.

Today's computers work so fast and can handle such large quantities of data that an expectant parent can see their future child moving on the screen (see Figure 4d.6).

It isn't just expectant mothers who are scanned using ultrasound. Anyone can be scanned if there are suspected problems with their internal organs – perhaps a twisted or blocked digestive tract, or a damaged kidney. Ultrasound is less damaging to living cells than X-rays. It is not thought to cause any harm to internal organs or to a growing baby, and this is a great advantage over X-rays. With even a small exposure to X-rays, there is a (very small) danger of cancer, so ultrasound scanning is greatly to be preferred.

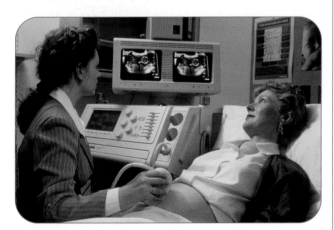

Figure 4d.5 In many countries, pre-natal scanning using ultrasound is now routine. The probe which is moved across the mother's body both emits the ultrasound and detects the reflected waves. A high-power computer then produces an image of the baby on the screen. For many young people, such an image is now to be found in the family photograph album.

Reflecting on ultrasound

When an expectant mother is to be scanned, her stomach is smeared with jelly before the probe is moved over it; this gives a good contact between the probe and the skin, so that the ultrasound waves enter the body rather than being reflected straight back into the probe.

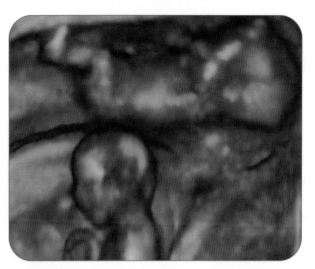

Figure 4d.6 An ultrasound scan of twins, 16 weeks into a pregnancy.

As the ultrasound waves pass down into the body, they meet layers of different materials – flesh, skin, bone and fluid. At each layer, the waves are partly reflected and partly transmitted (Figure 4d.7). This means that the probe receives a complicated mixture of reflected waves which the computer must disentangle to produce the image.

Because ultrasound is reflected when it passes from, say, flesh into skin, it can show up details of soft tissue which wouldn't show up in an X-ray image.

Medical ultrasound scanning uses frequencies of around 10 MHz (10 megahertz, or 10 million hertz). These have a wavelength of about 0.2 mm and this gives you an idea of the fine detail which can be seen in a scan.

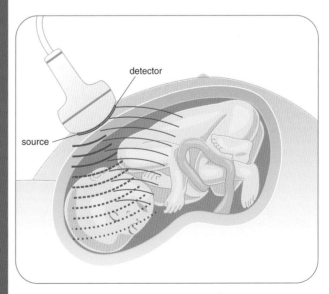

detector

source

Figure 4d.7 Ultrasound waves are partly reflected and partly transmitted as they pass through a body.

More medical uses

As well as body scanning, ultrasound has some other interesting uses in medicine.

Ultrasound can be used to break up dangerous accumulations of mineral material in the body, such as kidney stones. Rather than opening up a patient in an operation, a surgeon directs a focused beam of ultrasound into the body, aimed at one of the kidneys. The small stones vibrate violently and break up. The fine particles which remain are small enough to pass out of the body in the patient's urine.

Ultrasound is also used to measure the speed of blood flow. An ultrasound probe is placed on the patient's body, so that the ultrasound waves travel into a blood vessel. The waves reflect off the moving blood cells and are detected.

How does this tell us the speed of the blood? It's rather like throwing balls at a moving target: if the target is moving away, the balls bounce back at a slower speed. In the case of ultrasound, the waves' frequency is less when they bounce off cells which are moving away from the probe. By finding where blood flow is slowed down, doctors can identify veins or arteries which are 'furred up', a sign of circulatory disease.

SAQ

5 List *three* medical uses of ultrasound.

Summary

You should be able to:

◆ state that ultrasound travels as a longitudinal wave, a series of compressions and rarefactions

◆ state that the distance from compression to compression is the wavelength of the wave

◆ state that the greatest distance by which a particle moves from the undisturbed position is the amplitude of the wave

◆ state that the frequency of the wave is the number of waves per second

H ◆ describe how, in a transverse wave, the vibrations are at right angles to the direction of travel of the wave

◆ describe how ultrasound is used in medicine for body scanning, breaking up kidney stones and measuring blood flow

H ◆ explain how ultrasound can produce images of soft tissue and is less damaging to living cells than X-rays

Questions

1 Look at the diagram of a wave (Figure 4d.8).

   a What is the value of the amplitude of
   the wave?

   b What is its wavelength?

2 An ultrasound wave has a frequency of 50 kHz.
How many waves is that per second?

3 What is the difference between a compression
and a rarefaction in a sound wave? Illustrate
your answer with a sketch.

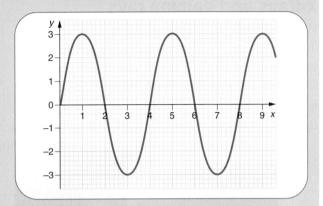

Figure 4d.8

4 Human beings can hear sounds over a wide range of frequencies.

   a Give approximate values for the highest and lowest frequencies which a young person can hear.

   b How does the upper limit of the audible range change as we get older?

   c What name is given to sounds whose frequency is higher than the upper limit?

   d Give *one* example of a way in which such inaudible sound is used.

H 5 Give *two* reasons why an ultrasound scan might be used in preference to making an X-ray of
a patient.

6 Figure 4d.9 shows how ultrasound can be used to check for defects in a metal rail. (These defects
are not visible from the outside, but could lead to the rail cracking when in use.) The arrows show
how ultrasound waves travel through the rail from a source to a detector. Explain how the signal
received by the probe could show whether or not there were defects in the rail.

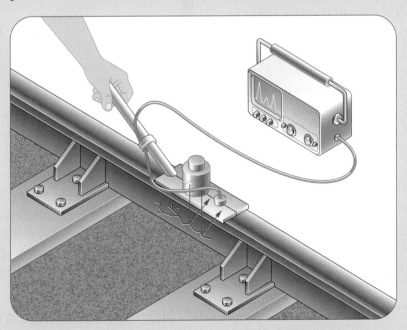

Figure 4d.9 Using ultrasound to look for defects inside a metal rail.

X-Rays and gamma rays

Most people have an X-ray at some time in their lives. Figure 4e.1 shows a dentist preparing a young patient for an X-ray of their teeth. Because X-rays can penetrate flesh and bone, the final image will enable the dentist to discover whether there is any decay inside the teeth.

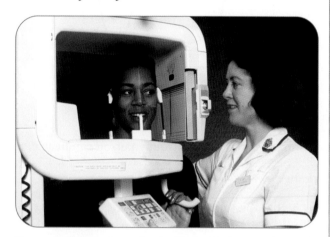

Figure 4e.1 The patient grips the film holder between her teeth as the dentist positions the X-ray machine.

Doctors also use X-rays for treatment. In this Item, we will look at how both **X-rays** and **gamma rays** are used in medicine.

X-Rays and gamma rays are similar.

- They are both forms of electromagnetic **radiation**.
- They both have short wavelengths.
- They are both highly penetrating, so they can get inside the body.

The difference is in how they are produced.

- X-Rays are made in X-ray machines.
- Gamma rays come from **radioactive substances**.

Gamma rays are produced by some radioactive substances as they decay. Because they come from the nucleus of the atom, they are described as **nuclear radiation**.

You should recall that there are three types of radiation which come from the nuclei of radioactive substances (Item P2d in Book 1). Alpha is the least penetrating and gamma the most. Only beta and gamma can pass through the skin.

SAQ

1 Put the three types of nuclear radiation in order, from most penetrating to least penetrating.

Producing X-rays

X-Rays are made in X-ray machines. Inside the machine, a beam of electrons is produced (Figure 4e.2). These are accelerated to very high speeds using very high voltages – thousands or even millions of volts. The electrons collide with a metal target and come rapidly to a halt. They lose their kinetic energy and this energy appears as X-rays. The tube is designed to produce a beam of X-rays.

X-rays have an important advantage over gamma rays. A radioactive substance emits radiation in all directions. A beam of X-rays can be directed exactly where it is needed.

This means that X-rays are easier to control than gamma rays. This is important because X-rays and gamma rays are both types of ionising radiation and can damage cells, so it is important that the radiation only reaches the parts of the body where it is wanted.

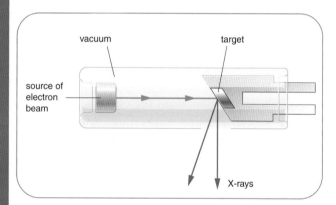

Figure 4e.2 An X-ray tube has a vacuum inside. The electron beam crashes into the metal target; it loses its energy as X-rays.

SAQ

2 Copy the sentences below, choosing the correct word from each italic pair.

If the accelerating voltage in an X-ray tube is increased, the electrons will move *faster/slower*. The electrons have more *kinetic/potential* energy when they strike the metal target. The X-rays produced will be *more/less* penetrating.

Nuclear radiation at work

Gamma rays are sometimes used to help destroy cancer cells – this is similar to the way in which

X-rays are being used in Figure 4e.3. The patient is X-rayed to establish the exact position of their tumour and then gamma rays are directed at the precise spot. The energy released kills the cancer cells. Figure 4e.4 shows one way in which this works, using a gamma knife. The patient, who has a brain tumour, wears a special helmet. Radioactive sources are placed in holes in the helmet and the radiation passes through the holes which are positioned so that the radiation strikes the tumour.

Because gamma rays are so good at killing cells, they are used for **sterilising** medical equipment. Items such as syringes and scalpels are packed in cellophane wrappers; they are then exposed to high intensity gamma rays, which penetrate the packaging and destroy any bacteria or other living cells. When the doctor or nurse opens the package, they can be sure that the item is entirely sterile.

A third use of nuclear radiation is in **tracers**. Suppose doctors think that their patient has a problem with the thyroid gland, which is in the neck. This gland uses iodine, so the patient is given a drink which contains radioactive iodine, which is the **radioactive tracer**. The substance spreads around the body and some reaches the gland. Gamma radiation coming from the tracer can be detected by placing a radiation detector on

Radiographers at work

In a hospital radiography department, **radiographers** work with several different techniques to diagnose and treat various ailments.

Figure 4e.3 shows two radiographers preparing a patient for treatment using X-rays. The patient has a tumour (cancer) on the skin of her cheek. The radiographers are adjusting the position of the X-ray tube so that the beam of X-rays will strike the affected area and, hopefully, kill off the cancer cells. By controlling the energy of the X-rays, the radiographers can ensure that the radiation does not penetrate too far into the patient's head.

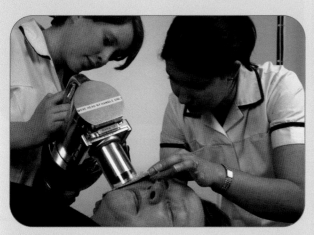

Figure 4e.3 This patient has skin cancer. The two radiographers are ensuring that the beam of X-rays strikes only the affected area. This treatment is usually very successful.

Radiographers must also know how to handle radioactive substances safely. A patient may be treated with beta or gamma radiation; they receive a reasonably large dose of radiation which may save their life. Radiographers come into contact with radiation every day, so it is important that they avoid exposing themselves to radiation or contaminating themselves with radioactive materials.

Radiographers must know how to interpret the images produced by different types of scans, such as X-ray CT scans and ultrasound scans. Because they look at images every day, they soon become expert in making sense of these images. Nowadays, scanning techniques produce electronic images (rather than images on photographic film). This means that the images can be stored easily on computers and can be transmitted rapidly to doctors and surgeons who may be working in different buildings or even different hospitals.

SAQ

3 Name the different types of radiation used by radiographers, as mentioned in the text.

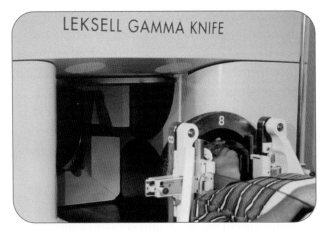

Figure 4e.4 This patient is suffering from brain cancer. The helmet of the gamma knife machine ensures that the gamma rays used to destroy the cancerous tissue are directed only at the tumour. The beam is sent in from different directions, so that the surrounding tissues receive a much weaker dose.

the patient's neck (see Figure 4e.5). If very little radiation appears, the doctors will know that the gland is not operating well. On the other hand, if there is a lot of radiation coming from the gland, this will indicate that it is overactive. In either case, treatment can cure the problem.

Radioactive tracers may be either beta or gamma emitters. They may be administered to the patient in the form of something to eat or drink, or they may be injected.

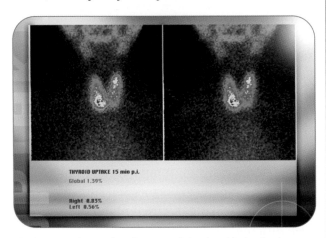

Figure 4e.5 The results of a scan of the thyroid gland using a radioactive iodine tracer. The yellow areas in the scan indicate a high level of gamma radiation coming from the tracer. This suggests that parts of the gland are overactive.

SAQ

4 Why would an alpha–emitting substance be of no use as a radioactive tracer?

Gamma rays treating cancer

Because X-rays and gamma rays are very energetic and penetrating, they have to be used with care. There is always a risk that a cancer might be started by radiation which is being used to diagnose or treat an illness.

Doctors must always balance this risk against the benefit to be gained from knowing more about the patient's condition and from treating it. The need to minimise the damage to healthy tissue whilst killing off cancerous tissue is illustrated by the way in which gamma rays and X-rays are used to treat cancer.

Figure 4e.6 shows a patient undergoing radiation treatment. The large machine rotates around the patient's body during the course of the treatment. Why is this?

● The beam of radiation is directed (focused) onto the tumour.
● Because the beam is fairly wide, it is relatively weak.
● As the beam is rotated around, only the tumour receives a full dose of radiation.
● Other, non-cancerous tissue receives only a limited dose.

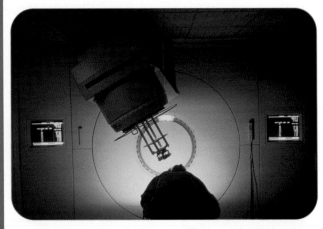

Figure 4e.6 This patient is being treated for cancer. Radiation (X-rays, in this case) is directed into their body from a source which rotates slowly around. This ensures that only the tumour receives a damaging dose.

SAQ

5 Suggest a reason why X-rays might be better than gamma rays for the method of treatment described here.

Summary

You should be able to:

- recall that nuclear radiation, including gamma rays, together with X-rays, can be used in medicine

H
- explain that X-rays are produced when fast-moving electrons hit a metal target

- explain that gamma rays come from the nuclei of radioactive substances

- recall that gamma rays and X-rays can be used to destroy cancer cells

- recall that, because nuclear radiation damages cells, it can be used to sterilise medical equipment

- describe that beta and gamma emitters can be used as tracers in the body

H
- explain how the radiation beam is directed at the tumour from different directions, to avoid damaging healthy tissue

- explain how the position of a tracer, as it moves around the body, can be detected from outside

Questions

1 In what ways are X-rays and gamma radiation similar to each other?

2 Here are four types of radiation: X-rays, alpha, beta, gamma.

   a Which *three* can be described as *nuclear radiation*?

   b Which *three* can pass through the skin?

   c Which *two* are emitted by radioactive tracers?

3 Explain why gamma radiation can be used to sterilise medical equipment.

H
4 X-rays can be used to treat cancer.

   a Why should a patient's exposure to X-rays be kept to a minimum?

   b Describe how it is possible to ensure that the cancer tissue receives a high dose of radiation while healthy tissue receives a much lower dose.

5 A substance called a *tracer* can be used to find whether a patient has an organ which is not functioning correctly.

   a How is the tracer administered to the patient?

   b How is the position of the tracer in the body detected?

Discovering radioactivity

Radioactivity was discovered by a French physicist, Henri Becquerel, in 1896. He had been investigating some phosphorescent rocks – rocks which glow for a while after they have been left under a bright light. His method was to leave a rock on his window sill in the light. Then he put it in a dark drawer on a piece of photographic film to record the light it gave out. He suspected that rocks containing uranium might be good for this but he discovered something even more dramatic: the photographic film was blackened even when the rock had not been exposed to light. He realised that some kind of invisible radiation was coming from the uranium. What was more, the longer he left it, the darker the photographic film became. Uranium gives out radiation all the time, without any obvious supply of energy.

Becquerel had discovered a way of revealing the presence of invisible radiation, using photographic film. This method is still used today. One of his first photographs of radiation is shown in Figure 4f.1.

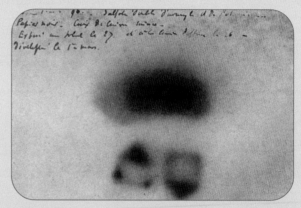

Figure 4f.1 One of Henri Becquerel's first photographic records of the radiation produced by uranium. The two black blobs are the outlines of two crystals containing uranium. To discover whether the radiation would pass through metal, he placed a copper cross between one of the crystals and the photographic film. You can see the 'shadow' of the cross on the photograph. The writing is Becquerel's; the last line says 'développé le 1er mars' – developed on the first of March (1896).

Radioactive substances

Radioactivity is a serious topic. You can tell that because people make lots of jokes about it.

- 'If you go on a school visit to a nuclear power station, you will come back with two heads.'
- 'If you have radiation treatment in hospital, you will glow in the dark.'

As with many jokes, there is a small element of truth here, and a great deal of fear of the unknown.

When radioactivity was first discovered, people became very excited by it. Some doctors claimed that it had great health-promoting effects. They sold radioactive water and added radioactive substances to chocolate, bread and toothpaste (see Figure 4f.2). There were radioactive cures for baldness, and contraceptive cream containing radium. This attitude still lingers on today, with some alpine spas offering residents the chance to breathe radioactive air in old mine tunnels!

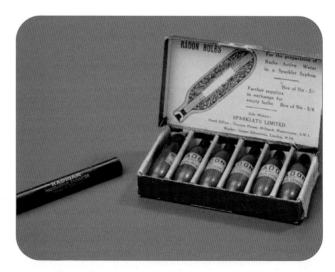

Figure 4f.2 In the 1930s, you could buy radioactive radon gas to dissolve in drinking water. An American called Ethan Byers drank a bottle a day for five years; he died of cancer of the jaw.

Measuring radioactivity

It takes a while to expose and develop a photographic film such as the one Becquerel used when he discovered radioactivity. For a quicker measurement of radiation, we can use a Geiger counter (Figure 4f.3). The detector is a Geiger–Müller tube, which is held close to a suspected source of radiation. The radiation enters the tube, which produces an electrical pulse every time it detects any radiation. The electronic counter adds up these pulses; it can give a click or beep for each pulse. In the photograph, a Geiger counter is being used to check that radiation levels close to a nuclear power station are safe.

Radioactive substances decay naturally – that is, they give out their radiation and eventually stop radiating. You can't switch them on or off – they just do it. You can't speed up their decay by heating them. Some radioactive substances have remained in the Earth ever since it formed, decaying slowly but still remaining radioactive today, after 4.5 billion years.

To see how radioactive a substance is, we simply measure the rate at which it emits radiation. If the Geiger counter gives many clicks or beeps per second, the substance is decaying rapidly. As it decays, it gradually becomes less and less radioactive.

SAQ

1 Which of the following is a correct explanation of how we measure radioactivity?

- The number of atoms in the sample.
- The total amount of radiation the sample gives out.
- The number of atoms which decay each second.
- The number of undecayed nuclei in the sample.

The pattern of decay

As time goes by, a radioactive substance becomes less and less radioactive. It gradually decays away. The amount of undecayed material, and therefore the amount of radiation, tails off more and more slowly. The graph of Figure 4f.4 shows this pattern. (There is more about this in Item P4g.)

Figure 4f.3 Using a Geiger counter to monitor radiation levels in the environment. This scientist has gathered vegetation in a National Park in France to check that it hasn't been contaminated by emissions from a nuclear power station in the valley below.

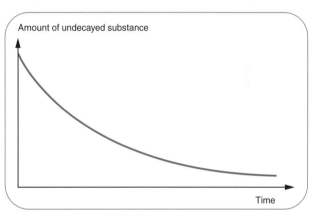

Figure 4f.4 The pattern of radioactive decay. The graph slopes down more and more slowly as time goes by.

Slower and slower

Henri Becquerel discovered the radioactivity of uranium. What surprised him was that uranium appears to be able to emit radiation endlessly, without ever running out of energy. This would go against the principle of conservation of energy. What he did not realise was that the uranium he used was undergoing very gradual decay. The problem was that uranium decays very slowly so that, even if Becquerel had carried on with his experiments for a thousand years, he would not have noticed any decrease in the activity of his samples. In fact, the uranium he was working with had been decaying gradually ever since the Earth formed, over 4.5 billion years ago.

Different radioactive substances decay at different rates, some much faster than others. However, they all decay with the same pattern, as

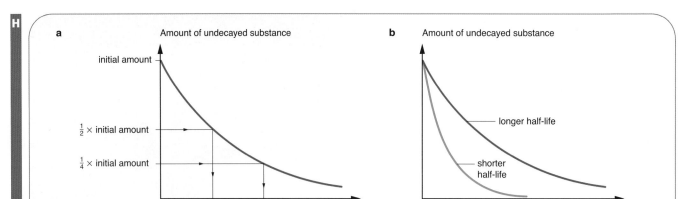

Figure 4f.5 a A decay graph for a radioactive substance. A curve of this shape is known as an exponential decay graph. **b** A steep graph shows that a substance has a short half-life.

shown in Figure 4f.5. The graph shows that the amount of a radioactive substance decreases rapidly at first, and then more and more slowly. In fact, because the graph tails off more and more slowly, we cannot say when the last atoms will decay.

Because we cannot say when the substance will have entirely decayed, we have to think of another way of describing the rate of decay. As shown on the graph in Figure 4f.5, we identify the half-life of the substance:

- the **half-life** of a radioactive substance is the average time taken for half of the atoms in a sample to decay.

Uranium decays slowly because it has a very long half-life; the radioactive samples used in schools usually have half-lives of a few years, so that they have to be replaced once in a while. Some radioactive substances have half-lives which are less than a microsecond. No sooner are they formed than they decay into something else.

Explaining half-life

After one half-life, half of the atoms in a radioactive sample have decayed. However, this does not mean that, after two half-lives, all of the atoms will have decayed. From the graph of Figure 4f.5, you can see that one quarter will still remain. Why is this?

Figure 4f.6 shows one way of picturing what is going on. We picture a sample of 100 undecayed atoms of a radioactive substance. They decay randomly; each one has a fifty–fifty chance of decaying in the course of one half-life.

- After one half-life, a random selection of 50 atoms has decayed.
- During the next half-life, a random selection of half of the remaining 50 atoms decays, leaving 25 undecayed.
- During the third half-life, half of the remaining atoms decay, leaving 12 or 13. (Of course, you can't have half an atom.)

So the number of undecayed atoms goes 100 – 50 – 25 – 12 and so on. It is because radioactive atoms decay in a random fashion that we get this pattern of decay. Notice that, just because one atom has not decayed in the first half-life does not mean that it is more likely to decay in the next half-life. It has no way of remembering its past.

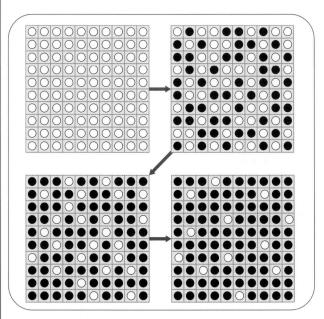

Figure 4f.6 The pattern of radioactive decay comes about because the decay of individual atoms is random. Half decay during each half-life, but we have no way of predicting which half.

Usually, we cannot measure the numbers of atoms in a sample. Instead, we measure the **count rate** using a Geiger counter or some other detector. The count rate decreases following the same pattern as the number of undecayed atoms.

Worked example 1

A sample of radioactive element X has an activity of 240 decays per second. If the half-life of X is 3 years, what will its activity be after 12 years?

Step 1: calculate the number of half-lives in 12 years.

12 years / 3 years = 4 half-lives

Hence we want to know the activity of the sample after 4 half-lives.

Step 2: calculate the activity after 1, 2, 3 and 4 half-lives. (Divide by 2 each time.)

initial activity   = 240

activity after 1 half-life = 120

activity after 2 half-lives = 60

activity after 3 half-lives = 30

activity after 4 half-lives = 15

So the activity of the sample has fallen to 15 decays per second after 12 years.

SAQ

2 A sample of a radioactive substance contains 8000 undecayed atoms. How many will remain after 6 half-lives (approximately)?

3 A radioactive substance has a half-life of 20 years. What fraction of its atoms will remain undecayed after 80 years?

Measuring a half-life

Figure 4f.7 shows how the half-life of a particular substance, protactinium-234, is measured in the laboratory. After the bottle has been shaken, the upper, oily layer of liquid contains protactinium, which emits beta radiation as it decays. Because its half-life is about a minute, the count rate decreases

quickly. The number of counts in successive intervals of 10 s is recorded and a graph plotted.

Figure 4f.8 shows how to deduce the half-life from the decay graph.

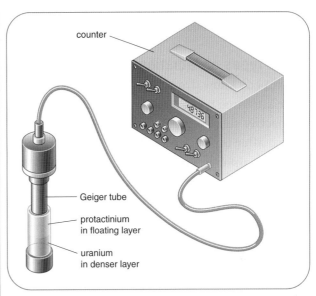

Figure 4f.7 A practical arrangement for measuring the half-life of the radioactive decay of protactinium-234.

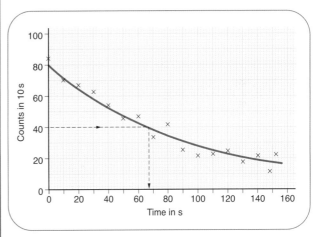

Figure 4f.8 The count rate for protactinium-234 decreases rapidly. The points show some experimental scatter, so a smooth curve is drawn. From this, the half-life can be deduced.

SAQ

4 From the graph of Figure 4f.8, what is the half-life of protactinium-234?

Radioactivity and atoms

To understand the nature of radioactivity, we need to picture what is going on at a microscopic level, on the level of atoms and nuclei.

● Why are some atoms radioactive and others not?

● What sort of radiation do they produce?

Radiation is emitted by the nucleus of an atom. We say that the nucleus is unstable; it emits radiation to become more stable. Fortunately, most of the atoms around us have stable nuclei. The unstable nuclei break up and emit radiation in an attempt to become more stable.

Three types of radiation

You already know (from Item P2d in Book 1) that there are three types of radiation emitted by radioactive substances (Table 4f.1). These are named after the first three letters of the Greek alphabet: alpha (α), beta (β) and gamma (γ). Alpha and beta are particles; gamma is a form of electromagnetic radiation.

Name	Symbol	What is it?
alpha	α	2 protons and 2 neutrons – the nucleus of a helium atom
beta	β	a fast-moving electron
gamma	γ	electromagnetic radiation

Table 4f.1 Three types of radiation produced by naturally occurring radioactive substances. To these we should add neutrons and positively charged beta radiation, produced by some artificial radioactive substances.

- An **alpha particle** is made up of two protons and two neutrons. (This is the same as the nucleus of a helium atom.) Because it contains protons, it is positively charged.
- A **beta particle** is an electron. It is not one of the electrons which orbit the nucleus; it comes from inside the nucleus. It is negatively charged and its mass is much less than that of an alpha particle.
- A **gamma ray** is a form of electromagnetic radiation, so it has no charge and no mass. We can think of it as a wave with a very short wavelength (similar to an X-ray).

An atom of a radioactive substance emits either an alpha particle or a beta particle; in addition, it may emit some energy in the form of a gamma ray. The gamma ray is usually emitted at the same time as the alpha or beta, but it may be emitted some time later. The radiation comes from the nucleus of the atom (Figure 4f.9).

The radiation carries away some of the excess energy which was making the nucleus unstable.

Alpha and beta particles have kinetic energy; a gamma ray transfers energy as electromagnetic radiation.

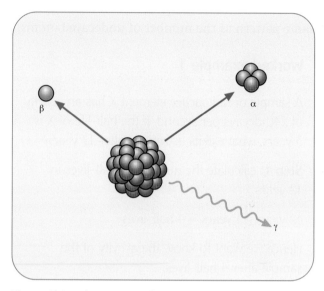

Figure 4f.9 Radiation comes from the nucleus of a radioactive atom.

SAQ

5 Which type of radiation from a radioactive substance:

 a has no mass?

 b is the same as a helium nucleus?

 c is a fast-moving electron?

Atoms and elements

The radiation from a radioactive substance comes from the atomic nucleus. The nucleus is made up of two types of particle, **protons** and **neutrons**. The protons carry the positive charge of the nucleus, whereas the neutrons are neutral. Together, protons and neutrons are known as **nucleons**.

Many different combinations of protons and neutrons are possible. Some combinations give stable atoms; others give unstable, radioactive atoms.

Each element has its own symbol, such as H for hydrogen. Sometimes an atom is represented by this symbol, with two numbers alongside:

$$^4_2\text{He}$$

This represents an atom of helium. The lower number tells us that there are two protons in the

H nucleus of the atom and the upper number tells us that there are a total of four nucleons. (From this, it is simple to work out that there must be two neutrons in the nucleus.)

In general, we can represent the nucleus of an atom of element E like this:

$$^A_Z E$$

Z is the **atomic number** – the number of protons in the nucleus.

A is the **mass number** – the number of nucleons (protons and neutrons) in the nucleus.

A neutral atom of element E will also have Z electrons orbiting the nucleus.

From Z and A you can work out the number of neutrons in the nucleus, $A–Z$.

SAQ

6 An atom of carbon is represented by $^{12}_6 C$.

   a What is its atomic number?

   b What is its mass number?

   c How many neutrons are there in its nucleus?

Nuclear equations - alpha decay

When an atom of a radioactive substance decays, it becomes an atom of another element. This is because, in alpha and beta decay, the number of protons in the nucleus changes. We can represent any radioactive decay by an equation using the notation explained above.

Here is an example equation for **alpha decay**:

$$^{241}_{95} Am \rightarrow ^{237}_{93} Np + ^4_2 He + energy$$

This represents the decay of americium-241, the radioactive substance used in smoke detectors. It emits an alpha particle (represented as a helium nucleus) and becomes an atom of neptunium.

Notice that the numbers in this equation must balance, because we cannot lose mass or charge:

   mass numbers: $241 \rightarrow 237 + 4$
   atomic numbers: $95 \rightarrow 93 + 2$

You can see that, in alpha decay:
- the nucleus loses two protons and two neutrons
- the mass number decreases by four and the atomic number by two.

H ## Nuclear equations - beta decay

Here is an example for **beta decay**:

$$^{14}_6 C \rightarrow ^{14}_7 N + ^0_{-1} e + energy$$

This is the decay which is used in **radiocarbon dating**. A carbon-14 nucleus decays to become a nitrogen-14 nucleus. (The beta particle, an electron, is represented by $^0_{-1}e$. Notice that the atomic number of the electron is −1, because its charge is negative.) If we could see inside the nucleus, we would see that a single neutron has decayed to become a proton:

$$^1_0 n \rightarrow ^1_1 p + ^0_{-1} e$$

For each of these two beta decay equations, you should be able to see that the mass numbers and atomic numbers are balanced. We say that, in radioactive decay, mass number and atomic number are *conserved*.

So, in beta decay:
- the nucleus loses one neutron and gains one proton;
- the mass number is unchanged and the atomic number increases by one.

SAQ

7 The equation below represents the decay of a uranium atom $^{235}_{92} U$.

$$^{235}_{92} U \rightarrow ^{231}_{90} Th + ^4_2 He$$

   a What type of particle does it emit?

   b Is the equation balanced?

Worked example 2

The polonium atom $^{214}_{84} Po$ decays by emitting an alpha particle. It becomes an atom of lead (Pb). Write down a balanced equation for this decay.

Step 1: write down an equation which includes the unknown atomic and mass numbers for the Pb atom.

$$^{214}_{84} Po \rightarrow ^A_Z Pb + ^4_2 He$$

continued on next page

Worked example 2 – *continued*

Step 2: knowing that mass number is conserved, calculate the mass number A.

$214 = A + 4$

so $A = 210$

Step 3: knowing that atomic number is conserved, calculate the atomic number Z.

$84 = Z + 2$

so $Z = 82$

Step 4: write down the complete equation.

$$^{214}_{84}\text{Po} \rightarrow {}^{210}_{82}\text{Pb} + {}^{4}_{2}\text{He}$$

SAQ

8 The strontium atom $^{90}_{38}\text{Sr}$ decays by emitting a beta particle to become an atom of yttrium (Y). Write a balanced equation for this decay.

Summary

You should be able to:

◆ state that an alpha particle is a helium nucleus

◆ state that a beta particle is a fast-moving electron

◆ state that radiation comes from the nucleus of a radioactive atom

◆ recognise that radioactivity is measured by the number of decays per second

◆ recognise that radioactivity decreases gradually with time

◆ explain and use the fact that the half-life of a radioactive substance is the time taken for half of its atoms to decay

◆ construct and use balanced equations to represent alpha and beta decay

Questions

1 Which particles make up the nucleus of an atom? Which orbit around the nucleus?

2 A sample of a radioactive substance contains 200 undecayed atoms. How many will remain undecayed after three half-lives?

3 The half-life of a radioactive substance X is 10 days. A sample gives an initial count rate of 440 counts per second. What will be the count rate after 30 days?

4 The radioactive substance E has a half-life of 2000 years. How long will it take the activity of a sample of E to decrease to one-eighth of its initial value?

5 An atom of a particular isotope of oxygen is written as $^{17}_{8}\text{O}$. What are its atomic number and mass number?

6 An atom of a particular isotope of lead (symbol Pb) contains 82 protons and 128 neutrons. Write down the symbol for this atom.

7 How many protons, neutrons and electrons are there in a neutral silver atom, $^{107}_{47}\text{Ag}$?

8 A radioactive decay is represented by the equation:

$$^{239}_{94}\text{Pu} \rightarrow {}^{A}_{Z}\text{U} + \alpha$$

Write this equation in balanced form, showing the alpha particle as a helium nucleus and giving the correct values of A and Z.

Radioactivity all around

We need to distinguish between two things: **radioactive substances** and the **radiation** which they give out. Many naturally occurring substances are radioactive; usually these are not very concentrated, so they don't cause a problem. There are two ways in which radioactive substances can cause us problems.

- If a radioactive substance gets inside us, its radiation can harm us. We say that we have been **contaminated**.
- If the radiation they produce hits our bodies, we say that we have received a dose of radiation – we have been **irradiated**.

In fact, we are exposed to low levels of radiation all the time; this is known as **background radiation**. In addition, we may be exposed to radiation from artificial sources, such as the radiation we receive if we have a medical X-ray.

Figure P2d.1 in Book 1 shows the different sources which contribute to the average dose of radiation received by people in the United Kingdom. It is divided into natural background radiation (about 87%) and radiation from artificial sources (about 13%). We will look at these different sources in turn.

Sources of background radiation

- Air is radioactive. It contains a radioactive gas called radon which seeps up to the Earth's surface from radioactive rocks underground. Because we breathe in air all the time, we are exposed to radiation from this substance. This contributes about half of our annual exposure. (This varies widely from one part of the country to another, depending on how much uranium there is in the underlying rocks.)
- The ground contains radioactive substances; we use materials from the ground to build our houses, so we are exposed to radiation from these.
- Our food and drink is also slightly radioactive. Living things grow by taking in materials from the air and the ground, so they are bound to be radioactive. Inside our bodies, our food then exposes us to radiation.

- Finally, radiation reaches us from space in the form of cosmic rays. Some of this comes from the Sun, some from further out in space. Most cosmic rays are stopped by the Earth's atmosphere; if you live up a mountain, you will be exposed to more radiation from this source. If you fly in an aircraft, you are high in the atmosphere. You are exposed to more cosmic rays. This is not serious for the occasional flier but airline crews have to keep a check on their exposure.

Because natural background radiation is around us all the time, we have to take account of it in experiments. It may be necessary to measure the background level and subtract it from experimental measurements.

SAQ

1 In space, astronauts are exposed to higher levels of radiation than on Earth. Which source of radiation is responsible?

2 Which give us a bigger dose of radiation, natural sources or artificial sources?

Sources of artificial radiation

- Most radiation from artificial sources comes from medical sources. This includes the use of X-rays and gamma rays for seeing inside the body and the use of radiation for destroying cancer cells. There is always a danger that exposure to such radiation may trigger cancer; medical physicists are always working to reduce the levels of radiation used in medical procedures. Overall, many more lives are saved than lost through this beneficial use of radiation.
- Today, most testing of nuclear weapons is done underground. In the past, bombs were detonated on land or in the air and this contributed much more to the radiation dose received by people around the world.
- Many people, such as medical radiographers and staff in a nuclear power station, work with radiation. Overall, this does not add much to the national average dose but, for individuals, it can increase their dose by up to 10%.

H
- Finally, small amounts of radioactive substances escape from the nuclear industry, which processes uranium for use as fuel in nuclear power stations and handles the highly radioactive spent fuel after it has been used.

SAQ

3 We are exposed to radiation from man-made sources. Give a single word which means the same as 'man-made'. Give a single word which is the opposite of 'man-made'.

Uses related to penetrating power

Radiation from radioactive materials can pass straight through solid matter. Different types of radiation have different penetrating powers. This is summed up in Figure 4g.1.

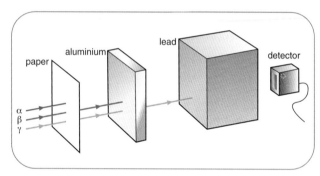

Figure 4g.1 How radiation is absorbed.

SAQ

4 Which type of radiation is the most penetrating? Which is the least penetrating?

Smoke detectors

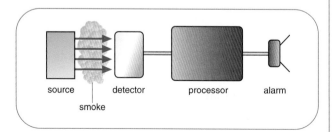

Figure 4g.2 A block diagram of the circuit which sounds the alarm when smoke absorbs the alpha radiation. Americium is an interesting element. It is not found in nature but must be made in nuclear reactors. It is element 95 in the Periodic Table and is described as a transuranic (beyond uranium).

These are often found in domestic kitchens and in public buildings such as offices and hotels. If you open a smoke detector to replace the battery, you may see a yellow and black radiation hazard warning sign (see Figure P2d.8 in Book 1). The radioactive material used is americium-241, a source of alpha radiation. Here is how it works (Figure 4g.2).

- Radiation from the source falls on a detector. Because alpha radiation is charged, a small current flows in the detector. The output from the processing circuit is OFF, so the alarm is silent.
- When smoke enters the gap between the source and the detector, it absorbs the radiation. Now no current flows in the detector and the processing circuit switches ON, sounding the alarm.

In this application, a source of alpha radiation is chosen because alpha radiation is easily absorbed. Smoke would not significantly absorb beta or gamma radiation.

Radioactive tracers

We have already seen (Item P4e) how a **radioactive tracer** can be used in medicine. A radioactive substance is put into a patient's body and its movement through the body can be traced by detecting the beta or gamma radiation which comes out of the body.

Engineers also use radioactive tracers. They may want to trace the flow of water in underground pipes – perhaps they suspect that one of the pipes is blocked, or it may be that the plans of underground piping have been lost.

Alternatively, they may want to track the movement of waste. For example, a waste dump may be suspected of leaking harmful chemicals into its surroundings or a pipe may be discharging waste into the sea. By adding a radioactive tracer to the waste, its movement can be monitored.

SAQ

5 The radiation coming from inside a pipe may have to pass through thick metal. What type of radiation would be best for this?

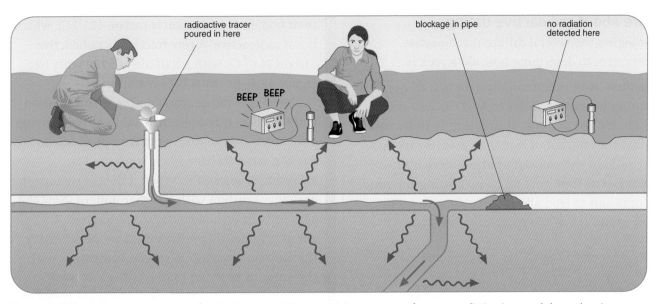

Figure 4g.3 Detecting the movement of water in pipes. Water containing a source of gamma radiation is passed down the pipes; radiation detectors reveal how the water moves through the pipes.

H Tracers in pipes

Here is how engineers use radioactive tracers to check pipes (Figure 4g.3).

- Under high pressure, they inject water containing a radioactive chemical into the pipe of interest.
- Then they move detectors over the surface of the pipe.
- The tracer is a gamma emitter, so the radiation can penetrate the walls of the pipe and be detected.
- At a point where no radiation is detected, this suggests that the pipe is blocked.

SAQ

6 What would you expect to find if a pipe was only partly blocked?

Dating using radioactivity

Radioactive substances decay all the time, quickly at first, and then more and more slowly. Because we can measure how fast they decay, we can use this information to find out how old things are. Here are two examples.

- Many rocks contain radioactive materials such as uranium. As time goes by, the amount of uranium in a rock decreases. So the rocks with the least amount of uranium are the oldest. Geologists have used this to find out the age of the Earth – it's about 4.5 billion years old.

- All living things contain carbon. One form of carbon, called carbon-14, is radioactive. When a living organism dies – perhaps a tree, or a person – the amount of radioactive carbon in their remains gradually decreases (Figure 4g.4). This can be used to find the age of the remains of anything which was once alive. The technique is known as **radiocarbon dating**.

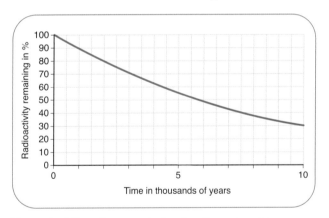

Figure 4g.4 How the amount of radioactive carbon-14 decreases with time. This is used to find out how long ago an organism was alive.

SAQ

7 Archaeologists are trying to date two pieces of wood: one from a weapon used at the Battle of Hastings (1066) and one from a sunken boat. They find that the wood from the boat contains a smaller fraction of carbon-14 than the wood from the weapon. What can you say about the age of the boat?

More about radioactive dating

Geologists use several different radioactive substances to determine how old a rock is.

Uranium–lead When uranium decays, its atoms eventually become atoms of lead. So, as a rock gets older, the amount of uranium in it decreases, whilst the amount of lead increases (Figure 4g.6). The ratio of the amount of uranium to lead decreases as time goes by, so measuring the ratio of uranium to lead allows a geologist to get an idea of the age of the rock.

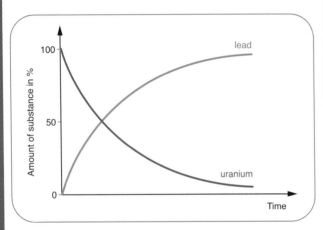

Figure 4g.6 Uranium turns to lead as a result of radioactive decay.

Radiocarbon dating Living organisms contain carbon. Plants get this from atmospheric carbon dioxide, which they use in photosynthesis. Animals get it from the plants they eat to build their bodies. Most carbon is carbon-12 ($^{12}_{6}C$), which is not radioactive. A tiny fraction is radioactive carbon-14 ($^{14}_{6}C$), with a half-life of 5730 years (see Figure 4g.4). (It emits beta radiation.)

The idea of radiocarbon dating is this: when a plant is alive, it is constantly taking in and giving out carbon (in the form of carbon dioxide). It thus has the same fraction of carbon-14 as the atmosphere around it. When it dies, it no longer exchanges carbon with its environment. Now the carbon-14 in its body decays. As time passes, the amount remaining decreases. If we can measure the amount remaining, we can compare it with the amount in living matter today and work out when the organism was alive. There are two ways to measure the amount of carbon-14 present in an object:

- by measuring the activity of the sample using a detector such as a Geiger counter
- by counting the number of carbon-14 atoms using a mass spectrometer.

Problems can arise with radiocarbon dating. It may be that the amount of carbon-14 present in the atmosphere was different in the past; certainly, nuclear weapons testing added extra carbon-14 to the atmosphere during the 1950s and 1960s; this means that living objects which died then have an excess of carbon-14, making them appear younger than they really are.

The Turin shroud

The Turin shroud (Figure 4g.5) is a religious relic which can be seen in a chapel at the cathedral in Turin, Italy. It carries markings which clearly show the figure of a bearded man, complete with injuries to his body. Since the 14th century, writers have claimed that this was the burial shroud of Christ, in which his body was wrapped after he was taken down from the cross.

In 1988, scientists were permitted to use a small fragment of the cloth to use in finding its age. The shroud was dated to AD 1325 (±33), showing that it did not date from biblical times.

The experimental result did not put an end to the debate about the shroud. Perhaps the cloth

Figure 4g.5 The Turin shroud was dated by radiocarbon dating. It was found to date from the 14th century, which matched the dates of the earliest historical records of its existence.

had been contaminated with more modern material, or perhaps the scientists were lying. An object like the shroud has a great significance to very many people, so no doubt the debate will continue for years to come.

H **SAQ**

8 Look at Figure 4g.4. Why would radiocarbon dating be unsuitable for a piece of wood from a tree which grew just 100 years ago? Why would it be unsuitable for dating an object which is 50 000 years old?

Summary

You should be able to:

- describe and recognise that we are continually exposed to background radiation from radioactive substances around us (e.g. in air, rocks, soil) and from cosmic rays

H - explain that some background radiation comes from artificial sources, e.g. radioactive waste from industry and hospitals and medical uses of radioactivity

- describe that smoke detectors make use of alpha-emitting sources

- recall that radioactive tracers can be used to track the dispersal of waste and to detect blockages in underground pipes

H - describe how radiation from a gamma source put into a pipe can reveal the movement of materials in the pipe

- recall that the amount of radioactivity in rocks and dead organic materials decreases with time; this can be used to date rocks and old materials

H - explain how rocks are dated using the uranium–lead ratio

H - explain how radiocarbon dating uses the decay of carbon-14 in material which was once living

Questions

1 Which type of radiation is made use of in a smoke detector? Explain this choice.

2 What is meant by *background radiation*? List the main sources of natural background radiation.

3 At a factory, engineers suspect that water is leaking from an underground pipe. How could they test this idea using a radioactive tracer?

H 4 In 1991, the dehydrated body of a man who lived in the late Stone Age was found in the Alps. Samples were taken for radiocarbon dating.

a Which form of carbon is used in this technique?

b It was found that 46% of this carbon had decayed since the man's death. Use the graph in Figure 4g.4 to estimate the time which had passed since he died.

Nuclear power

The electricity generated by a nuclear power station is exactly the same as the electricity from any other power station. Figure 4h.1 compares a nuclear station with a coal-fired one. The principal difference is at the left-hand side.

- In a nuclear power station, heat is generated in a nuclear **reactor**, which uses uranium as its fuel.
- This energy is transferred by a **coolant** (such as carbon dioxide gas) to a boiler which generates steam.

  How does this compare when coal is used as a fuel?

- In a coal-fired station, burning coal produces heat.
- This heats the water in the boiler to make steam. After that, the process is the same in both cases.
- High-pressure steam from the boiler is used to turn turbines.
- The turbines then turn generators.

  So you can see that the main difference is that a nuclear power station has a reactor in which energy is released from uranium fuel; this happens by a process called **nuclear fission**. In this Item, you will learn more about this process.

SAQ

1 Plutonium is one element which can be used as a fuel in nuclear power stations. Name another.

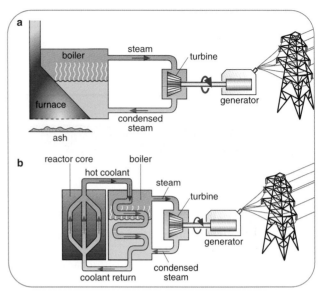

Figure 4h.1 These block diagrams show the similarities between coal-fired (**a**) and nuclear (**b**) power stations. The main difference is in the way heat is produced and transferred to the boiler.

The process of fission

In Item P4f, we saw why radioactive decay occurs: an unstable nucleus becomes more stable by emitting a small particle (alpha or beta) and usually some electromagnetic radiation (a gamma ray). Nuclear fission is another way in which an unstable nucleus can become stable.

- In nuclear fission, a large nucleus splits in two to form two smaller nuclei. In the process, some neutrons are also released. (The word *fission* means 'splitting'. Think of a fissure, which is a large crack in a rock.)

The history of nuclear power

The nuclear power industry has had a troubled history. Nuclear power stations make use of nuclear fission, the splitting of uranium atoms. In its early days, half a century ago, nuclear power was seen as a cheap, clean and safe way of providing electricity on a large scale. Today, a significant proportion of the world's energy resources come from nuclear power – roughly 20% of the UK's electricity, more than 75% of France's.

The first countries to develop nuclear power were those which had developed nuclear weapons during the Second World War – the USA, the UK and the Soviet Union (now Russia). Some people opposed this development because early reactors were often designed to produce plutonium for use in nuclear weapons, as well as electricity. There is still concern that, when nuclear power technology is sold to other countries, they are being given the means to produce nuclear weapons.

H There are two ways in which fission can occur. Sometimes, a large unstable nucleus will split spontaneously; this is *spontaneous fission*. In the other kind of fission, a neutron collides with a large nucleus, making it even more unstable and causing it to split; this is *neutron-induced fission*. We will concentrate on this second mechanism, because it is the one which is made use of in nuclear power stations and in nuclear bombs.

Figure 4h.2 shows a uranium nucleus undergoing fission. A neutron hits the uranium nucleus and is absorbed. The nucleus is now very unstable and starts to deform. Soon, it splits completely in two, to form two 'daughter' nuclei. At the same time, two or three neutrons are released.

SAQ

2 Look at Figure 4h.2. What can you say about the masses of the two daughter nuclei which result from the fission process?

Releasing energy

This process releases energy which was previously stored in the uranium nucleus.

This energy takes several forms.

- The daughter nuclei move apart; they have kinetic energy.
- The neutrons also fly off at high speed; they too have kinetic energy.
- Gamma rays are released; this is energy in the form of photons of electromagnetic radiation.

H (Sometimes, people talk about 'nuclear energy'. We might use this name for the energy stored by the uranium nucleus. This energy is converted to the kinetic energy of the particles and the energy of the gamma radiation.)

When a single uranium nucleus splits like this, the energy released is about 10^{-11} J. This is a tiny amount but it is much more than the energy released in a chemical reaction. For example, when two hydrogen atoms and an oxygen atom bond together to form a molecule of water, the energy released is less than one-millionth of that released by the fission of a uranium nucleus. Hence, we can think of uranium as a highly concentrated store of energy, roughly one million times as concentrated as an equal mass of coal and oxygen. That is what attracted physicists to the idea of using uranium as a fuel for electricity generation.

SAQ

3 In a nuclear power station, the energy released in fission is used to boil water. What energy transformation has taken place?

Chain reaction

In the process of fission, the nucleus of a uranium atom splits in two and energy is released. In the process, two or more neutrons are released. These fly off at high speed – perhaps 10^7 m/s. Each of these neutrons may collide with another uranium

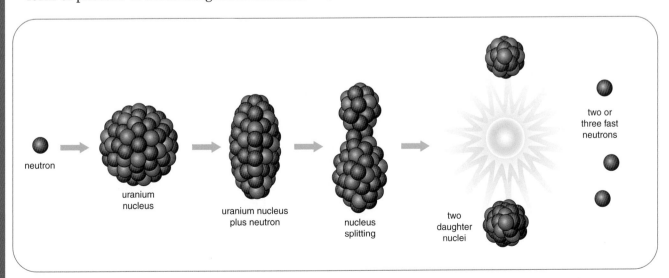

Figure 4h.2 A uranium nucleus undergoing neutron-induced fission. In the process, a lot of energy is released. Two daughter nuclei are formed and two or three neutrons are released. These neutrons may go on to cause more uranium nuclei to split.

neutron

uranium nucleus

uranium nucleus plus neutron

nucleus splitting

two daughter nuclei

two or three fast neutrons

nucleus and cause it to split. The result can be a **chain reaction** (see Figure 4h.3).

- In a nuclear power station, the chain reaction is kept under control. The reactor is designed so that just one of the neutrons goes on to cause another nucleus to split. The other neutrons escape from the uranium and are absorbed by the surrounding concrete.

- In a nuclear bomb, the idea is to get as many uranium nuclei as possible to undergo fission in a very short time; from the first fission, perhaps two neutrons go on to cause uranium nuclei to split, releasing four neutrons, which cause fission, then 8, 16 and so on. The nuclear reaction escalates out of control and vast amounts of energy are released in a fraction of a second.

The reactor in a nuclear power station is designed so that an uncontrolled chain reaction cannot happen. The biggest danger is that the reaction may go so fast that the uranium overheats. It may melt and the reactor may burst into flames.

In a nuclear bomb (Figure 4h.4), a chain reaction has been deliberately allowed to go out of control. Several kilograms of uranium may give up their stored energy in a fraction of a second.

SAQ

4 What is the main difference between a chain reaction in a nuclear power station and in a nuclear bomb?

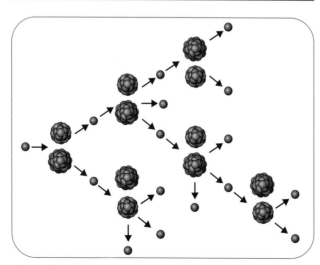

Figure 4h.3 A nuclear chain reaction. It does not really happen in straight lines like this; the neutrons released fly off in all directions and they may travel some distance through the uranium before they strike another nucleus and cause it to split.

Figure 4h.4 A nuclear bomb – a chain reaction in uranium has been allowed to go out of control.

Under control

There are several different designs of nuclear reactor in use around the world. Figure 4h.5 shows a 'Magnox' reactor, a type which has been in use in the United Kingdom and elsewhere for 40 years. Notice the following features.

- The fuel is contained in rods; inside each **fuel rod**, the fuel is in the form of pellets. A spent rod can be removed and replaced while the reactor is running.

- Between the fuel rods are blocks of graphite (carbon), which form the **moderator**. Without the moderator, the neutrons would tend to fly straight out of the reactor. The neutrons are slowed down as they pass through the graphite, making them more likely to interact with uranium nuclei.

- Also passing through the graphite are **control rods**. These contain a material (boron) which is good at absorbing neutrons. They can be lowered into the reactor to slow or stop the reaction; they can be removed to speed it up. They are designed to drop in automatically if the reactor overheats. The idea is that, to keep the reaction going at a steady rate, an average of just one neutron from each nucleus that splits should go on to cause another atom to split.

- The **coolant** is carbon dioxide gas. It is pumped along pipes in the hot core of the reactor, where it heats up. The hot gas then passes down through pipes in the boiler, resulting in high-pressure steam.

H

boiler　　reactor core　　boiler

steam out

hotter CO$_2$

uranium fuel rods

cool CO$_2$

cold water in

gas circulator

Figure 4h.5 The core of a Magnox-type reactor. The chain reaction occurs in the uranium fuel; its heat is carried away by the coolant.

These four parts make up the core of the reactor. The coolant passes through pipes in the core so that it isn't contaminated by the core's radioactivity. The core is surrounded by thick layers of concrete, which absorb neutrons that escape from it; this means that it is safe to work on top of the core even when it is operating.

SAQ

5　Explain why the chain reaction in a reactor comes to a halt when the control rods are fully lowered.

Radioactive waste

A big coal-fired power station needs a lot of coal. It may need a trainload every hour – over a thousand tonnes. Nuclear fuel is a much more concentrated source of energy, so a nuclear power station may need just one truckload of fuel each week to produce the same amount of electricity.

Coal-fired stations produce ash, carbon dioxide and other waste gases. Nuclear power stations produce radioactive waste.

There are two reasons why the waste from a nuclear power station is radioactive.

- When the nucleus of a uranium nucleus splits, two smaller atoms are formed. These are unstable, so they are radioactive.
- The neutrons which are released by the splitting nucleus may be absorbed by other materials in the reactor, such as the steel container. These materials then become radioactive.

When spent fuel rods are removed from the reactor core, they must first be stored in deep tanks of water. This removes the heat generated by the radioactive decay of the daughter products. After a few weeks, the spent rods are packed into a strong metal container and transported to a reprocessing plant.

SAQ

6　Put these sources of energy in order, from most concentrated to least concentrated.

wind　　　　uranium　　　　coal

What's to be done?

Radioactive waste from nuclear power stations has to be dealt with. Some of the radioisotopes are highly radioactive, but that can be a good thing. After a few years, their activity will have decreased to a safe level. Others are less radioactive and this means they must be stored safely for a long time, perhaps centuries, until they are safe to release to the environment. Figure 4h.6 shows how the activity of nuclear waste decreases over the years.

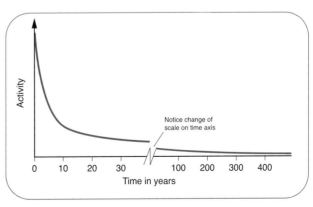

Figure 4h.6 Nuclear waste is a mixture of radioactive substances, some with short half-lives, others with long half-lives. Its activity decreases rapidly at first, but the longer-lived isotopes may be active for thousands of years.

Figure 4h.7 This nuclear reprocessing plant at Cap de la Hague in northern France receives shipments of spent uranium fuel from many countries. It separates out the radioactive daughter products and reconcentrates the uranium for reuse in power stations.

It is difficult to decide the best way to deal with these dangerous materials. Is it best to concentrate them so that only a small volume of highly dangerous material must be stored, or is it best to keep them in a more dilute form?

A second problem is that the chain reaction in uranium usually comes to a halt long before all of the uranium nuclei have been split. This is because it has become too dilute; there are no longer enough uranium nuclei present to ensure that at least one neutron from each fission event goes on to cause another fission. The spent fuel must be processed to reconcentrate the uranium and to remove the hazardous daughter products. Such reprocessing requires great technical skill to avoid releasing hazardous materials to the environment. Because there are only a handful of reprocessing plants in the world (Figure 4h.7), spent fuel must be transported over large distances and this is another cause for concern. An accident might result in the release of radioactive material; a terrorist attack might put bomb-making material in the hands of dangerous groups.

Nuclear power and the environment

Any method of generating electricity is bound to damage the environment. Burning fossil fuels produces carbon dioxide and contributes to the greenhouse effect. A dam for a hydroelectric scheme can flood useful farmland and displace local people. Nuclear power stations produce radioactive waste. This is the major environmental concern about nuclear power. How should the waste be handled? Should it be dumped underground or stored above ground, where its condition can be monitored for decades into the future?

Cost is another concern with nuclear power. All power stations are expensive to build and the operator gets no return on investment until electricity starts to flow. This may be as long as ten years from the start of planning and construction. Dealing with the radioactive waste produced is expensive. Finally, at the end of its life, a nuclear power station must be dismantled and made safe (Figure 4h.8). The site must be

Figure 4h.8 Berkeley in the west of England was one of the UK's first nuclear power stations. At the end of its working life, much of it was dismantled. Here you can see the large rusty-red boilers being laid down on the ground. Eventually levels of radioactivity will be low enough for the whole site to be grassed over.

looked after for up to 100 years. All this adds to the cost, so that the original idea of electricity 'so cheap that it would not be worth the expense of metering' proved to be an over-optimistic dream.

Summary

You should be able to:

♦ recall that uranium is the fuel used in a nuclear power station

♦ recall that a nuclear reaction releases energy by the process of fission

H ♦ describe how, in the fission process, a uranium nucleus is split when it is hit by a neutron and energy is released

♦ describe how this heat is used to produce steam, which turns a turbine, which turns an electricity generator

♦ recall that a chain reaction is needed to release large amounts of energy

♦ recall that a nuclear explosion is an uncontrolled chain reaction

H ♦ explain how, in a chain reaction, the neutrons released by one uranium atom go on to split others

♦ explain how control rods absorb neutrons and ensure that just enough neutrons go on to maintain the chain reaction

Questions

1 Put the following items in order, to describe how a nuclear power station works.

- heat released
- nuclear reaction
- turbine turned
- electricity generated
- steam produced

2 What is the name of the process by which energy is released in a nuclear reactor?

3 Neutrons are released by uranium in a nuclear reactor. These neutrons may be absorbed by other materials in the reactor. What is the result of this?

H 4 What particle can induce the fission of a uranium nucleus?

5 Look at Figure 4h.3. How can you tell that the daughter nuclei are not uranium nuclei but the nuclei of other elements?

6 What is the difference between a controlled chain reaction (as in a nuclear reactor) and an uncontrolled chain reaction (as in a nuclear bomb)?

7 Here is an equation which shows one way in which a uranium nucleus may split:

$$^{235}_{92}U + ^{1}_{0}n \longrightarrow ^{141}_{56}Ba + ^{92}_{36}Kr + 3^{1}_{0}n + energy$$

a How many neutrons are produced when the uranium nucleus splits?

b How many of these must go on to cause another nucleus to split, if the chain reaction is to continue at a steady rate?

c Show that the equation is balanced.

Answers to SAQs: Biology

Item B3a Molecules of life

1 Yes, plant cells do contain mitochondria. All living organisms, including plants, respire.

2 a 40 (because A always pairs with T)
   b 60 (because all the other base pairs must be G–C)

3 The bases are 'read' in groups of three, each group coding for one amino acid. If an extra base is inserted, then all the groups of three from then onwards will be different. So all the amino acids from then on in the protein will be different, making an entirely different protein.

4 The suspect has only his own blood on him. No one has the victim's blood. So it does not help us at all to decide who killed the victim. It neither disproves nor supports the hypothesis.

5 a 7
   b Pepsin has a much lower optimum pH than most enzymes. This fits in with the environment in which it functions – in the hydrochloric acid in the stomach, where the pH can be as low as 2.

Item B3b Diffusion

1 The particles in a gas are moving fast and freely, scarcely interacting with each other at all. The particles in a liquid are also in constant motion, but they are still interacting with each other and are less free to move around. So diffusion happens faster in a gas than in a liquid.

2 a Glucose is likely to be used as a respiratory substrate and to be broken down to release energy.
   b Amino acids are likely to be used to make proteins, for growth and repair of cells and also making enzymes, haemoglobin, antibodies, keratin and many other proteins.

3 a The placenta has many folds, increasing the surface area of contact between the mother and foetus.
   b i The concentration gradient for glucose is from the mother's blood to the foetus's blood. This happens because the foetus's cells are constantly respiring and using glucose, which is resupplied from the mother's digestive system or from glucose stores in her liver.
      ii The concentration gradient for carbon dioxide is from the foetus's blood to the mother's blood. This is because the foetus's cells are constantly respiring and producing carbon dioxide, which is lost from the mother's body when she breathes out.
   c The foetus's blood and mother's blood are separated by very thin walls of tissue.

4 a 0.02 mm per second (2×10^{-5} m/s)
   b The transmitter substance slots into receptors in the cell membrane of the next neurone, and starts up an action potential there.

5 The air is saturated with water vapour. The concentration of water vapour inside the leaf will not be any greater than it is in the air. Thus, there is no concentration gradient for water vapour and it does not diffuse out of the leaf.

Item B3c Keep it moving

1 Respiration is constantly taking place, using up oxygen.

2 The haemoglobin in the red blood cells combines with oxygen in the air, becoming bright red.

3 a Insulin enters the blood from the pancreas.
   b Water enters the blood from the alimentary canal (especially the small intestine and the colon).
   c Carbon dioxide enters the blood from all the respiring tissues, all over the body.

4 a hepatic portal vein, liver, hepatic vein, vena cava, right side of heart, pulmonary artery, lungs, pulmonary vein, left side of heart, aorta, carotid artery, brain
   b iliac vein, vena cava, right side of heart, pulmonary artery, lungs

5 The walls of the heart are made of muscle and the more force the muscle needs to produce, the thicker it grows. The muscles in the atria need only produce enough force to squeeze the blood down into the ventricles. The muscles in the ventricles must produce enough force to push the blood all around the body.

6 Advantages: can be done even when the person's own heart is very badly damaged; the heart should work on its own and not need a power supply as an articifial pacemaker does; there will be no problem of clotting around a replacement valve so anti-clotting drugs need not be taken.
Disadvantages: the operation is very serious; there is a shortage of donors so the person may have to wait much longer than if they just need to have a valve replaced; their immune system may reject the heart unless they take immunosuppressant drugs for the rest of their life.

Item B3d Divide and rule

1 a 2   b 4   c 8
2 a DNA
  b It is replicating, forming exact copies of itself – the two chromatids in each chromosome.
3 a 4   b 8
4 The first, when the paired chromosomes separate from each other.
5 No, because there is only one set of chromosomes so they would not have partners to pair up with.

Item B3e Growing up

1 Mitosis.
2 Chloroplasts contain chlorophyll, which absorbs light for photosynthesis. Onion bulbs grow underground, where there is no light.
3 a The root cap protects the root as it pushes through the soil.
  b In the region a little way back from the tip, where some of the cells are changing into xylem and phloem cells.

4 Most body cells are specialised and cannot become any other kind of cell. Stem cells are undifferentiated (unspecialised) and can divide to form other kinds of cells.
5 a elephant, rhino, cow, horse, lion, goat, cat, squirrel, chipmunk, mouse
  b In general, the larger the animal, the longer its gestation period. However, there are some exceptions; for example, a cow has a greater body mass than a horse but a shorter gestation period.
  c All the organisms begin as a single cell. Many more cell divisions are needed to make enough cells to produce an elephant than are needed to produce a mouse, so the process takes longer.
6 a The head. Perhaps it is important for the brain to develop early on, because it may influence the growth of the rest of the body.
  b The reproductive organs will not be needed until the person is older, so there is no point in them growing early on.
7 a 32 kg
  b 71 − 32 = 39 kg
  c 11−13 (this is when the curve is steepest)
  d Adolescence.
  e The graph has still not levelled off, so he could still be growing.
8

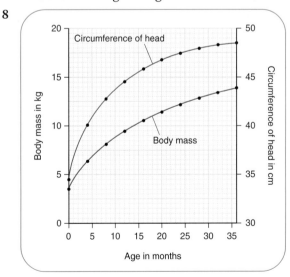

b 0−4 years
c 0−4 years

Item B3f Controlling plant growth

1 a i Same kinds of seedlings; same age of seedlings; same cotton wool; same amount of water; same size of box.

ii Any of them could affect the rate or direction of growth of the seedlings.

b Whether or not there is light and the direction from which it comes. It has been done by having different openings, or no opening, in the dark boxes.

c The seedlings in the dark were yellow, whereas those that had light were greener. In the dark, plants do not make chlorophyll; it would be useless because there is no light for it to absorb.

2 The results show that the tip senses the direction of light. If the tip is removed or covered, the shoot doesn't grow at all or grows straight upwards, even if light is coming from one side. When the tip is cut off and placed on jelly, as in D, and the jelly is replaced on the cut shoot, it grows towards the light just like the intact shoot B. This suggests that a chemical in the tip soaked into the jelly and this made shoot D grow towards the light.

3 a Shoots should grow upwards, away from gravity, so that flowers are held in the air for pollination and leaves are held in the light for photosynthesis.

b Roots should grow downwards, towards gravity, so that they can anchor the plant in the soil and absorb water and mineral ions.

4 You could place the clinostats so they point upwards rather than sideways. Pin germinating seedlings onto them. Put both clinostats into a box with light from one side. Switch one on so that it turns around all the time and leave the other one switched off so that it stays still. Compare the direction of growth of the shoots. The ones that were turning round effectively had light coming from all directions, and the others had light coming from one side.

5 It gives time for picking, packaging, transporting to the shop or market stall and then waiting for people to buy it.

Item B3g New genes for old

1 a

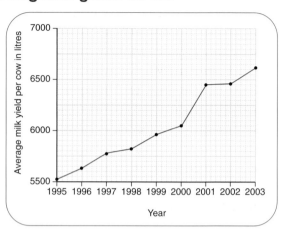

b 1097 litres

c Only cows with the highest milk yields would be chosen to breed. Bulls would be chosen whose mothers, sisters or daughters had high milk yields. This would be repeated in each generation.

d Perhaps, as time went on, the cows were kept in better conditions, with more food or a better diet.

2 If two dogs are closely related, they could both have the genotype Hh. When they breed together, there is a one in four chance that a puppy will have the genotype hh and have the hip weakness.

	H	h
H	HH normal	Hh normal
h	Hh normal	hh weak hips

The genetic diagram should show and label clearly: the genotypes of the parents; the genotypes of the gametes; the possible genotypes and phenotypes of the offspring.

Item B3h More of the same

1 a When the embryos are split; each embryo that is made by splitting one original embryo will be genetically identical to the other.

b No. Each of the original embryos will be genetically different because they have been produced by the fusion of different eggs and sperms.

2 a The Finn Dorset.

b She is genetically identical with her Finn Dorset parent.

3 There are no right or wrong answers to this question. Your answer should contain some science to back up your opinions or arguments.

Item B4a Who planted that there?

1 $6CO_2 + 6H_2O \rightarrow C_6H_{12}O_6 + 6O_2$

or

carbon dioxide + water $\rightarrow$ glucose + oxygen

2 cell membrane, cell wall, nucleus and chromosomes

(However, xylem cells are dead and only have a cell wall.)

3 Photosynthesis uses carbon dioxide, so its concentration in the leaf is very low. It is even lower than the concentration outside the leaf, so carbon dioxide diffuses into the leaf down its concentration gradient.

4 a Water vapour.

b On very hot days, a lot of water vapour diffuses out of the leaves. Closing the stomata would mean that they lose less water vapour from their leaves and so are less likely to become dehydrated.

Item B4b Water, water everywhere

1 Any three of: high temperature, large concentration gradient, thin membrane, large surface area of membrane

2 Space X is filled with whatever is outside the cell, in this case the concentrated solution. All the particles in the solution can move freely through the cell wall.

Item B4c Transport in plants

1 They are in the veins.

2 Sugar would be needed as an energy source, for growth. It could be used to make starch for storage and to make cellulose for new cell walls. A flower would need sugar to make

nectar, to attract insects to pollinate it. A fruit would need sugar to make it taste sweet, so animals would eat it and disperse the seeds.

3 At higher temperatures, particles move more quickly.

4 a There is likely to be less transpiration in a dark cupboard because:
   ● the water vapour might build up in the air in the cupboard, so there would be less of a concentration gradient between the leaves and the air;
   ● the plant might close its stomata in the cupboard, because it has no light and so does not need to allow carbon dioxide to diffuse into the leaves for photosynthesis.

b There is likely to be more transpiration if a fan is blown onto the plant because the moving air will take away the water vapour, making sure that there is still a concentration gradient for water vapour from the leaf into the air.

5 a Use the same shoot all the time, keeping all the conditions the same except temperature. Take a series of readings in each of at least five different temperatures. Repeat the experiment three times at each temperature and plot the average readings against time for each temperature.

b In the same way, place the apparatus and shoot in different humidities and take readings as described for temperature.

6 You would need to know the diameter of the bore of the tube. Once you know this, you can calculate the volume of water in a given length of tube like this:

volume =

cross-sectional area of tube × length of tube

= πr^2 × length of tube

So you take readings in the usual way, but convert the distance reading into a volume reading using the equation above.

7 Light makes the guard cells more turgid. They pull away from each other, opening the stoma between them. This allows carbon dioxide to diffuse into the leaf, so the cells can use the light for photosynthesis.

Item B4d Plants need minerals too

1 GPS can ensure that he does not put fertiliser on parts of the field that don't need it, and that he puts enough on the parts that do need it. In this way, he can save money by not wasting fertiliser and also by making sure that his growing crops have enough fertiliser to make them grow well.

2 This technology could reduce the quantity of fertiliser that is put onto the land, which might wash off into waterways and cause algal blooms and eutrophication.

3 a To prevent light getting in. Light might stimulate the growth of algae (microscopic plants) in the culture solution, and they might use up some of the minerals.

   b The roots require oxygen for respiration. Air contains oxygen.

Item B4e Energy flow

1 a One guinea pig needs three grass plants.

   b One anaconda needs nine grass plants.

2

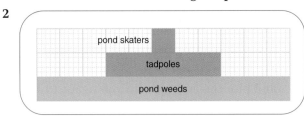

3 a The energy is transferred when the cattle eat the maize, in the form of chemical potential energy in the maize.

   b One-tenth.

   c They do not eat absolutely all of it. They cannot digest all of it and so some is lost in their faeces.

   d The cattle lose some of the energy themselves, and it is lost as heat from their bodies. The people do not eat all of the cattle – for example their hooves or skin. They do not digest all of the meat or milk and so some is lost in their faeces.

4 efficiency $= \dfrac{50\,\text{kJ}}{1000\,\text{kJ}} \times 100\% = 5\%$

5 If an organism is very large (such as tree) then just one may be able to support a large number of smaller organisms that feed on it.

6 There are many possible answers to the first part. The rarest animals are the ones at the end of the chain – the top predators. There is little energy left for them, because so much is lost at each step in the chain.

7 Transporting the biomass to the power station will generate carbon dioxide by burning petrol or diesel in the engine of the vehicles.

8 When biogas or bioethanol is burnt, it does not release sulfur dioxide, which is a toxic gas and can also produce acid rain. Also, the plants that are grown to produce bioethanol took the carbon dioxide from the air only a short time ago (not millions of years ago as with fossil fuels), so the carbon dioxide that goes back into the air when the bioethanol is burnt is balanced by the carbon dioxide that the plants took in.

9 It depends on how good our digestive systems are at digesting the maize. We are not as good as cows at doing this because we don't have a cellulase enzyme to digest plant cell walls. If we could digest the maize then it would be better to eat that, because a lot of energy is lost when the cattle eat the maize. Cooking the maize makes it easier to digest, so if we do that then eating maize would give us more energy than eating meat and milk from the cattle.

Item B4f Farming

1 a The wild oats compete with the wheat for light, water and mineral ions from the soil. So the wheat plants get less of these requirements and do not grow so large.

   b If the crop grows well, he sells it for £500. If there are 35 wild oats per m², he gets only £182. So he loses £318.

   c If he sprays it with herbicide, he spends £20. He then gets £500 for the crop, instead of £182. He gains £318 for his spend of £20. So his return is £15.90 for each £1 that he spends.

2 The glasshouse is enclosed, so you have a small area that you are trying to deal with. You can target pesticides or herbicides very precisely, rather than having to apply them over a wide area. Once you have got rid of pests, it is less easy for them to get back inside the glasshouse.

Once you have got rid of weeds, it is difficult for new seeds to get into the glasshouse.

3 **a** $4.48\,\text{kg} \div 9.9 = 0.45\,\text{kg}$

   b $25\,°\text{C}$

   c At $23\,°\text{C}$, the metabolic reactions in the cucumbers, including photosynthesis, took place more slowly than at $25\,°\text{C}$, because molecules would not have as much kinetic energy and were moving around more slowly. So collisions between reactants, and between enzymes and their substrates, would happen less often.

   At $27\,°\text{C}$, enzyme molecules begin to be denatured (plant enzymes tend to have lower optimum temperatures than mammalian enzymes), so the reactions that they catalyse happen more slowly. Another possible reason is that the plants were losing more water from their leaves at this temperature and so they closed their stomata to conserve water. This would mean that less carbon dioxide was getting into their leaves, so photosynthesis slowed down.

   In both cases, photosynthesis would happen more slowly than at $25\,°\text{C}$, and if photosynthesis is slow then growth is also slow.

4 **a** Carbohydrates and fats, and also proteins (to a lesser extent).

   b The food that the cattle eat contains chemical potential energy. If they do not digest some of the carbohydrates, fats or proteins, these are lost in their faeces. Urine contains urea, which also contains chemical potential energy.

   c Energy taken in = 500 MJ per day.
   Energy lost = 115 + 20 + 170 + 120 = 425 MJ.
   So energy available for producing new tissues = 500 − 425 = 75 MJ per day.

   d If kept warm and with restricted movement, the cattle will lose less energy as heat produced in their bodies. There will more energy available for making milk or producing new tissues.

5 Fewer insects are killed by pesticides, so there is more food for spiders, birds and bats. Also, there is a greater range of weeds growing, which can provide a wider range of habitats for insects and other animals.

6 **a** Both wheat and oats have higher yields per hectare when grown conventionally than when grown organically. The difference is greater for wheat than for oats. For wheat, the yield is 2.82 tonnes per hectare more, whereas for oats it is 1.45 tonnes per hectare more.

   b There may be more insects feeding on the organically grown crop. There may be more weeds competing with the organically grown crop for water, mineral ions and light.

   c There is no difference in yield for beans when grown organically or conventionally. Perhaps beans are not troubled by insect pests. In fact, bean flowers are pollinated by insects, so not using insecticides might actually increase the pollination success rate, so more beans are produced from the pollinated flowers. This might balance out any loss of yield caused by more competition from weeds.

Item B4g Decay

1 Banana skin, brown paper bag, toe nail clippings.

2 When the food is thawed and comes to room temperature, bacteria can grow in it. If you refreeze it, these bacteria will not be killed but just go into suspended animation. When the food is thawed again, there could be a large number of bacteria already growing in it.

3 Protease − substrate is proteins, product is amino acids.
   Lipase − substrate is fats, product is fatty acids and glycerol.
   Amylase − substrate is starch, product is maltose.

Item B4h Recycling

1 The plant could die and sink into a swamp. It does not fully decay but gets squashed and compressed as more and more material accumulates on top of it. Over millions of years, it turns into coal.

   The coal is mined and burnt. Carbon dioxide is released from it and goes into the air. A plant takes in the carbon dioxide and uses it in photosynthesis to make starch.

You eat the seeds of the plant, containing starch. You digest the starch to maltose and then glucose, which is absorbed into your bloodstream. The glucose is absorbed into your brain cells.

2 You could add another arrow at the top right, leading to 'phytoplankton' and labelled 'photosynthesis'. Then show zooplankton feeding on the phytoplankton. Then show the zooplankton dying, with an arrow going to 'limestone forming on the sea floor'; then an arrow from the limestone to carbon dioxide in the air, labelled 'uplift and weathering'.

3 If the bacterium is a nitrifying bacterium then the nitrogen could become part of an ammonium or nitrate ion. This could be taken up by a plant and used to make proteins. You eat the plant and digest the proteins to produce amino acids. These are absorbed into your blood and then taken up by a cell and used to make proteins.

4 If you have eaten anything containing protein then, yes, you have eaten something containing nitrogen. The nitrogen got into your food because plants used nitrates or ammonium ions to make the proteins that you ate. If you ate something that came from an animal, that animal must have eaten plants or other animals containing proteins.

Item C3a What are atoms like?

1 a over 100
   b A substance that cannot be broken down chemically. A substance made of only one type of atom.
2 a The atoms in each diagram are all the same.
   b The atoms of different metals are shown different sizes.
3 a Silver, nitrogen and oxygen.
   b 5
4 a 6   b 6   c 6   d nucleus   e shells
5 a 3   b 3   c third   d Li
6 a 9   b 9   c ninth   d F
7 a 7   b 19
8 a 1, 1 and 1
   b first, first and first
   c 1, 2 and 3
9 a Column 4: 10, 17, 17, 18, 19, 28, 53
      Column 5: 10, 18, 20, 22, 21, 30, 74
      Column 6: 10, 17, 17, 18, 19, 28, 53
   b Chlorine-35 and chlorine-37 are isotopes.
10 a $^{9}_{4}$Be
   b It has four protons and four electrons. The total charge on one atom is therefore +4 −4 = 0.
11 a 2, 8 and 14
   b 2, 8 and 14
   c

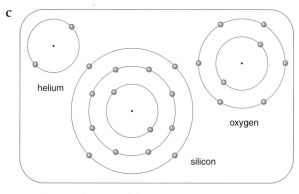

   d He is 2; O is 2.6; Si is 2.8.4.

Item C3b How atoms combine – ionic bonding

1 a They lose one electron each.
   b Positive ions.
   c They gain one electron each.
   d Negative ions.

2 The positive sodium ions and the negative chloride ions attract each other.
3 a Na^+ or Cl^- b Na (or Cl)   c Cl_2
4 a Sodium chloride has a high melting point because the sodium ions and the chloride ions attract each other strongly.
   b The sodium ions and chloride ions are not free to move in solid sodium chloride.
   c The sodium ions and chloride ions are free to move in sodium chloride solution.
   d The sodium ions and chloride ions are free to move in molten sodium chloride.
5 a They lose two electrons.
   b 2+ because two electrons are lost.
   c They gain two electrons.
   d 2− because two electrons are gained.
6 a Magnesium oxide has a higher melting point because 2+ and 2− ions attract each other more strongly than 1+ and 1− ions.
   b The magnesium ions and oxide ions are not free to move in solid magnesium oxide.
   c The magnesium ions and oxide ions are free to move in molten magnesium oxide.
7 a Cl^-: 2.8.8; Mg^{2+}: 2.8; O^{2-}: 2.8
   b i 11, 12, 10; ii 12, 12, 10;
      iii 8, 8, 10; iv 17, 18, 18.
8 a Magnesium atoms lose two electrons when they react (they have two outer electrons). Chlorine atoms gain one electron when they react (they have one gap in their outer shell). One chlorine atom cannot gain two electrons from one magnesium atom.
      Alternatively: Mg forms 2+ ions and Cl forms 1− ions, so an MgCl formula unit wouldn't have a total charge of zero.
   b Sodium atoms lose one electron when they react (they have one outer electron). Oxygen atoms gain two electrons when they react (they have two gaps in their outer shell). One sodium atom cannot transfer two electrons to one oxygen atom.
      Alternatively: Na forms 1+ ions and O forms 2− ions, so an NaO formula unit wouldn't have a total charge of zero.

9 Row 2: Mg^{2+}, F^-; MgF_2
Row 3: Na^+, S^{2-}; Na_2S
Row 4: Mg^{2+}, S^{2-}; MgS
Row 5: Ag^+, O^{2-}; Ag_2O
Row 6: Al^{3+}, O^{2-}; Al_2O_3

Item C3c Covalent bonding and the structure of the Periodic Table

1 a 1   b 2   c 3
  d covalent bonds
  e Water molecules are only attracted to neighbouring water molecules by weak forces.
  f Because water molecules are electrically neutral.
2 a Na and Mg   b Na^+ and O^{2-}
  c H_2O and CO_2 d NaCl and MgO
3 a 7   b 1   c 8
  d They do not take part in bonding.
4 a 4   b 1
  c When it makes four bonds the carbon atom has a full outer shell.
5 a 6   b 1
  c When it makes two bonds the oxygen atom has a full outer shell.
6 Each oxygen atom needs to receive two electrons to have a full outer shell. The carbon atom needs to receive four electrons to have a full outer shell. It does this by receiving two electrons from each oxygen. Each double bond therefore involves two electrons from the carbon and two electrons from one of the oxygen atoms.
7 a e.g. oxygen, polonium
  b e.g. sodium, argon
8 a 5   b none
  c Very unreactive, gas.
9 a 2   b 3   c 1   d 4
10 a Group 5, Period 2
  b Group 2, Period 4
  c Group 1, Period 3
  d Group 3, Period 2

Item C3d The Group 1 elements

1 Bright and shiny, good conductors of electricity.
2 a sodium + water →
               sodium hydroxide + hydrogen
  b $2Na + 2H_2O \rightarrow 2NaOH + H_2$

3 a potassium + water →
               potassium hydroxide + hydrogen
  b $2K + 2H_2O \rightarrow 2KOH + H_2$
4 a If similar sized pieces are used, the sodium will finish reacting before the lithium. The sodium gets hot enough to melt during the reaction, the lithium does not.
  b When the potassium touches the water, a flame is seen; sodium doesn't do this. The reaction with potassium is very rapid and usually ends with a small explosion; sodium doesn't do this.
5 a $2Rb + 2H_2O \rightarrow 2RbOH + H_2$
  b Alkaline, soluble in water.
  c $2Cs + 2H_2O \rightarrow 2CsOH + H_2$
  d Metal, bright and shiny, conducts electricity, very soft, stored under oil, floats on water, reacts with water, products of reaction with water are an alkali and hydrogen, extremely reactive with water.
6 a $Na \rightarrow Na^+ + e^-$
  b
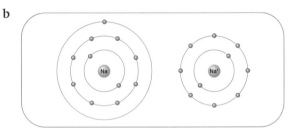
  c Potassium atoms are larger than sodium atoms. When these atoms react with water, they each lose their single outer electron. It is easier for the larger atoms (potassium in this case) to lose their outer-shell electron. The outer-shell electron is lost more easily because it is further from the nucleus.
7 a potassium   b sodium   c lithium

Item C3e The Group 7 elements

1 a Water treatment, making pesticides, making PVC.
  b Preserving food, flavour enhancer, making chlorine.
2 a caesium + bromine → caesium bromide
  b potassium + fluorine → potassium fluoride
  c lithium + chlorine → lithium chloride
  d rubidium + iodine → rubidium iodide
3 They all have atoms with seven electrons in the outer shell.

4 a $2Rb + F_2 \rightarrow 2RbF$

b $2K + I_2 \rightarrow 2KI$

c $2Li + Br_2 \rightarrow 2LiBr$

d $2Cs + Cl_2 \rightarrow 2CsCl$

5 Each lithium atom loses one electron.

6 a sodium + fluorine → sodium fluoride

sodium + chlorine → sodium chloride

sodium + bromine → sodium bromide

sodium + iodine → sodium iodide

b $2Na + F_2 \rightarrow 2NaF$

$2Na + Cl_2 \rightarrow 2NaCl$

$2Na + Br_2 \rightarrow 2NaBr$

$2Na + I_2 \rightarrow 2NaI$

7 a sodium iodide + chlorine →

sodium chloride + iodine

b Brown colour of iodine is seen.

c Chlorine is more reactive than iodine.

d No change / stays brown. Iodine cannot displace chlorine from sodium chloride because chlorine is more reactive than iodine.

8 a $Cl_2 + 2NaBr \rightarrow Br_2 + 2NaCl$

b $Cl_2 + 2NaI \rightarrow I_2 + 2NaCl$

c $Br_2 + 2NaI \rightarrow I_2 + 2NaBr$

9 a $F + e^- \rightarrow F^-$

b

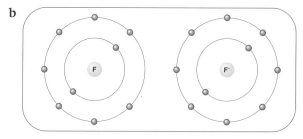

c When fluorine and chlorine atoms react, they gain electrons. Fluorine atoms are smaller than chlorine atoms. The smaller fluorine atoms are more reactive because the gap in their outer shell is nearer the nucleus. The nucleus therefore exerts a stronger attractive force on an electron which approaches the gap. Smaller atoms gain electrons more easily than larger atoms.

10 a Fluorine is very pale yellow or green. The point is that the colours get paler up the group. Astatine is very dark – possibly black. The point is that the colours get darker down the group.

b For fluorine, something below −35 °C; −135 °C would fit in with the trend; the actual boiling

point is −188 °C. For astatine, something above 170 °C; 280 °C would fit in with the trend; the actual boiling point is 337 °C.

Item C3f Electrolysis

1 a yes   **b** no

2 a Negative ion.   **b** Positive electrode.

c Positive ion.

d A liquid that conducts electricity and is permanently changed by a DC current.

e Negative electrode.

f The permanent changing of a compound owing to the passing of a DC electric current.

3 a Bauxite.

b aluminium oxide → aluminium + oxygen

c Because of the high cost of the electricity needed to make it.

4 a Aluminium oxide doesn't conduct electricity when solid.

b Graphite.

c They burn away.

5 a Oxide ions lose electrons.

b 2

6 Aluminium ions gain electrons.

7 a At the anode.

b At the cathode.

c Oxygen relights a glowing splint, hydrogen burns with a pop.

d Dilute sulfuric acid.

Item C3g Transition elements

1 26 and 79 (iron and gold)

2 Tungsten, W. Good conductor of heat and electricity, bright and shiny appearance, malleable and ductile.

3 a green

b orange or brown

c blue or greeny-blue

4 A substance that speeds up a reaction without being changed itself.

5 Having coloured compounds and catalytic activity.

6 Carbonate is an anion and zinc is a cation.

7 **a** manganese(II) carbonate →

manganese(II) oxide + carbon dioxide

b zinc carbonate →

zinc oxide + carbon dioxide

8 a $MnCO_3 \rightarrow MnO + CO_2$

  b $ZnCO_3 \rightarrow ZnO + CO_2$

9 a Fe^{3+} b Fe^{2+}

10 a $Fe^{2+} + 2OH^- \rightarrow Fe(OH)_2$

   b $Fe^{3+} + 3OH^- \rightarrow Fe(OH)_3$

Item C3h Metal structure and properties

1 a Hard, high density, lustrous, high tensile strength, high melting and boiling points, good conductors of heat and electricity.

  b Anvils, weights, jewellery, cables, light bulbs, saucepans, wires.

2 a Copper and zinc.

  b Strong, malleable, cheap.

  c Excellent electrical conductivity.

3 a covalent b ionic   c metallic

4 a Free to move.

  b Its outer shell electrons.

  c The ions are positive, the electrons are negative.

  d Metallic bonding is strong.

5 a Its electrical resistance.

  b Sufficiently low temperature.

  c Power transmission, electronics, electromagnets.

Item C4a Acids and bases

1 a red, orange or yellow

  b blue or purple

  c green

2 a 7; b 5; c 4; d 1 or 2; e 10; f 13 or 14

3 lemonade

4 a 1 or 2, 5, 7 or 8, 10, 13 or 14

  b It goes up.

5 a 13 or 14, 10, 7 or 8, 5, 1 or 2

  b It goes down.

6 carbon dioxide

7 base, alkali, alkali, base, acid

8 a sulfuric acid + copper(II) oxide $\rightarrow$
                            copper(II) sulfate + water

  The product, copper(II) sulfate, is blue.

9 a potassium chloride

  b sodium sulfate

  c calcium nitrate

10 Hydroxide ions.

11 The carbonate ion has a 2– charge. The sodium ion has a 1+ charge. Two sodium ions and one carbonate ion give a formula unit in which the total charge is zero: (1+) + (1+) + (2–) = 0.

  The carbonate ion has a 2– charge. The calcium ion has a 2+ charge. One calcium ions and one carbonate ion give a formula unit in which the total charge is zero: (2+) + (2–) = 0

12 a hydrochloric acid + sodium carbonate $\rightarrow$ carbon dioxide + water + sodium chloride

   $2HCl + Na_2CO_3 \rightarrow CO_2 + H_2O + 2NaCl$

  b sulfuric acid + sodium hydroxide $\rightarrow$
                          water + sodium sulfate

   $H_2SO_4 + 2NaOH \rightarrow 2H_2O + Na_2SO_4$

  c nitric acid + copper oxide $\rightarrow$
                          water + copper nitrate

   $2HNO_3 + CuO \rightarrow H_2O + Cu(NO_3)_2$

  d nitric acid + ammonium hydroxide $\rightarrow$
                      water + ammonium nitrate

   $HNO_3 + NH_4OH \rightarrow H_2O + NH_4NO_3$

Item C4b Reacting masses

1 a 8   b 16   c 35

2 a 16   b 32   c 80

3 a Scale **a** is balanced – there is a mass of 32 on each side. Scale **d** is balanced – there is a mass of 80 on each side.

  b In scale **b**, the bromine atom is heavier than the two sulfur atoms. In scale **c**, the three sulfur atoms are heavier than the bromine atom.

4 a 18   b 160   c 100

5 ten

6 The copper(II) ion has a 2+ charge. The nitrate ion has a 1– charge. One copper(II) ion and two nitrate ions give a formula unit in which the total charge is zero: (2+) + (1–) + (1–) = 0.

7 188

8 The iron(III) ion has a 3+ charge. The hydroxide ion has a 1– charge. One iron(III) ion and three hydroxide ions give a formula unit in which the total charge is zero (3+) + (1–) + (1–) + (1–) = 0).

9 107

10 a 74   b 132

11 40 g

12 32 g

13 a 12 g   b 140 kg

14 a 15 g   b 250 g

15 a copper(II) sulfate and water

  b copper(II) oxide and sulfuric acid

16 a 160 g   **b** 18 g

17 a 196 g   **b** 36 g

18 800 g

19 a 3.2 g   **b** 8.0 g

20 a 2.00 g (method: (3.00 g/48) × 32 = 2.00)

   b 5.00 g (method: 3.00 g + 2.00 g = 5.00 g)

   c 0.42 g (method: (0.63 g/48) × 32 = 0.42 g)

   d 1.05 g (method: 0.63 g + 0.42 g = 1.05 g)

21 a 50%   **b** 80%   **c** 70%   **d** 40%   **e** 85%

22 a D A C B E

Item C4c Fertilisers and crop yield

1 a nitrogen

   b nitrogen, phosphorus

   c nitrogen

   d potassium, nitrogen

2 a 80   **b** 149   **c** 132   **d** 101

3 The crops will remove nutrients from the soil very rapidly. Very little natural decay processes replacing these nutrients. Soil becomes poor in nutrients.

4 Fertiliser washes into river → algal bloom → light blocked → plants dies → plants rot → oxygen depleted → fish die.

5 a Ammonium nitrate is made with ammonia and nitric acid. Ammonium phosphate is made with ammonia and phosphoric acid. Ammonium sulfate is made with ammonia and sulfuric acid.

   b Ammonium nitrate contains nitrogen. Ammonium phosphate contains nitrogen and phosphorus. Ammonium sulfate contains nitrogen.

6 A = litmus, B = burette, C = funnel, D = measuring cylinder, E = conical flask, F = graduated pipette.

7 a 20.8% (method: (31/149) × 100% = 20.8%)

   b 28.2% (method: (42/149) × 100% = 28.2%)

   c 13.9% (method: (14/101) × 100% = 13.9%)

   d 38.6% (method: (39/101) × 100% = 38.6%)

8 The ammonium nitrate is 35% nitrogen, 0% phosphorus and 0% potassium.

Item C4d Making ammonia – Haber process and costs

1 Cost of raw materials, low rate of reaction, labour costs, cost of plant, energy costs.

2 a NH_3

   b The Haber process.

   c Nitrogen and hydrogen.

   d The air, natural gas/cracking of crude oil.

   e Making fertilisers, making nitric acid, making cleaning agents.

3 a The reaction is reversible.

   b High pressure involves a plant that is expensive to build and maintain. High temperatures can involve high energy costs.

   c Automation results in lower labour costs.

4 $N_2 + 3H_2 \rightleftharpoons 2NH_3$

5 a High pressure gives a higher yield and a faster rate but it is costly.

   b High temperature gives a faster rate but it can increase energy costs and it gives a lower yield.

6 The iron catalyst increases the rate of the reaction. It has no effect on yield, is not expensive to buy and is not used up.

Item C4e Detergents

1 a To digest food stains.

   b To remove dirt from clothes.

   c To make white clothes look whiter.

   d To remove coloured stains.

   e To remove the chemicals that cause hardness.

2 Wash at 40 °C, do not bleach, use a hot iron, do not dry clean, can be tumble dried.

3 Washing above 30 °C, ironing, bleaching, dry cleaning.

4 a Water-hating and water-loving.

   b Oil doesn't dissolve in water.

   c The detergent molecules make the oil soluble in water.

5 Less likely to damage certain clothes

6 a Water molecules attract each other by polar forces. Oil molecules attract each other by non-polar forces.

   b The molecules of oily dirt and the molecules of dry cleaning solvent attract each other by non-polar forces.

7 a The detergent and dirt would dry on the washing up if it was left to drain.

   b It wouldn't wash the food off the plates and cutlery.

   c The washing-up liquid would be difficult to squirt into the water.

   d The washing-up liquid would be less appealing.

Item C4f Batch or continuous?

1 a A process in which reactants are continually being fed in and converted into products and products are continually being removed.

  b Makes large amounts of product, minimises start-up time, reduces energy costs.

2 a Advantages – easy to match supply to demand, good way to make small amounts of product, easier to ensure consistent high quality of product. Disadvantages – labour intensive, higher energy costs, small scale.

  b Demand for ammonia is high and predictable.

  c Pharmaceuticals are often made in smaller amounts, high level of quality control is essential.

3 High labour costs, high energy costs, research costs, expensive raw materials, marketing costs, legislative demands, e.g. need to test thoroughly.

4 a To break open the plant cells.

  b To dissolve the desired chemical, the solution can then be separated from solid material by filtering.

  c To separate the desired chemical from other dissolved substances.

5 a Increases profits, shortens pay-back time if successful.

  b More likely to develop a drug with the desired properties.

  c To ensure there are no undesirable side effects.

Item C4g Nanochemistry

1 a diamond, graphite, buckminsterfullerene

  b buckminsterfullerene

  c graphite

  d diamond

2 a Lustrous, colourless, transparent, insoluble in water, non-conductor of electricity, high melting point, hard.

  b Lustrous, colourless, transparent, expensive.

  c High melting point, hard.

3 a All the carbon atoms are bonded to their neighbours by strong covalent bonds.

  b There are no delocalised electrons (electrons are not free to move).

4 a Dark black, lustrous, opaque, insoluble in water, conducts electricity, high melting point, slippery.

  b Dark black, opaque, slippery.

  c Slippery.

5 a The layers of carbon atoms are only held together by weak forces. Because of this, the layers of carbon atoms can slide over each other easily.

  b The graphite structure contains delocalised electrons.

6 a C_{60}

  b red or purple

7 a Tubular structures made of carbon atoms, only a few nanometres in diameter.

  b Electronic circuits, strengthening tennis rackets, support surface for catalyst particles.

8 Adding atoms or molecules to a small starter molecule until it has been built up to a molecule of the desired shape.
Removing atoms or molecules from a larger starter molecule until it has been slimmed down to a molecule of the desired shape.

9 The sizes are conveniently measured in nanometres. One nanometre is one billionth of a metre.

Item C4h How pure is our water?

1 a lakes, rivers, aquifers, reservoirs

  b reservoirs

  c Reducing the amount of water used at each toilet flush, not watering lawns.

2 Lead compounds dissolve out of lead piping; pesticides sprayed on crops get washed into water supplies by rain.

3 a Fertiliser applied to agricultural land gets washed into water supplies by rain.

  b Ammonium nitrate dissolves in water and does not react with any water treatment substances to give an insoluble sediment that can be filtered out.

4 The middle-eastern country has a more plentiful, cheaper supply of energy than the UK. The UK can rely more on rainfall than the middle-eastern country.

5 silver ions + bromide ions → silver bromide

6 silver ions + iodide ions → silver iodide

7 a $Ag^+ + Br^- \rightarrow AgBr$

  b $Ag^+ + I^- \rightarrow AgI$

Item P3a Speed

1 Inches per minute.
2 s/m, ms
3 Fastest: Car C. Slowest: Car B.
4 250 m/s
5 75 km/h
6 1 728 000 km
7 3 h 20 min
8 1.6 m/s
9

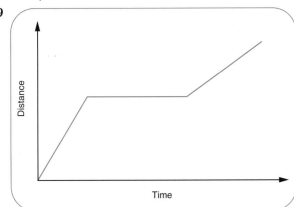

10 Fastest during first segment of journey.
11

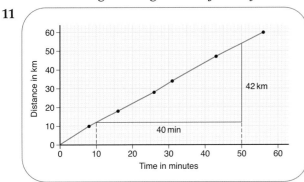

Average speed = 63 km/h

Item P3b Changing speed

1 Decelerating (deceleration).
2 km/s
3 1.5 m/s^2
4 0.2 m/s^2
5 56 m/s^2
6 −280 m/s^2
7 25 m/s
8 11 s

9

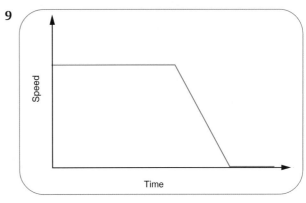

10 a A, C, G
   b F
   c E
   d B, D

11 a

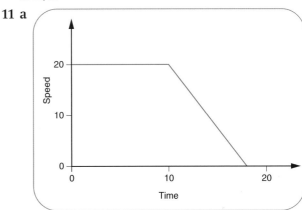

   b 2.5 m/s^2
   c 280 m

Item P3c Forces and motion

1 a John – greater force gives greater acceleration.
   b Increase the mass of John's ball.
2 1500 N
3 1.96 N
4 25 m/s^2
5 800 kg

6

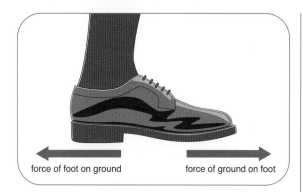

force of foot on ground   force of ground on foot

The force of the ground on your foot causes you to accelerate.

7 a Braking distance is greater.

b At 20 mph.

8 Slippery road affects braking distance (can't apply such a great braking force). Alcohol and drugs slow reaction time, so thinking distance increases; and affect ability to judge suitable braking force, so braking distance increases.

Item P3d Work and power

1 100 J
2 300 J
3 300 J
4 400 N
5 50 000 W (50 kW)
6 8 W

Item P3e Energy on the move

1 a KE increasing.
  b KE constant.
  c KE decreasing.
2 Less KE on return because less mass (but same speed).
3 1600 J
4 800 000 J
5 Three times the speed; nine times the KE; braking force must do nine times as much work, so takes nine times the distance (assuming the force is the same).
6 Best fuel economy: Herbie. Worst fuel economy: Batmobile.
7 The original source of the energy is sunlight.

Item P3f Crumple zones

1 The time of stopping is decreased because there is no 'give', so the force is greater.
2 Instead of the car stopping in a short distance because the body does not collapse, it stops over a longer distance as the crumple zone crushes. The force is less because it does work to stop the car over a greater distance.
3 Traction control.

Item P3g Falling safely

1 Where the images get further apart, it is accelerating. Where they get closer together, it is decelerating.
2 The forces are balanced.
3 a Accelerating; weight is greater.
  b Air resistance is greater, so forces are unbalanced; this causes the parachutist to decelerate. Eventually, forces are balanced again and the parachutist stops decelerating.
4

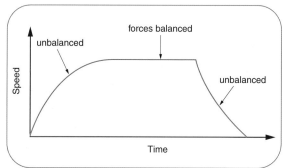

Item P3h The energy of games and theme rides

1 a GPE decreasing.
  b GPE constant.
  c GPE increasing.
2 Greatest GPE: Jane's rock. Least GPE: Jack's rock.
3 Weakest g: Mars (the Moon is not a planet). Strongest g: Jupiter.
4 a 5 N
  b 800 N
  c 12 000 N
5 150 kg
6 1000 J
7 100 m

8 a GPE
  b GPE changes to KE
  c KE
  d KE changed to heat by brakes/friction.

Item P4a Electrostatics – sparks

1 Friction.
2 Repel.
3 a Negative.
  b Repel.
4 a Positive.
  b Electrons are transferred from the rod to the cloth, so that the rod loses negative charge and becomes positive. The cloth gains the negative charge of the electrons.
  c No charge has been lost to the surroundings, or gained from them.
5 a Positive.
  b Custard particles are flammable when mixed with air. A spark jumped between the charged pipe and earth, igniting the mixture.
6 a Friction between oil and pipe causes oil to gain negative charge. This spreads round walls of tank. The pipe is left with positive charge. Sparks may jump between pipe and walls, igniting mixture of oil vapour and air.
  b Charge cannot accumulate in oil as it flows through pipe; electrons will flow from oil to pipe.

Item P4b Uses of electrostatics

1 Insulating handles; standing back.
2 Negative charge.
3 a Negative charge.
  b Opposite charges attract.
  c Dust particles would stick to wires.
4 Positive charge.
5 They have the same charge (positive), so they repel each other. They spread all round the negatively charged object.

Item P4c Safe electricals

1 a Increase.
  b Current will decrease.
  c Lamp will get dimmer.
2 4 Ω
3 100 V
4 360 V
5 0.24 A
6 Neutral and earth wires.
7 The earth wire.
8 The current flowing through the fuse wire causes the wire to become so hot that it melts.
9 The fuses will allow several amps to flow before they blow; an RCD trips when a much smaller current is flowing to earth.

Item P4d Ultrasound

1 a A compression.
  b Particles farther apart; a rarefaction.
2 a X is transverse, Y is longitudinal.
  b X: from side to side. Y: forwards and backwards, along line of spring.
3

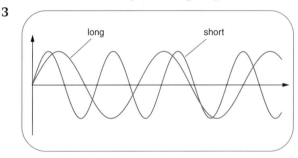

4 20 Hz
5 Body scanning, breaking up kidney stones, measuring speed of blood flow.

Item P4e Treatment

1 (most penetrating) gamma, beta, alpha (least penetrating)
2 If the accelerating voltage in an X-ray tube is increased, the electrons will move *faster*. The electrons have more *kinetic* energy when they strike the metal target. The X-rays produced will be *more* penetrating.
3 X-rays, beta, gamma
4 Alpha radiation would be absorbed by the body, so could not be detected outside the body.
5 We can control the energy (penetrating power) of X-rays; can produce a finer beam.

Item P4f What is radioactivity?

1 Radioactivity is the number of atoms which decay each second.
2 125
3 1/16
4 67 s (approx)
5 a gamma
  b alpha
  c beta
6 a 6
  b 12
  c 6
7 a alpha
  b Yes (A: 235 = 231 + 4; Z: 92 = 90 + 2)
8 $^{90}_{38}\text{Sr} \rightarrow \,^{90}_{39}\text{Y} + \,^{0}_{-1}\text{e}$

Item P4g Uses of radioisotopes

1 Cosmic rays.
2 Natural sources.
3 Artificial; natural.
4 Most penetrating: gamma. Least penetrating: alpha.
5 Gamma.
6 Weaker radiation beyond the point of blockage.
7 The wood from which the boat was made grew before 1066.
8 100 years: little measurable decrease in radioactivity.
  50 000 years: very little radioactive carbon will be left.

Item P4h Fission

1 Uranium.
2 Their masses are smaller.
3 Nuclear (potential) energy transformed to heat energy.
4 In a bomb, the chain reaction happens very quickly, in an uncontrolled way.
5 Neutrons are absorbed so that few can go on to cause further fission events; the chain reaction stops.
6 (most concentrated) uranium, coal, wind (least concentrated)

Glossary: Biology

active site the part of an enzyme molecule where the substrate temporarily binds

active transport the movement of substances against their concentration gradient, using energy provided by the cell

adolescence the period between about 10–11 years old and about 17 years old

aerobic respiration the release of energy from glucose by combining it with oxygen

alveolus a tiny air sac in the lungs, where gaseous exchange takes place (*plural*: alveoli)

amino acid a molecule containing C, H, O and N; there are 20 different kinds of amino acids which link together in long chains to form proteins

amylase an enzyme that catalyses the conversion of starch to maltose

antibodies chemicals secreted by lymphocytes that are specific for a particular antigen; they help to destroy the antigen

antitrypsin an enzyme that can help to treat cystic fibrosis

aorta the large artery that takes oxygenated blood from the left ventricle of the heart and distributes it to other parts of the body

arteries thick-walled, elastic blood vessels, that carry blood away from the heart

aseptic technique taking care not to allow micro-organisms to contaminate apparatus and materials as you work

atrium one of the upper chambers of the heart, which receive blood (*plural*: atria)

auxin a plant hormone that is involved in the growth of a shoot towards the light

bases parts of a DNA molecule; there are four different bases and they pair up with one another to hold the two strands of the molecule together

biconcave disc the shape of a red blood cell, like a disc with a dent in each surface

biodegradable can be broken down by micro-organisms

bioethanol ethanol that has been produced by the fermentation of biomass, for example sugar cane or maize

biogas a mixture of carbon dioxide and methane, produced by the fermentation of biological waste

biological control using a parasite or predator to control the population of a pest

biomass the mass of living material

blood plasma the liquid part of blood

bronchiole a small tube leading from the bronchus to the alveoli in the lungs

callus a small lump of undifferentiated cells formed when explants are grown in tissue culture

capillaries tiny blood vessels, with walls only one cell thick, that distribute blood into the body tissues

carbon neutral using a resource in such a way that the carbon dioxide given out is equal to the carbon dioxide taken in

cell differentiation the change of a general purpose cell into one specialised for a particular function

cell sap the sugary fluid inside the large vacuole in a plant cell

cell surface membrane the membrane covering the surface of a cell, which controls the entry and exit of substances to and from the cell

cell wall a layer of criss-crossing cellulose fibres, which covers the outside of a plant cell; bacteria also have cell walls, but not made of cellulose

childhood the period between about 3 years old and 10–11 years old

chlorophyll a green pigment found in chloroplasts in plant cells, which absorbs energy from sunlight

chloroplast an organelle containing chlorophyll, found in some plant cells, in which photosynthesis takes place

cholesterol a fat-like substance; too much in the blood can lead to the formation of plaques in arteries and possibly a heart attack

chromosome a long thread in the nucleus of a cell, made of DNA and containing many genes

clones genetically identical organisms

cloning producing genetically identical copies of an organism

coleoptile a young shoot of wheat, barley or other cereal plant

complementary base pairing the way in which bases on one strand of a DNA molecule pair up with bases on the other strand; A always pairs with T, and C always pairs with G.

concentration gradient a difference in concentration of a particular substance between two regions

consumer an organism that has to take in organic substances made by other organisms; animals and decomposers are consumers

crenation the shrinkage and deformation of animal cells owing to water loss by osmosis

crop rotation growing different crops in a piece of ground each year, over a period of four or more years

cuticle a waterproofing layer of wax on the surface of a leaf

cuttings pieces of stem, leaf or root taken from a plant, which can grow into an entire new plant

cytoplasm a jelly-like material in a cell, in which enzyme-controlled reactions take place

decomposers micro-organisms that digest and break down waste material from animals and plants

denatured loss of shape; enzyme molecules are denatured at high temperatures

denitrifying bacteria bacteria that convert nitrate ions into nitrogen gas

detritivores animals that feed on dead or decaying waste from animals and plants, for example earthworms

differential growth the growth of different parts of the body at different rates

differentiation the way in which cells specialise to perform different functions

diffusion the net movement of particles from a region of high concentration to a region of lower concentration

digester a container within which biological materials are fermented by micro-organisms, for example to produce biogas

diploid containing two complete sets of chromosomes

DNA the chemical found in chromosomes, which carries instructions about which proteins to make

DNA fingerprinting making a record of the base sequence in part of someone's DNA

DNA replication copying a DNA molecule to make a new, identical one

dormancy a state of suspended animation, for example in seeds, which rest and do not grow through the winter

double circulatory system a circulatory system like that of humans, where blood goes to the lungs and back to the heart, and then around the body and back to the heart, in order to make one complete circuit

double helix the shape of a DNA molecule; there are two long strands coiled around one another

egestion getting rid of undigested food materials, in the form of faeces

embryo transplant producing embryos by *in vitro* fertilisation, and then placing them into the uterus of a different female to develop normally

enzymes proteins that act as biological catalysts

epidermis a layer of cells covering the surface of a plant organ, for example a leaf or a root

ethene a plant hormone that stimulates fruit to ripen

explant a small piece of a plant that can be used in tissue culture to grow new plants

fermentation the use of a substance as an energy source by micro-organisms, producing new materials

fertiliser a substance containing mineral ions, such as nitrate or phosphate, that is added to the soil to help plants to grow

flaccid soft and floppy; a flaccid plant cell has lost water

fungicide a chemical used to kill fungi

gamete a sex cell, such as an egg or sperm, that will fuse with another gamete to form a zygote

gaseous exchange the movement of gases into and out of an organism, for example oxygen and carbon dioxide move into and out of the blood in the lungs

gene a section of a DNA molecule carrying instructions for making a particular protein

gene pool all the alleles of all the genes in a population

gene technology altering the genes in an organism, so that the organism has characteristics that are useful to us

genetic code the sequence of bases in a DNA molecule, which determines the sequence of amino acids in the protein that is made

genetic engineering altering the genes in an organism, so that the organism has characteristics that are useful to us

genetically modified (GM) having altered genes, or possessing genes that have come from another organism

geotropism growth towards or away from gravity

gestation period the time during which a foetus develops in the uterus

gibberellin a plant hormone that stimulates seeds to germinate

guard cells a pair of sausage-shaped cells surrounding a stoma, which can change shape to open or close it

haemoglobin a red pigment found in red blood cells that combines temporarily with oxygen and so transports it around the body; haemoglobin is a protein containing iron

haploid containing one complete set of chromosomes

heart transplant the replacement of a faulty heart with a heart taken from another person's body

herbicide a chemical used to kill weeds

hormones chemicals that are made in endocrine glands are transported in the blood and that have an effect on target organs

hydroponics growing plants in a culture solution, without soil

hypha a long thread that makes up the body of a fungus (*plural*: hyphae)

immunosuppressant drugs drugs given after a transplant operation, to prevent the recipient's immune system from attacking and destroying the transplanted organ

inbreeding breeding between closely related animals or plants; it increases the risk of two harmful recessive alleles coming together

infancy the period between birth and about two years old

insecticides chemicals that kill insects

intensive farming using high levels of inputs (for example fertilisers) to get high outputs from a given area of land

lignin a tough, waterproof chemical found in the cell walls of xylem vessels

limestone rock made of calcium carbonate, usually from the shells of tiny animals

lumen the space in the centre of a hollow tube

lysis the destruction of a cell by bursting

maturity fully grown and fully developed

medium a nutrient source in which an organism can grow, for example agar jelly for bacteria

meiosis a type of cell division in which a diploid cell divides to form four genetically different haploid cells

mesophyll the tissues in the middle of a leaf; the palisade layer and the spongy layer

metabolic reactions chemical reactions that take place in a living organism

micropropagation growing new organisms from a tiny group of cells, for example growing plants from tissue cultures

microvillus tiny projections on the outer surfaces of the cells covering the villi (*plural*: microvilli)

mitochondrion an organelle where aerobic respiration takes place, releasing energy for the cell to use (*plural*: mitochondria)

mitosis a type of cell division in which a cell divides to form two genetically identical daughter cells

multicellular made of many cells

mutation a random and unpredictable change in a gene or a chromosome

negative phototropism growth away from light

neurotransmitter a substance which diffuses across a synapse from one nerve cell to another, transferring the impulse to the next nerve cell

nitrifying bacteria bacteria that convert ammonia into nitrate ions

nitrogen-fixing bacteria bacteria that convert nitrogen gas into ammonium ions; many of them live in root nodules in plants of the pea and bean family

nodules little swellings; root nodules of some plants contain nitrogen-fixing bacteria

NPK fertiliser a fertiliser containing nitrogen (in the form of ammonium ions or nitrate ions), phosphate and potassium

nucleus the part of a cell that contains chromosomes

nutrients substances needed by living organisms to stay alive

optimum best; for example, the optimum temperature for an enzyme is the temperature at which it works fastest

osmosis the diffusion of water molecules from an area of high water concentration to an area of low water concentration, through a partially permeable membrane

oxyhaemoglobin the compound formed when haemoglobin combines with oxygen

pacemaker a patch of tissue in the right atrium of the heart, which sets the pace for the heart muscle to contract and relax

palisade layer a layer of cells near the upper surface of a leaf; they contain many chloroplasts and photosynthesise

partially permeable membrane a membrane that allows some substances to pass through but not others

persistent does not break down readily, and so stays in the environment and can accumulate as it passes along a food chain

pesticide a chemical used to kill pests

phagocytosis taking bacteria or other substances into a vacuole in a cell

phloem tubes long tubes made of living cells with cellulose cell walls, which transport sucrose in plants

phototropism growth towards or away from light

phytoplankton tiny plants that float in the sea

placenta the organ in which the blood of a foetus and its mother are brought close together and can exchange materials

plaque a build-up of cholesterol in the wall of an artery

plasmolysed a plant cell that has lost so much water that the cell membrane has shrunk away from the cell wall

platelets tiny cell fragments found in the blood that help in clotting

positive phototropism growth towards light

potometer an instrument for measuring the rate at which a shoot takes up water

producer an organism that uses inorganic materials and an energy source to make organic substances that can be eaten by consumers; plants are producers

product a substance that is made as a result of a chemical reaction

protein synthesis making proteins inside a cell, by linking amino acids together

puberty the time at which a person becomes sexually mature

pulmonary artery an artery that carries deoxygenated blood from the right ventricle of the heart to the lungs

pulmonary vein a vein that carries oxygenated blood from the lungs to the left atrium of the heart

pyramid of biomass a diagram showing the mass of organisms at each trophic level in a food chain

pyramid of numbers a diagram showing the numbers of organisms at each trophic level in a food chain

red blood cells cells that contain the red pigment haemoglobin and transport oxygen from the lungs to the rest of the body

renewable energy source a source of energy that is replenished, so that we can continue to use it without using it up

respiration the release of energy by the breakdown of glucose (see **aerobic respiration**, **anaerobic respiration**)

root hair part of a cell in the epidermis of a root, which extends in between soil particles and has a large surface area for water uptake

rooting powder a powder containing auxin, which stimulates cuttings to grow roots

saprophyte an organism that feeds by secreting enzymes onto its food, then absorbing the digested nutrients

selective breeding choosing animals or plants with desired characteristics for breeding; it is usually carried on for many generations

selective weedkiller a chemical that kills some plants but not others

septum the wall of tissue that separates the left and right sides of the heart

small intestine the part of the alimentary canal where the absorption of digested food takes place

specificity the ability of an enzyme to act on only one kind of substrate, and not others

spongy layer a layer of cells near the lower surface of a leaf; the cells contain chloroplasts and can photosynthesise, and have large air spaces between them through which gases can easily diffuse

stem cells undifferentiated animal cells that are able to divide and produce cells that can differentiate into various kinds of specialised cells

stoma a tiny gap in the lower epidermis of a leaf, through which carbon dioxide, oxygen and water vapour can diffuse (*plural*: stomata)

substrate a substance that is changed by the activity of an enzyme

symbiosis organisms of two different species living very closely together and both benefiting from the relationship

synapse a tiny gap between two nerve cells (neurones)

tissue culture the growth of new plants from tiny pieces cut from another plant

transamination changing one kind of amino acid into a different kind

translocation the transport of products of photosynthesis, for example sucrose, through a plant, in phloem tubes

transmitter substance a substance which diffuses across a synapse from one nerve cell to another, transferring the impulse to the next nerve cell

transpiration the loss of water vapour from a plant's leaves

triplet a sequence of three bases in a DNA molecule, coding for one amino acid

trophic level the position at which an organism feeds in a food chain

tropism a growth response, in which the growth is towards or away from the stimulus

turgid a turgid plant cell is full of water and is stiff and firm

turgor pressure the outwards pressure on a plant cells wall when it is **turgid**

vacuole an area in a cell containing a liquid, for example cell sap

valves structures that allow a fluid to flow in one direction, but not the other

vascular bundles bundles of phloem tubes and xylem vessels running alongside each other in a plant

vein (in plants) a group of phloem and xylem vessels, through which nutrients and water are transported

veins (in animals) thin-walled blood vessels, containing valves, that carry blood back to the heart

vena cava a large vein that brings deoxygenated blood back to the right atrium of the heart (*plural*: venae cavae)

ventricle one of the lower chambers of the heart, which pump blood out into the arteries

villus a tiny finger-like projection on the inner surface of the small intestine (*plural*: villi)

white blood cells cells found in the blood that help to defend the body against pathogens

xylem vessels long tubes made of dead, empty cells with lignified walls, which transport water in plants

zooplankton tiny animals that float in the sea

zygote the cell formed when two gametes fuse together at fertilisation

acid a substance that dissolves in water to give a solution with a pH of below 7; a substance that dissolves in water to give a solution containing raised levels of H^+ ions

alkali a soluble base; a substance that dissolves in water to give a solution with a pH of above 7; a substance that dissolves in water to give a solution containing raised levels of OH^- ions

alkali metals the elements in Group 1 of the Periodic Table: lithium, sodium, potassium, rubidium, caesium and francium

allotropes different forms of one element; the allotropes of carbon are diamond, graphite and buckminsterfullerene

anion a negative ion (which is therefore attracted to an anode)

anode a positive electrode

aquifer a usable underground water source

atom one of the tiny particles that all matter is made of, consisting of a positive nucleus (made of protons and neutrons) orbited by one or more negative electrons; every element has its own type of atom

atomic number the number of protons in the nucleus of each atom of an element; the position of the element in the Periodic Table.

base a substance that can neutralise an acid, e.g. metal oxides, metal hydroxides and metal carbonates

batch process an industrial manufacturing process in which a fixed amount of reactants are made into products; the process then stops and the products are removed; many pharmaceuticals are made by batch processes (see **continuous process**)

bauxite impure aluminium oxide, an important ore of aluminium

bleach a chemical agent that reacts with colours to give a colourless product; washing powders contain bleaches which remove coloured stains

burette a graduated tube used to measure a volume of liquid accurately when titrating

catalyst a substance that speeds up a chemical reaction while remaining unchanged itself

cathode a negative electrode

cation a positive ion (which is therefore attracted to a cathode)

chlorination adding chlorine to domestic water supplies in order to kill micro-organisms

chromatography a means of separating a mixture; chromatography usually involves a solvent moving slowly through a stationary solid material; the solvent carries different substances with it at different speeds

compound a substance made from more than one type of atom (and therefore more than one type of element) chemically bonded together; it can be decomposed

continuous process an industrial manufacturing process in which new reactants are constantly fed in and products are constantly removed, e.g. the Haber process (see **batch process**)

covalent bond a shared pair of electrons

covalent bonding a way of holding atoms together that involves the sharing of pairs of electrons; this type of bonding is strong

delocalised free to move; a substance with delocalised electrons will conduct electric current

detergent an active agent contained in washing powders and washing-up liquids that helps to remove dirt, especially oily dirt

displacement reaction a reaction in which an element in a compound is displaced (or replaced) by another element

distillation a way of purifying a liquid by boiling it, then cooling and condensing its vapour

ductile can be drawn out easily into wires

electrolysis the decomposition of a molten ionic substance, or of an aqueous solution of an ionic substance, caused by the passing of a DC electric current

electrolyte an ionic compound, either molten or dissolved in water, that conducts electricity and is permanently changed by a DC electric current

electron particle found within every atom that has a mass of 0.0005 units and an electric charge of -1; constantly moving on a path called a shell

electronic structure the number of electrons on each shell in an atom; can be shown diagrammatically or with numbers, e.g. 2.8.7

electrostatic force the force of attraction or repulsion between two electrically charged objects, e.g. the force of attraction between a sodium ion and a chloride ion

element a substance that cannot be decomposed (broken down) chemically; a substance made of only one type of atom

enzyme a biological catalyst that is added to washing powders to digest food stains

eutrophication a progressive process started by fertiliser run-off, resulting in low dissolved oxygen levels in a waterway, and causing death of water animals

fertiliser a chemical compound containing one or more essential elements needed by crops

filtration the removal of undissolved solids from a liquid

flame test identification of the metal present in a compound by looking at the colour change the compound causes in a flame

formula unit the group of atoms represented by the formula of a compound, e.g. NaCl; this term is most often used when the bonding of the compound is ionic

fullerene a spherical or tube-shaped molecule made of carbon atoms

giant covalent structure a highly regular, repeating pattern consisting of a very large number of atoms joined by covalent bonds, e.g. diamond, graphite

giant ionic lattice the highly regular, repeating pattern of positive and negative ions found in a crystalline piece of an ionic solid

good conductor of electricity electric current flows easily through this substance

good conductor of heat heat flows quickly through this substance

graduated pipette a very accurate device for measuring a fixed volume of a liquid and used when performing a titration

group a column of the Periodic Table (alternatively the names of the elements in a column of the Periodic Table)

Haber process an industrial process in which nitrogen and hydrogen react together to make ammonia

half-equation an equation that shows the formation of an ion from an atom, e.g. $Li \rightarrow Li^+ + e^-$, or the formation of an atom from an ion, e.g. $Al^{3+} + 3e^- \rightarrow Al$; also called **ionic half-equation**

halide ion negative ion with a one minus charge formed when a halogen atom gains one electron, e.g. F^-, Cl^-, Br^- and I^-

halogens the elements in Group 7 of the Periodic Table: fluorine, chlorine, bromine, iodine and astatine

hard (of metal) does not scratch easily

hard (of water) water that contains dissolved chemicals that make it difficult for detergents to clean clothes properly

high density one cubic centimetre of this material has a high mass (compared with other materials)

high melting and boiling points this material will not melt or boil unless it is heated to a high temperature (compared with many other materials)

high tensile strength when a piece of this material is pulled it will not snap unless a high force is used (compared with other materials)

hydrophilic water-loving, i.e. mixes with or dissolves in water

hydrophobic water-hating, i.e. won't mix with or dissolve in water but will mix with oil

insoluble will not dissolve (in a particular solvent), e.g. oily dirt is insoluble in water

intermolecular forces the bonds that hold molecules together; they are never strong but can vary between weak and very weak

ion an electrically charged atom or group of atoms

ionic bond the force of attraction between a positive ion and a negative ion

ionic bonding a way of holding the atoms in a compound together that involves the attraction between positively charged ions and negatively charged ions; this type of bonding is strong

ionic compound a compound consisting of positive and negative ions, many ionic compounds consist of metal (positive ions) combined with non-metal (negative ions)

ionic half-equation an equation that shows the formation of an ion from an atom, e.g. $Li \rightarrow Li^+ + e^-$, or the formation of an atom from an ion, e.g. $Al^{3+} + 3e^- \rightarrow Al$; also called **half-equation**

isotopes atoms with the same atomic numbers but with different mass numbers

labour intensive used to describe a process that requires a large workforce

legislative demands if said about a new medicine, this term refers to laws that cause the product to cost more to develop, e.g. testing stages required by law

limescale a build-up of calcium carbonate due to dissolved chemicals in hard water; it is commonly seen on the heating elements of washing machines, in kettles, and inside central heating systems

lustrous bright and shiny

malleable can be moulded into new shapes without breaking

mass number the total number of protons and neutrons in an atom

metal halide an ionic compound consisting of metal ions and halide ions

metal hydroxide an ionic compound in which the positive ion is a metal and the negative ion is the hydroxide ion, OH^-; it may be formed as a precipitate

metallic bond the strong bond between atoms in a metal

molecular manufacturing producing a desired product molecule in one of two ways: a small molecule can be gradually added to until the desired product molecule is achieved; a larger molecule can be gradually subtracted from until the desired product molecule is achieved

molecule the group of atoms represented by the formula of a compound, e.g. H_2O, or element, e.g. O_2; this term is most often used when the bonding in the compound or element is covalent

nanometre a unit of length equal to one billionth of a metre (1×10^{-9} m)

nanoparticle a molecule whose size is conveniently measured in nanometres

nanotube a tube-shaped molecule made of carbon atoms. Typically nanotubes are between 1 and 20 nanometres in thickness

negative ion an ion with a negative electric charge caused when one or more electrons were gained; a negative ion has more electrons than protons

neutral when used to describe solutions neutral means neither acidic nor alkaline; a neutral solution has a pH of 7

neutralisation a reaction in which an acid and a base produce a neutral solution

neutron particle found within every atom (except hydrogen atoms) that has a mass of 1 unit and no electric charge; it is found in the atom's nucleus

nucleus tightly packed bundle of protons and neutrons in the centre of an atom

optical brightener a constituent of some washing powders that causes white clothes to shine and look especially clean; this effect is produced by the optical brightener together with the ultraviolet light from the Sun

oxidation a description of what happens to a substance if it loses electrons during a reaction; can also be applied to a substance if it gains oxygen during a reaction

pay-back time the time taken for a new product to pay for the costs incurred during its development period

period a row of the Periodic Table (alternatively the names of the elements in a row of the Periodic Table)

Periodic Table the known elements listed in order of increasing atomic number

pesticides chemical agents used on agricultural land to kill pests

pH scale a numerical scale used to describe acidic and alkaline solutions; an acidic solution has a pH of below 7; a neutral solution has a pH of 7; an alkaline solution has a pH of above 7

pharmaceuticals legal drugs and medicines used for treating illnesses

plant an industrial installation, usually consisting of a large amount of heavy duty equipment

positive ion an ion with a positive electric charge caused when one or more electrons were lost; a positive ion has more protons than electrons

precipitate a solid product formed in a solution

precipitation reaction a chemical reaction in which two solutions react to give an insoluble product; the insoluble product is called a precipitate

proton particle found within every atom that has a mass of 1 unit and an electric charge of +1; found in the atom's nucleus

reduction a description of what happens to a substance if it gains electrons during a reaction; can also be applied to a substance if it loses oxygen during a reaction

relative atomic mass how heavy the atoms of an element are, relative to a carbon atom (that has six protons and six neutrons)

relative formula mass the relative mass of one molecule or formula unit of a substance

reservoir a man-made lake produced by damming a river valley

reversible reaction a reaction that can proceed in both direction, i.e. the reactants can react to make products but these products can also react together, re-forming the original reactants

salt a neutral compound produced when an acid is neutralised by a base; an ionic compound formed when the hydrogen ions in an acid are replaced by metal ions

sedimentation a way of removing dissolved substances from domestic water supplies; the solutes are made to form an insoluble sediment that can be filtered out

shell the path in an atom followed by an electron; a specific number of electrons can move around each shell

simple molecules molecules that are held to their neighbours by weak intermolecular forces

soluble will dissolve (in a particular solvent), e.g. oily dirt is soluble in dry cleaning fluid

solute a substance that can dissolve in a liquid called a solvent, e.g. when dry cleaning clothes the dirt is the solute

solution a liquid mixture that will not separate out, made from a liquid called a solvent with a substance called a solute dissolved in it, e.g. when dry cleaning clothes, the dry cleaning fluid plus dirt is the solution

solvent a liquid that other substances may dissolve in, e.g. when dry cleaning clothes the dry cleaning fluid is the solvent

superconductivity a superconducting material does not have any electrical resistance

synthetic man-made, artificial

therapeutic activity the desired effect of a drug or medicine; the benefit given to the patient

titration a technique in which the volumes of two solutions which will react together are measured exactly

transition elements one of the elements found in the central block of the Periodic Table; all transition elements are metals

trend gradual change in property; usually used to describe the gradual change in a property seen from one element to the next down a group in the Periodic Table, e.g. each Group 1 metal is more reactive with water than the one above it

water softener a chemical that removes the dissolved chemicals in water that cause hardness and limescale; washing powders may contain water softeners

yield the amount of product obtained from a chemical reaction, usually calculated as a percentage:

$$\text{percentage yield} = \frac{\text{actual yield}}{\text{expected yield}} \times 100\%$$

Glossary: Physics

ABS brakes a braking system which automatically prevents skidding during braking

acceleration the rate of change of an object's velocity

active safety features features of a vehicle which automatically take action to control the vehicle

air bags energy-absorbing bags which inflate to protect the occupants of a car during an impact

air resistance the frictional force when an object moves through air

alpha decay the decay of a radioactive nucleus by emission of an alpha particle

alpha particle a particle of two protons and two neutrons emitted during radioactive decay

ammeter a meter for measuring electrical current

amplitude the greatest height of a wave above its undisturbed level

atomic number the number of protons in an atomic nucleus

average speed speed calculated from total distance travelled divided by total time taken

background radiation the radiation from the environment to which we are exposed all the time

beta decay the decay of a radioactive nucleus by emission of a beta particle

beta particle a particle (an electron) emitted during radioactive decay

braking distance the distance travelled between a driver operating the brakes and the vehicle stopping

chain reaction when neutrons released in the course of nuclear fission go on to cause further atomic nuclei to split

circuit breaker a safety device which automatically switches off a circuit when the current becomes too high

compression region of a sound wave where the particles are pushed close together

contaminated when an object has acquired some unwanted radioactive substance

control rods rods used to control the rate of fission in a reactor by absorbing neutrons

coolant a substance which flows through a reactor to transfer heat to the boiler

count rate the number of decaying radioactive atoms detected each second (or hour etc.)

crest the highest point of a wave

cruise control an automatic system which keeps a vehicle travelling at a steady speed

crumple zones the parts of a car at front and rear, designed to collapse in an impact, absorbing energy

defibrillator a device which uses static electricity to attempt to restart a patient's heart

doing work transferring energy by means of a force

double insulated when an electrical appliance is so well insulated that no earth connection is needed

drag the frictional force when an object moves through a fluid (liquid or gas)

driving force the forward force provided by a vehicle's engine

earth wire the wire in an electrical supply which is connected directly to earth

earthed when the case of an electrical appliance is connected to the earth wire

electrical resistance a measure of the difficulty of making an electric current flow through a device

electrons negatively charged particles, smaller than an atom

electrostatic smoke precipitator a device which uses static electricity to remove dust or smoke particles from the air

energy the capacity to do work

frequency the number of waves per second

friction the force which acts when two surfaces rub over one another

fuel rods the fuel used in a nuclear reactor

fuse a device in a circuit that prevents excessive current flowing; if it does, the fuse wire melts, breaking the circuit

gamma ray electromagnetic radiation emitted during radioactive decay

gravitational field strength a measure of the strength of gravity at a particular place; the gravitational force acting on each kilogram of mass

gravitational potential energy (GPE) the energy of an object raised up against the force of gravity

half-life the time taken for half the atoms in a sample of a radioactive material to decay

infrasound sound waves whose frequency is so low that they cannot be heard by humans

insulator a material which does not conduct electricity

interrupt card a piece of card which breaks the light beam of a light gate

irradiated when an object has been exposed to radiation

kinetic energy (KE) the energy of a moving object

light gates devices for recording the passage of a moving object when it breaks a light beam

live wire the wire in an electrical supply whose voltage varies from positive to negative relative to earth

longitudinal wave a wave whose vibrations are at right angles to the direction of travel

lubrication using a liquid such as oil to reduce friction

mass number the number of protons and neutrons in an atomic nucleus

model a way of representing a system in order to understand its functioning; usually mathematical

moderator a substance which slows down neutrons in a nuclear reactor, making it more likely that they will cause fission

negative charge one type of electric charge

neutral having no overall positive or negative electric charge

neutral wire the wire in an electrical supply whose voltage remains close to earth

neutrons electrically neutral particles found in the atomic nucleus

Newton's third law of motion when two objects interact, they exert equal and opposite forces on each other

nuclear fission the process by which massive nuclei split, releasing energy

nuclear radiation radiation coming from the nucleus of an atom

nucleons particles found in the atomic nucleus; protons and neutrons

passive safety features features of a vehicle which contribute to the occupants' safety

positive charge one type of electric charge

potential difference (p.d.) another name for the voltage between two points

potential energy (PE) energy stored in an object because of its position

power the rate at which energy is transferred or work is done

protons positively charged particles found in the atomic nucleus

radiation any particles or energy spreading out from a source

radioactive substances substances which decay by emitting radiation from their atomic nuclei

radioactive tracer a radioactive substance used to trace the flow of liquid or gas, or to find the position of cancerous tissue in the body

radiocarbon dating a technique which uses the known rate of decay of radioactive carbon-14 to find the approximate age of an object made from dead organic material

radiographers people who work with radiation (usually medical staff)

radioisotope a radioactive isotope

rarefaction region of a sound wave where the particles are further apart

reactor a place where nuclear fission happens at a controlled rate

residual current device (RCD) a safety device which detects when the live and neutral currents are different, and switches off the circuit

ring main circuit the arrangement of wires running around a house, bringing electricity to the different rooms

safety cage the rigid central section of a car which protects the occupants during an impact

seat belts straps which restrain the occupants of a car during an impact

sound wave a wave which carries sound from place to place

stability control a braking system which automatically prevents a car from yawing (twisting round) during travel

static electricity electrical charge held by a charged insulator

sterilising killing any micro-organisms which are contaminating an object

stopping distance the distance travelled between a driver detecting a hazard and the vehicle stopping

streamlining shaping an object to reduce drag

terminal speed the greatest speed reached by an object when moving through a fluid

thinking distance the distance travelled between a driver detecting a hazard and operating the brakes

thinking time the time between a driver detecting a hazard and operating the brakes

tracer a substance, often radioactive, used to trace the flow of liquid or gas, or to find the position of cancerous tissue in the body

traction control a braking system which automatically prevents skidding during acceleration

transverse wave a wave whose vibrations are along the direction of travel

trough the lowest point of a wave

ultrasound sound waves whose frequency is so high that they cannot be heard by humans

upper threshold of hearing the highest frequency of sound which can just be heard by humans

variable resistor a resistor whose resistance can be changed, e.g. by turning a knob

voltmeter a meter for measuring the potential difference (voltage) between two points

wavelength the distance between adjacent crests (or troughs) of a wave

weight the downward force of gravity acting on a body

work done the amount of energy transferred when one body exerts a force on another

Periodic Table

Key

| relative atomic mass |
| **atomic symbol** |
| name |
| atomic (proton) number |

1	2												3	4	5	6	7	8
																		4 **He** helium 2
7 **Li** lithium 3	9 **Be** beryllium 4					1 **H** hydrogen 1							11 **B** boron 5	12 **C** carbon 6	14 **N** nitrogen 7	16 **O** oxygen 8	19 **F** fluorine 9	20 **Ne** neon 10
23 **Na** sodium 11	24 **Mg** magnesium 12												27 **Al** aluminium 13	28 **Si** silicon 14	31 **P** phosphorus 15	32 **S** sulfur 16	35.5 **Cl** chlorine 17	40 **Ar** argon 18
39 **K** potassium 19	40 **Ca** calcium 20	45 **Sc** scandium 21	48 **Ti** titanium 22	51 **V** vanadium 23	52 **Cr** chromium 24	55 **Mn** manganese 25	56 **Fe** iron 26	59 **Co** cobalt 27	59 **Ni** nickel 28	63.5 **Cu** copper 29	65 **Zn** zinc 30		70 **Ga** gallium 31	73 **Ge** germanium 32	75 **As** arsenic 33	79 **Se** selenium 34	80 **Br** bromine 35	84 **Kr** krypton 36
85 **Rb** rubidium 37	88 **Sr** strontium 38	89 **Y** yttrium 39	91 **Zr** zirconium 40	93 **Nb** niobium 41	96 **Mo** molybdenum 42	[98] **Tc** technetium 43	101 **Ru** ruthenium 44	103 **Rh** rhodium 45	106 **Pd** palladium 46	108 **Ag** silver 47	112 **Cd** cadmium 48		115 **In** indium 49	119 **Sn** tin 50	122 **Sb** antimony 51	128 **Te** tellurium 52	127 **I** iodine 53	131 **Xe** xenon 54
133 **Cs** caesium 55	137 **Ba** barium 56	139 **La*** lanthanum 57	178 **Hf** hafnium 72	181 **Ta** tantalum 73	184 **W** tungsten 74	186 **Re** rhenium 75	190 **Os** osmium 76	192 **Ir** iridium 77	195 **Pt** platinum 78	197 **Au** gold 79	201 **Hg** mercury 80		204 **Tl** thallium 81	207 **Pb** lead 82	209 **Bi** bismuth 83	[209] **Po** polonium 84	[210] **At** astatine 85	[222] **Rn** radon 86
[223] **Fr** francium 87	[226] **Ra** radium 88	[227] **Ac*** actinium 89	[261] **Rf** rutherfordium 104	[262] **Db** dubnium 105	[266] **Sg** seaborgium 106	[264] **Bh** bohrium 107	[277] **Hs** hassium 108	[268] **Mt** meitnerium 109	[271] **Ds** darmstadtium 110	[272] **Rg** roentgenium 111								

Elements with atomic numbers 112–116 have been reported but not fully authenticated

*The Lanthanides (atomic numbers 58–71) and the Actinides (atomic numbers 90–103) have been omitted

Cu and Cl have not been rounded to the nearest whole number

Group 8 is usually called Group 0; see *Additional Science Class Book*, Item C3c

Physics formulae

Module P3 Forces for transport

You should be able to state and use the following equations:

◆ For a moving object:

$$\text{speed} = \frac{\text{distance}}{\text{time}}$$ (page 216)

$$\text{acceleration} = \frac{\text{change in speed}}{\text{time taken}}$$ (page 226)

◆ Acceleration produced by a force:

force = mass × acceleration   (page 235)

◆ Energy transfer by a force:

work done = force × distance   (page 243)

$$\text{power} = \frac{\text{work done}}{\text{time}}$$ (page 246)

H ◆ Energy of a moving object:

kinetic energy = $\frac{1}{2}$ × mass × speed^2

$KE = \frac{1}{2}mv^2$ (page 248)

◆ An object in a gravitational field:

gravitational potential energy =
    mass × gravitational field strength × height

$GPE = mgh$ (page 264)

weight = mass × gravitational field strength

weight = mg (page 263)

Module P4 Radiation for life

◆ You should be able to state and use the formula for electrical resistance:

$$\text{resistance} = \frac{\text{voltage}}{\text{current}}$$ (page 281)

◆ You should be able to construct and balance simple formulae in terms of mass numbers and atomic numbers to represent alpha and beta decay (pages 303–304).

Index

Bold page references are to Higher material.

ABS brakes, 255, **256**
absorption of digested food, 11–12
acceleration, **223**, 225–9, **229**, 234–5
acids and bases, 161–5, **165–7**, 168
action and reaction, **236**
active safety features, 255
active sites of enzymes, 6
active transport, 77
adjustable seating, 255
adolescence, 34
aerobic respiration, 1
air bags, 253, 254
air resistance, 258
air spaces in leaves, **62**
albumen, 3
alcohol, 82
alkali metals, 129, 132–7
alkalis, 132, 161–2, 164
allotropes, 203
alpha decay, **303**
alpha particles, 302
aluminium, 144–5
aluminium oxide, 144–5, **146**
alveoli, 11
amber, 270
amino acids, 3
ammeter, 280
ammonia, 179, 185–8, **188–9**
ammonium hydroxide, **166**
ammonium nitrate, 179, 209, 210
ammonium phosphate, 179
ammonium sulfate, 168, 179, 181–2
amoebas, 26
amplitude, 289–90
amylase, 5
animal cells, 1, 32, 64
animal cloning, **57**
anions, 144
anodes, 144, **146**, **148**
anti-lock braking system, 255, **256**
antibodies, 3, 18
antitrypsin, **49**
aorta, 20–1
appliance insulation, **284**
aquifers, 208
arteries, 19–22
artificial radiation, 305–6
aseptic technique, **57**
asexual reproduction, 51, 55
atomic mass, 175
atomic numbers, 110–11, **112**, **303**
atomic structure, 110
atomic theory, 108
atoms, 108–15, 124
atrium, 20
auxin, 40–1
average speed, 216

background radiation, 305
bacteria, 91, 96, **105–6**
balanced forces, 259, **260**
balanced symbol equations, **139**, **152**, 166–7
banana ripening, 42
barium chloride solution, 211, **213**
barium ions, **213**
bases (chemistry), 163–4, **166**
bases (DNA), 2
batch processes, 198–9
bats, 290
battery farming, 90
bauxite, 144
Becquerel, Henri, 298, **299**
beta decay, **303**
beta particles, 302
biconcave disc, 17
biodegradable detergents, 193
biodegradable pesticides, 88
biodegradable substances, 96
bioethanol, 82
biogas, 82–3
biological control, 93
biological washing powders, 192–3
biomass, 79–82, **83**
bird droppings, 76
bleach, 191
blood, 17
blood plasma, 17–18
blood vessels, 18–19, **20**
braking distance, 237–9
brass, 157
breathing, 11
bromine, 130, 138, **141**
bronchioles, 11
buckminsterfullerene, 203, 205–6
burette, 181
buses, 225

callus, **56**
cancer, **296**
cane toads, 93–4
canning food, 97
Cannizzaro, Stanislao, 175
capillaries, 19, **20**
car collisions, 253
car safety features, **254**
carbohydrates, 79
carbon, 110, 203
carbon cycle, 103–4
carbon dioxide, 9–11, 14, 61, 79, 124, 127–9
carbon neutral, **82**
carbonates, **120**, 136, 151, 163
carnivorous plants, 106
cars of the future, 250–1
catalysts, 150, 186
catalytic converters, 153
cathodes, 144, **146**, **148**
cations, 144
Cavendish, Henry, 147

cell differentiation, 26
cell sap, 32
cell surface membranes, 9, 32
cell wall, 32, 80
cells (biology), 1–2, 26–31
   in animals, 1, 32
   diffusion into and out of, 9–10
   and osmosis, 64–5
   in plants, 32
cellulose, 80
centre of gravity, 264
chain reaction, 311–12
changing direction, **228**
charging up objects, 270–1
chemical formulae with brackets, 171
chemicals, extraction from plants, 199–200
childhood, 34
chloride ions, 116–17, **141**, **213**
chlorination, 210
chlorine, 114, **125–6**, 130, 138
chlorophyll, 32, 59, 61
chloroplasts, 32, 60
cholesterol, 22
chromatography, 200
chromosomes, 2, 27, **28**
cinchona bark, 201
circuit breakers, 285
circuits, 279–80
circulatory system, 18–21
clones, 51
cloning, 51
   of animals, **57**
   of cows, 51
   of humans, 54
   of plants, 54–5, **57**
coleoptiles, 40
collagen, 3
collisions, 253
coloured compounds, 150
complementary base pairing, **2**
compost making, **98–9**
compounds, 109
compressions, 288
concentrated solutions (effect of), 63
concentration gradient, 9, 63
conduction of electricity, 156, 158
conduction of heat, 156
conjoined twins, separation of, 23–4
consumers, 79
contamination from radioactive substances, 305
continuous processes, 198
control rods, **312**
coolants, 310, **312**, 313
cooling food, 97
copper, 157
copper(II) carbonate, 151
copper(II) compounds in general, 150
copper(II) hydroxide, 154
copper(II) nitrate, 171
copper(II) oxide, 151

copper(II) sulfate, 164
count rate, **301**
covalent bonding, 124
covalent bonds, 124
crash barriers, **254**
crenation, 64
crests of waves, 289
crop rotation, 91–2
crops in glasshouses, 88–9
cruise control, 255
crumple zones, 253, **254**
current, **282**
cuticle in leaves, 59
cuttings of plants, 42, 54
cytoplasm, 1, 32

Dalton, John, 108, 175
damaged valves, 22
DDT, 88
decay, 96–102
deceleration, 225, 234
decomposers, 98, 103, **106**
defibrillators, 275–6
dehydration, 136
delocalised electrons, **157**, 204
denatured proteins, 6
denitrifying bacteria, **106**
detergents, 191–7
   how they work, **194**
detritivores, 98, 103
diamond, 203–4
   structure and properties, **204**
differential growth, 36
differentiation, 33–4
diffusion, 9–16, 63
   factors affecting rate of, **10**
   into and out of cells, 9–10
digester (biogas), **82**, 83
dilute solutions (effect of), 63
diploid cells, 27
diploid zygote, 29
displacement reactions, 140
dissolved substances, 209
distance, 216, **217**, 229, **230–1**
distance–time data, 221–2
distance–time graphs, 220–1
distillation, **210**
DNA, 2–5
   fingerprinting, 4–5
   profiling, **4**
   replication, 3
doing work, 242–3
Dolly the cloned sheep, 51, **52–3**
dopamine, 35
dormancy, 42
dot and cross diagrams, **119–20**, **125–9**
double circulatory system, **21**
double helix, 2
double insulated appliances, 284
drag, 258–60
drinking water from sea water, **210**
driving force, **260**
drowning, 65–6
drug therapeutic activity, 205–6
drugs, making new, **200–1**
dry cleaning, 194
dry ice, **128**
drying food, 97
ductile metals, 150

earth connection, **286**
earth wire, 283, 284
earthing, 284
efficiency of energy transfer, **81**
egestion, 80
egg cells, 30–1
electric circuits, 279–80
electric windows, 255
electrical resistance, 158, 281
electricity, 81–2, 144, 279
electrolysis, 144–9
   of dilute sulfuric acid, 146, **148**
   of molten aluminium oxide, 144–5, **146**
electrolytes, 144
electromagnetic waves, **289**
electron gain, **140–1**
electronic structure, **112–14**, **119**, 134
electrons, 109–10, **112–14**, **121**, **157**,
   270, **271**
electrostatic forces, **122**
electrostatic smoke precipitators,
   276–7
elements, 108
elephants, 290
embryo transplanting, 51
Encarsia wasp, 93
energy, 242, 279
   from fuel, **250**
   from uranium, **311**
energy calculations, **266–7**
energy flow, 79–85
energy losses in food chains, 80
energy losses and pyramids, **81**
energy transfer efficiency, **81**
entomology, 99
enzymes, 1–3, 5, **98**, 192
   and pH, 7
   and temperature, 6
   how they work, **5–6**
epidermis (in leaves), 59, **62**
ethene, 42
eutrophication, **179**, 180, 193
explant, **56**
express trains, 225
extensive farming, 86

falling through the air, 258–9
farming, 75, 86–95
fermentation, **82**
fertilisers, 75, 178
   effect of, **178**
   making by neutralisation, 181–2
   percentages of essential elements
      in, **182–3**
filtration, 200, 209
firebricks, **120**
fish farming, 91
fission (nuclear), 310–16
flaccid cells, **66**
flame tests, 135
fluorine, 138, 142
fluoroapatite, 142
food chains, 79
food preservation, 97
force, 233–7
   doing no work, 244–5
   doing work, 242–3
forces, pairs of, **235–6**
forensic entomology, 99

formula unit, 109, 125
formulae
   of acids and bases, 165–6
   of compounds, 109
   of ionic compounds, 119, **122**
fossil fuels, **82**, 249
Franklin, Benjamin, 269
free-fall parachutists, 258–9
freezing food, 97
frequency, 288
friction, **236**, 255, 257–8, 270–1
friction reduction, 259–60
fuel consumption figures, 249–50
fuel rods, **312**
full outer shells of electrons, **121**
fullerene cages, **205–6**
fullerenes, 205
fungi, 87, 96, **98**
fungicide, 87
fuses, 284–5, **286**

Galileo, 266
gametes, 29
gamma rays, 294, **296**, 302
gas exchange in leaves, 14
gaseous exchange in the lungs, 10–11
Geiger counter, 299
Geiger–Müller tube, 299
gene pool, **45–6**
gene technology, 46–7
genes, 2
genetic code, 3, **4**
genetic engineering, 46, **47–8**
genetically modified (GM) organisms,
   46, 48, **49**
geodesic domes, 206
geotropism, 40–1
gestation periods, 34–5
giant covalent structure, **204**
giant ionic lattice, **122**
gibberellin, 42
glasshouses, 88–9
GPS (global positioning system)
   technology, 75
graduated pipette, 181
graphite, 203–4
   structure and properties, **204–5**
gravitation potential energy (GPE),
   242, 262, **263–4**
gravitational field strength, 263
gravity, **236**, 257, 266
gravity strength, 263
Group 0 elements, 130
Group 1 elements, 129, 132–7
Group 7 elements, 130, 138–43
groups (chemistry), 129
growth, 32, 36
   in animals and plants, 33
   in humans, 34
   in the uterus, 34
growth data, **37**
guano, 76
guard cells, 60

H_2O, 109
Haber process, 150, 186–8, 198
   conditions, **188–9**
   making a profit from, **187**
haemoglobin, 3, 17, **18**

half-equations, **134**, **141**
half-life measurement, **301**
half-life of radioactive substances, **300-1**
halide ions, 211
halogens, 130, 138-43
 properties of the, 141
 reactions with alkali metals, 138-9
haploid cells, 29
hard water, 191
hearing, 288
heart, 18, 20-1
heart disease, 21-4
heart transplants, 23
height, **264**
herbicides, 86-7
high pressure, 186-7
high tensile strength, 156
hormones, 39
human alimentary canal, 12
human growth data, **37**
hydrochloric acid, **165**
hydrogen, **125**, 147, 148
hydrogen cars, 251
hydrogen ions, **165**
hydrophilic, **194**
hydrophobic, **194**
hydroponics, 89-90, **90**
hydroxide ions, **165**
hyphae, **98**

identification of hydrogen, 148
identification of oxygen, 147
immunosuppressant drugs, 23
inbreeding, **45-6**
indicator, 181
infancy, 34
information, 279
infra-red gates, 220
infrasound frequencies, 290
insecticides, 87
insoluble dirt, 194
insulators, 270
intensive farming, 86, **88**
intermolecular forces, **195**
internal surface area of a leaf, **62**
interrupt card, 219-20
iodine, 130, 138, **141**
ionic bonding, 116-23
ionic bonds, 116
ionic compounds, 119, **122**, 171
ionic equations, **213**
ionic half-equations, **134**, **141**
ions, 116, 124
iridium, 153
iron, 150, 156, 186
iron(II) carbonate, 151
iron(II) compounds in general, 150
iron(III) compounds, 150
iron(III) hydroxide, 171
irradiation, 305
isotopes, 111

jewellery, 153

keratin, 3
kinetic energy, 248
 calculation, **248-9**
 decreasing, **249**
kinetic-energy–gravitational-potential-
 energy transformations, 264-5

labour-intensive processes, 199
lakes and rivers, 208
Large Electron–Positron collider (LEP), 113
lattice, **122**
lead compounds, 209
lead water pipes, 209
leaves, 14-15, 59-60, **62**
legislative demands, 199
lethal injection, 136
levodopa, 35
lifting an object, 242
light gates, 219-20
lightning, **105**, 269
lignin, 69
limescale, 191
limestone, **104**
linen, 70
liquid oxygen, 147
lithium, 132, 135
lithium carbonate, 136
lithium fluoride, 138-9
lithium hydroxide, **134**
live wire, 282
longitudinal waves, 289
lubrication, 259
lumen, 19
lungs, gaseous exchange in the, 10-11
lustrous, 203
lysis, 64

magnesia, **120**
magnesite, **120**
magnesium, **75**, **120**
magnesium carbonate, **120**
magnesium chloride, **121**
magnesium oxide, 118-19, **120-1**
magnetic forces, **236**
Magnox nuclear reactor, 312-13
malaria, 201
malleable, 150
maltose, 5
manic depression, 136
margarine, 150
mass, 235-7
mass conservation, **172-3**
mass numbers, 111, **112**, 303
maternal and foetal rights, 14
maturity and old age, 34
medical scans, 158
medicines, 199
medium (tissue culture), **56**
meiosis, 29-30
membranes, 63
meniscus, 71
mesophyll, 59
metabolic reactions, 1-2, 79
metal carbonate compounds, 163
metal halides, 138
metal hydroxides, **154**, 164
metal oxides, 163
metallic bonds, 157

metals, 156-60
 properties and uses of, 156-7
methane, **126**
micro-organisms, 96
 survival and growth, **96-7**
micropropagation, **55**
microvilli in the small intestine, **12**
milk of magnesia, **120**
mineral absorption in plants, 77
mineral deficiencies in plants, 76
minerals, 75
Mississippi river, 180
mitochondria, 1, 32
mitosis, 27-8
models, 289
moderators, **312**
molecular manufacturing, 206
molecules, 109, 124-5
molybdenum, 153
motorway crash barriers, **254**
MRI scans, 158
Mucor (fungus), **98**
multicellular animals, 26
mutation, **4**, 44

nanochemistry, 203-7
nanometre, 203
nanoparticles, 205
nanotubes, 205
natural chemicals, 199
negative charge, 270
negative ions, 116, 119
negative phototropism, 39
neurones (nerve cells), **13**
neurotransmitters, **13**
neutral atoms, **271**
neutral substances, 162
neutral wire, 282
neutralisation, 162-3
neutralisation reactions, **166-7**
neutron-induced fission, **311**
neutrons, 109-10, **302**
newtons (N), 233-4
Newton's third law of motion, **236**
nickel, 150
niobium, 153
nitrate ions, 105, **213**
nitrates, **75**
nitrification, **106**
nitrifying bacteria, **106**
nitrogen, 178
nitrogen cycle, 104-5
nitrogen fixation, **105**
nitrogen-fixing bacteria, 91, **105**
nitrogen-fixing crops, 91
nitrogenous fertilisers, **178**, 180
noble gases, 130
nodules, **105-6**
non-biological washing powder, 193
non-polar forces, **195**
NPK fertilisers, 75
nuclear bomb, 312
nuclear equations, **303**
nuclear fission, **310-11**
nuclear power, 310
 and the environment, 314
nuclear power stations, 312
nuclear radiation, 294-6
nuclear reactors, 310, **312**

nucleons, **302**
nucleus (atom), 110
nucleus (cell), 2, 32
nutrient solutions, 89
nutrients, 18

ohms (|), **282**
oil-tanker hold safety, **273**
oily dirt, **194**, 195
onion cells, 32–3
Onnes, Heike, 158
optical brighteners, 191, 193
optimum temperature, 6
organic farming, 91, **92**
organs for human transplants, 53
osmosis, **40**, 63
   in animal cells, 64
   in plant cells, 65
oxidation, **135**
oxygen, 9–11, 14, 17, **18**, 147
oxyhaemoglobin, **18**

pacemakers, 22–3
paddle controls, 256
paint spraying, 277
palisade cells, 60
palisade layer, 59, **62**
palladium, 153
parachutists, 258–9
Parkinson's disease, 35
partially permeable membranes, 63
passive safety features, 255–6
pay-back time, 201
penetrating power of radiation, 306
percentage yield, 174, **187**
percentages of essential elements in
   fertilisers, **182–3**
Periodic Table, 109, 129, 170
   groups, 129–30
   periods, 129–30
persistent insecticides, 88
pesticides, 87–8, 209
pH, **7**
pH scale, 162
phagocytosis, 18
pharmaceuticals, 199
phloem tubes, 60, 69–70
phosphates, **75**, 193
phosphorus, **75**, 178
photosynthesis, 14, 60–1, 79
phototropism, 39
phytoplankton, **104**
pitcher plants, 106
placenta, 12–13
plant cells, 32–3
   and osmosis, 65
plant cloning, **57**
plant fibres, 70
plant fossils, 61
plant growth (biology), 39–43
plant hormones, 39, 41–2
plant transport system, 68
plants (biology), 66
   absorption of minerals, 77
   essential elements, 178
   extracting chemicals from, 199–200
   mineral deficiencies, 76
   reducing water loss in, 72–3

plants (chemical), 185
plaque in artery walls, 22
plasmolysed plant cells, 66
platelets, 17
plugs, 283–4
polar forces, **195**
polyvinylchloride (PVC), 138
positive charge, 270
positive ions, 116, 119
positive phototropism, 39
potassium, **75**, 133–5, 178
potassium bromide, 139
potassium chloride, 136
potassium hydroxide, **166**
potassium nitrate, 179
potential difference (p.d.), 281
potential energy (PE), 262
potometer, 71–2
power, 245–6
precipitates, 152, 211
precipitation reactions, 152–3, 211
preserving food, 97
pressure, 186–7
Priestley, Joseph, 147
producers, 79
product, 5
product loss in chemical reactions,
   174–5
productivity of plants, 89
protein synthesis, 3
proteins, 105
protons, 109–10, **302**
puberty, 34
pulmonary artery, 21
pulmonary vein, 21
PVC, 138
pyramid of numbers, 79–80
pyramids of biomass, 80, **81**

quinine, 201

racing car top speed, **260**
radiation, 294, 302, 305, **305–6**
radioactive dating, 307, **308**
radioactive decay, 299
radioactive decay equations, **303**
radioactive substance decay graph, **300**
radioactive substances, 294, 298, 305
radioactive tracers, 295–6, 306, **307**
radioactive waste, 313–14
radioactivity, 298–9, **299–301**, 301–2,
   302–4
radiocarbon dating, **303**, 307, **308**
radiographers, 295–6
radioisotopes, 305–9
rare breeds, 46
rarefactions, 288
rate of diffusion, **10**
rate of reaction, **187**
rate of transpiration, 71
ratios of reactant to product, 172
raw materials, 185
reacting masses, 171
reaction of potassium with water, 133
reactivity of halogens, 139, **141**
reactors, 310
recycling, 103–7
red blood cells, 17, **18**
reduction, **141**

relative atomic mass, 170, 175
relative formula mass, 170–1
renewable energy, 81
replacement organs, 53
reservoirs, 209
residual current device (RCD), **286**
resistance, 158, 281–2
resistors, 280–1
respiration, 1, 9, 14, 80, 103
respiratory surface area, **27**
reversible reactions, 186
rheostat, 280–1
ring main circuit, 283
road traffic accidents, 240
roller-coaster rides, 233, 265
root hairs, 66
rooting powder, 41
roots, **33**, 66, 68
rubidium, **133–4**, 135
running uphill, 242
Rutland Water, 212

safety cages, 253
salts, 138, 162, 164
saprophytes, **98**
Scheele, Carl, 147
seat belts, 253–4
sedimentation, 209
selective breeding, 44–5
selective weedkillers, 41, 86
septum, 20
sewage treatment, 99–100
sexual reproduction, 29–31
shells in atoms, 110
SI units, 216
silver, 153
silver chloride, **122**
silver iodide, 212
silver ions, **213**
silver nitrate solution, 211–12, **213**
silver sulfide, **122**
single-celled organisms, 26
sisal, 70
skidding, 255, **256**
slope, **222–3**
small intestine, 11, **12**
smoke detectors, 306
smoke precipitators, 276–7
Snuppy the cloned dog, 53
sodium, 132–5
sodium chloride, 116–18, **119–20**, 136,
   140–1, 164
sodium ions, 116–17
sodium oxide, **121–2**
soluble dirt, 194
solutes, 194
solutions, 63, 194
solvents, 194
sound models, 289
sound waves, 288–9
space rocket lift-off, 233
specificity of enzymes, 5
spectator ions, **213**
speed, 215–22, **222–3**, 225–9, **229–31**
speed calculation, **222–3**
speed cameras, 219
speed control, **259**
speed measurement, 215–16, 218

speed–time graphs, 228–9
    calculations using, **229**
    shapes of, 228
    use of, **231**
sperm cells, 30
spongy layer in leaves, 60, **62**
spontaneous fission, **311**
stability control, 255
starch, 5
static electricity, 269–73, **273–4**
steel, 156
stem cells, 33–4, **57**
sterilisation of medical equipment, 295
stomata, 14, 60–1, **73**
stopping distance, 237–8
streamlining, 260
substrate of enzymes, 5
sulfate ions, **213**
sulfur, 168
sulfuric acid, 146, **148**, 168
sundew plants, 106
superconductivity, 158–9
support in plants, 66
surface area to volume ratio, **26–7**
symbiosis, **105**
symbol equations, **139**, **152**, **166–7**
synapses, **13**
synthetic chemicals, 199

tantalum, 153
temperature, 6
tensile strength, 156
terminal speed, 259, **266**
therapeutic activity of drugs, 205–6
thermal decomposition reactions, 151
thinking distance, 237–9
thinking time, 238
time, 216, **217**
tissue culture, **55–7**
titration, 181
toads, 93–4
toothpaste, 142
top speed, 260, **260**
tracers, 295

traction control, 255
transamination, 3
transition element carbonates, 151
transition elements, 150–5
transition metal hydroxides, **154**
transition metals, 150–5
translocation, 69
transmitter substances, **13**
transpiration, 15, 70–3
Transport Research Laboratory
    (TRL), 253
transverse waves, **289**
treatment for damaged hearts, 22–3
trends, 133
trip switches, 285
triplets (DNA), **4**
triploid zygotes, 29
trophic levels, 79
tropism, 39
troughs of waves, 289
tungsten, 153
tuning fork vibrations, 288
turgid cells, **66**
turgor pressure, **66**
Turin shroud, 308
twins, separation of conjoined, 23–4

ultrasound, 288–93
ultrasound scans, 291
unbalanced forces, 259
units
    of acceleration, 226
    of power, 246
universal indicator, 161–3
upper threshold of hearing, 288
uranium, **299**, **311**
uranium–lead ratio, **308**
urea, **106**, 179

vacuoles, 32
valves in veins, **20**, 22
Van de Graaff generator, 272
variable resistors, 280
vascular bundles, 70
veins (in animals), 19, **19–20**

veins (in plants), 59–61
venae cavae, 20–1
ventricles, 20
Venus flytraps, 106
vibrations, 288
villi in the small intestine, **12**
volcanic eruptions, **104**
voltage, 281, **282**, 285
voltmeter, 282

washing labels, 192
washing powder, 191
washing-up liquid, 195
waste products, 18
water, 18, 124, **127–9**
water conservation, 209
water molecule, 109
water softener, 191
water supplies, 208
water testing, 211–12, **213**
water transport in plants, 68
water treatment, 209–10
water uptake of roots, 66
watts (W), 246
wavelength, 289–90
waves, 288–9
waxy cuticles in plants, 72
weedkillers, 41, 86
weight, 257, **263**
wheat, 86
white blood cells, 17–18
wire insulation, **284**
wiring, 282–3
wood, 82
work, 242–5

X-rays, 291, 294
xylem vessels, 60, 68–70

yeasts, 26
yield, 174, **187**

zinc carbonate, 151
zooplankton, **104**
zygote, 29

Acknowledgements

Cover images, Digital Vision; **B3a.2**, Bruce Iverson / SPL; **B3a.4**, **B4a.3**, Andrew Syred / SPL; **B3a.10**, Alfred Pasieka / SPL; B3b.9, Steve Gschmeissner / SPL; **B3c.1**, Ed Reschke, Peter Arnold Inc. / SPL; **B3c.12**, **C3e.4**, **C4e.2**, Sheila Terry / SPL; **B3c.14**, Alexander Tsiaras / SPL; **B3d.1**, **B4g.2**, Astrid & Hanns-Frieder Michler / SPL; **B3d.7**, Maurice Nimmo / SPL; **B3d.13**, Eye Of Science / SPL; **B3e.4**, Ron Sachs / Corbis; **B3f.6a**, David Askham / Garden Picture Library; **B3f.6b**, CuboImages srl / Alamy; **B3f.8**, Pablo Corral V / Corbis; **B3f.9**, **P3b.5**, Erich Schrempp / SPL; **B3f.10**, Catherine Karnow / Corbis; **B3g.1**, Gregory K. Scott / SPL; **B3g.2**, Primrose Peacock / Holt / FLPA; **B3g.3**, David Aubrey / SPL; **B3g.4a**, **B3g.4b**, **B4d.4a–d**, Nigel Cattlin / FLPA; **B3g.5**, Bjorn Svensson / SPL; **B3g.6**, Gary Cook / Alamy; **B3g.7**, AJ / IRRI / Corbis; **B3g.9**, Greenpeace / Cobbing; **B3g.11**, John Carnemolla / Australian Picture Library / Corbis; **B3h.1**, **B4f.13**, **C4c.1**, **C4c.2**, Holt Studios International Ltd / Alamy; **B3h.2**, Helen Mcardle / SPL; **B3h.4a**, Agripicture Images / Alamy; **B3h.4b**, Tim Thompson / Corbis; **B3h.5**, Getty Images; **B3h.5a**, **C3f.11**, Phototake Inc. / Alamy; **B3h.6**, Seoul National University / Getty Images; **B3h.7**, Angelo Christo / zefa / Corbis; **B3h.10**, **B3h.12**, Pierre Vauthey / Corbis Sygma; **B4a.4**, **B4b.8**, **B4h.6**, **C3b.14**, Dr Jeremy Burgess / SPL; **B4c.6**, **B4g.6a**, **B4g.6b**, **B4h.7**, **B4h.8**, Eleanor Jones; **B4c.7a**, **B4c.7c**, **B4c.7e**, **B4c.13**, Geoff Jones; **B4d.1**, David R. Frazier Photolibrary, Inc. / Alamy; **B4d.3**, **P3c.2**, Corbis; **B4e.6**, **C3h.10**, **C4h.5**, Martin Bond / SPL; **B4e.9**, Patrick Ward / Alamy; **B4f.1**, David Aubrey / SPL; **B4f.2**, Russell Graves / Agstock / SPL; **B4f.3**, Skyscan / SPL; **B4f.4**, Francoise Sauze / SPL; **B4f.5**, Claude Nuridsany & Marie Perennou / SPL; **B4f.7**, Hank Morgan / SPL; **B4f.9**, David Reed / Corbis; **B4f.10**, **B4g.10**, Simon Fraser / SPL; **B4f.11**, Jim Gipe / Agstock / SPL; **B4f.15**, Kathie Atkinson / Oxford Scientific; **B4g.1**, C. James Webb; **B4g.4**, Nordicphotos / Alamy; **B4g.5**, Owen Franken / Corbis; **B4g.9**, Thomas Wright / University of Florida / IFAS; **B4h.2**, Martin Dohrn / SPL; **B4h.3**, Krafft / Explorer / SPL; **C3a.1a**, Andrew Palmer / Alamy; **C3a.1b**, Susumu Nishinaga / SPL; **C3a.2a**, Alan King / Alamy; **C3a.2c**, **C3d.2**, **C3g.2c**, **C4e.8**, Martyn F. Chillmaid / SPL; **C3a.9**, **C3h.13**, David Parker / SPL; **C3a.10**, CERN / SPL; **C3b.1a**, **C3c.13**, **C3d.4**, **C3d.5**, **C3d.6**, **C3d.9**, **C3d.10**, **C3d.11**, **C3f.9**, **C3g.2b**, Andrew Lambert Photography / SPL; **C3b.1b**, Leslie Garland Picture Library / Alamy; **C3b.1c**, **C3b.1d**, **P4c.2**, **P4c.3a**, **P4c.4**, **P4c.10**, **P4c.11**, **P4c.13**, Andrew Lambert; **C3b.4a–d**, **C3e.11a,b**, **C3e.12a,b**, **C3f.13**, **C3g.2a**, **C3g.3a–d**, **C3g.7a,b**, **C3g.8a,b**, **C3g.9a,b**, **C4a.1a,b**, **C4a.2a–f**, **C4a.3a–d**, **C4a.5a–e**, **C4a.6a–e**, **C4a.7a–e**, **C4a8ai,aii,b**, **C4b.5a,b**, **C4b.9a,b**, **C4c.8**, **C4c.9**, **C4c.10**, **C4c.11**, **C4c.12**, **C4c.13**, **C4c.14**, **C4h8ai,aii,bi,bii**, **C4h.9a,b**, **C4h.10a,b**, **C4h.11a,b**, **C4h.12a,b**, David Acaster; **C3b.8**, Roberto De Gugliemo / SPL; **C3b.9**, Kevin Fleming / Corbis; **C3b.10**, Danita Delimont / Alamy; **C3c.11**, **P4a.6**, Roger Ressmeyer / Corbis; **C3c.15**, sciencephotos / Alamy; **C3d.12**, Liba Taylor / Corbis; **C3e.2**, Darren Matthews / Alamy; **C3e.3**, Garry D. McMichael / SPL; **C3e.5**, Mike Devlin / SPL; **C3e.6**, PhotoCuisine / Corbis; **C3e.7**, BananaStock / Alamy; **C3e.8**, James Holmes, Hays Chemicals / SPL; **C3e.14**, **C4f.9**, Josh Sher / SPL; **C3f.2**, Howard Davies / Corbis; **C3f.3a,b**, Alcan; **C3f.4**, James L. Amos / Corbis; **C3f.10**, **C3f.12**, **C4d.7b**, **P4e.1**, SPL; **C3g.10**, **C3h.4**, Chris Knapton / SPL; **C3g.11**, Silverlon; **C3h.1**, Jeff Greenberg / SPL; **C3h.2**, Robert Fried / Alamy; **C3h.3**, Simon Lewis / SPL; **C3h.5**, Daniel Templeton / Alamy; **C3h.6**, Powered by Light / Alan Spencer / Alamy; **C3h.7**, **P4c.1**, Maximilian Stock Ltd / SPL; **C3h.9**, WoodyStock / Alamy; **C3h.11**, Geoff Tompkinson / SPL; **C3h.12**, Photofusion Picture Library / Alamy; **C4a.15**, Ace Stock Limited / Alamy; **C4b.13**, John McLean / SPL; **C4c.3**, Chris Westwood; **C4c.4**, Nick Hawkes / Ecoscene; **C4c.6**, Richard Hamilton Smith / Corbis; **C4c.7a**, Tim Wimborne / Reuters / Corbis; **C4c.7b**, **C4e.1a,b**, **C4e.3a,b**, **C4e.4**, **C4e.7**, **C4e.10**, Vanessa Miles; **C4c.7c**, Minnesota Historical Society / Corbis; **C4d.2**, VStock / Alamy; **C4d.3**, **P3e.3**, Time Life Pictures / Getty Images; **C4d.7a**, Emilio Segre Visual Archives / American Institute Of Physics / SPL; **C4e.5**, Grace / zefa / Corbis; **C4e.9**, Chinch Gryniewicz / Ecoscene; **C4e.16**, C. Schmidt / zefa / Corbis; **C4f.1**, Visuals Unlimited / Corbis; **C4f.2**, **P4d.5**, Saturn Stills / SPL; **C4f.3a**, Geoff Kidd / SPL; **C4f.3b**, Rosal / SPL; **C4f.7**, Tek Image / SPL; **C4f.8**, Dr Morley Read / SPL; **C4g.1**, Cinemaphoto / Corbis; **C4g.2**, Maximilian Weinzierl / Alamy; **C4g.4**, David J. Green / Alamy; **C4g.9**, Kathy deWitt / Alamy; **C4h.1a,b**, World Vision; **C4h.2**, Paul Hardy / Paul Hardy / Corbis; **C4h.3**, Mark Boulton / Alamy; **C4h.4**, Christine Strover / Alamy; **C4h.13**, Arthur Morris / Corbis; **P3a.1**, **P3f.1**, TRL Ltd / SPL; **P3a.2**, Iain Masterton / Alamy; **P3a.3**, Nigel Luckhurst; **P3a.6**, Cordelia Molloy / SPL; **P3a.7**, **P3a.8**, Michael Donne / SPL; **P3a.9**, Instruments Direct; **P3a.12a**, Milepost 92_; **P3b.1**, Bryan F. Peterson / CORBIS; **P3b.2**, Jeremy Hoare / Alamy; **P3c.1**, Nelson Jeans / CORBIS; **P3c.10**, Ashley Cooper / Corbis; **P3c.11**, Robert Brook / SPL; **P3c.12**, Jim Varney / SPL; **P3d.1**, Brian Shuel / Collections; **P3d.7**, Purestock / Alamy; **P3e.1**, Roger G. Howard Photography; **P3e.4**, AFP / Getty Images; **P3e.5**, Reuters / Corbis; **P3f.4**, Vic Pigula / Alamy; **P3g.1**, Hermann Eisenbeiss / SPL; **P3g.4**, StockShot / Alamy; **P3g.7**, Christie & Cole / Corbis; **P3h.1**, Noah K. Murray / Star Ledger / Corbis; **P3h.3**, NASA / Corbis; **P3h.5**, Jessica Rinaldi / Reuters / Corbis; **P3h.6**, Royalty-free / Corbis; **P3h.9**, Sonny Meddle / Rex Features; **P3h.10**, National Maritime Museum Picture Library, London; **P3h.11**, Sergio Pitamitz / zefa / Corbis; **P4a.1**, Museum of the City of New York / Corbis; **P4a.2**, The Library Company of Philadelphia; **P4a.5**, Richard Levine / Alamy; **P4b.1**, Adam Hart-Davis / SPL; **P4b.3**, Eberhard Streichan / zefa / Corbis; **P4c.14**, **P4c.15**, Sheila Terry / SPL; **P4c.5**, Ian M. Butterfield / Alamy; **P4d.4**, Dietmar Nill / naturepl.com; **P4d.6**, Dr Najeeb Layyous / SPL; **P4e.3**, Simon Fraser / NCCT, Freeman Trust, Newcastle-Upon-Tyne / SPL; **P4e.4**, CC Studio / SPL; **P4e.5**, Miriam Maslo / SPL; **P4e.6**, BSIP LECA / SPL; **P4f.1**, Bibliothèque Centrale de l'Ecole Polytechnique, Paris, France; **P4f.2**, National Radiological Protection Board; **P4f.3**, Pascal Goetgheluck / SPL; **P4g.7**, Reuters / photo by Claudio Papi; **P4h.4**, US Department Of Energy / SPL; **P4h.7**, Michael Marten / SPL; **P4h.8**, BNFL.

Abbreviations: FLPA, Frank Lane Picture Agency; SPL, Science Photo Library.
Picture research: Vanessa Miles.